FIBER IN
HUMAN NUTRITION

FIBER IN
HUMAN NUTRITION

Edited by

Gene A. Spiller
and Ronald J. Amen

Institute of Agriscience and Nutrition
Syntex Research
Palo Alto, California

PLENUM PRESS · NEW YORK AND LONDON

Library of Congress Cataloging in Publication Data

Main entry under title:

Fiber in human nutrition.

Includes bibliographies and index.
1. High-fiber diet. 2. Fiber deficiency diseases. 3. Digestive organs — Diseases —
Nutritional aspects. I. Spiller, Gene A. II. Amen, Ronald J.
RM237.6.F54   612'.39   76-20767
ISBN 0-306-30919-X

The value of what one knows is doubled if one confesses to not knowing what one does not know. What one knows is then raised beyond the suspicion to which it is exposed when one claims to know what one does not know.

SCHOPENHAUER

The problem of our mistaken view as to the nature of the link between cause and effect lies in the propensity to attribute necessity of connection between the members of certain sequences of ideas. Now the linking together of ideas arises from association promoted by the three relations of resemblance, contiguity in space and time, and cause and effect.

HUME

Selected by contributors

Jenny Eastwood
Martin A. Eastwood
M. Ward

Contributors

George M. Briggs, Ph.D.

Professor of Nutrition, Department of Nutritional Sciences, University of California, Berkeley, California

John H. Cummings, M.B., M.Sc., M.R.C.P.

Clinical Scientific Staff, Medical Research Council Gastroenterology Unit, Central Middlesex Hospital, London, England

Jenny Eastwood, M.B., Dip. Soc. Med.

Community Medicine Medical Officer, Lothian Health Board, Edinburgh, Scotland

Martin A. Eastwood, M.B., M.Sc., F.R.C.P.

Consultant Physician, Wolfson Gastrointestinal Laboratories, Department of Medicine, Western General Hospital, Edinburgh, Scotland

Robert E. Hungate, Ph.D.

Professor Emeritus of Bacteriology, Department of Bacteriology, University of California, Davis, California

David Kritchevsky, Ph.D.

Associate Director, Wistar Institute of Anatomy and Biology; Wistar Professor of Biochemistry, University of Pennsylvania, School of Veterinary Medicine, Philadelphia, Pennsylvania

John A. Lang, Ph.D.

Postdoctoral Fellow, Department of Nutritional Sciences, University of California, Berkeley, California

Ian Macdonald, M.D., D.Sc.

Professor of Applied Physiology, Department of Physiology, Guy's Hospital Medical School, London, England

W.D. Mitchell, Ph.D., F.R.I.C.

Wolfson Gastrointestinal Laboratories, Western General Hospital, Edinburgh, Scotland

David A.T. Southgate, Ph.D.

Dunn Nutritional Laboratory, Medical Research Council, Cambridge, England

Jon A. Story, Ph.D.

Research Scientist, Wistar Institute of Anatomy and Biology, Philadelphia, Pennsylvania

Michael Ward, M.R.C.P.

Lecturer in Medicine, Wolfson Gastrointestinal Laboratories, Western General Hospital, Edinburgh, Scotland

Alexander R.P. Walker, D.Sc.

Medical Research Council, Human Biochemistry Research Unit, South African Institute for Medical Research, Johannesburg, South Africa

Preface

The editors have designed this book to serve both as a textbook on fiber in nutrition and, we hope, as the first complete reference on the subject. For the past 25 years, the study of plant fibers and their effect on human physiology has generally been relegated to a low-priority status. Recently, however, this area of research has enjoyed a renaissance unparalleled in the history of the food and nutritional sciences, a reawakening which has occurred primarily as a result of epidemiology reports that suggested a positive relationship between plant fiber ingestion and health.

As interest among the scientific community increased and new research programs were initiated to test objectively the epidemiological hypotheses, major gaps in the fundamental pool of knowledge became apparent. To compound the difficulty, scientists often did not agree upon what "fiber" is. Some investigators restricted their definition to the structural polymers of the plant, while others expanded theirs to include the entire plant cell wall with all its fibrous and associated nonfibrous substances.

As a result, research that was performed and reported frequently only obscured the issue still further; at best it exposed whole new areas of ignorance in a field once considered too uninteresting to pursue. Despite voluminous research, scientists generally have still not been able to identify with certainty the specific component(s) of the plant cell-wall system that causes the various observed physiological effects. In fact, they do not yet agree upon the nomenclatures involved.

In the editors' belief, the present volume is the first publication consolidating information gained by the various disciplines involved in plant fiber research. We think this book represents the most up-to-date knowledge pertaining to the chemistry, analysis, physicochemical properties, animal and human experiments, and clinical and epidemiological studies of fiber and related substances. We were exceptionally fortunate in obtaining the cooperation of leading experts on fiber in nutrition.

The book comprises all facets of the study of fiber in nutrition, from an overall view of fiber (Chapter 1) through detailed chemistry (Chapter 2), analytical methodology (Chapter 3), physical properties (Chapter 4), microbiological digestion (Chapter 5), animal experimentation (Chapter 6), and physiological, clinical, and epidemiological aspects (Chapters 7–11).

Since this area of research still remains controversial and relatively new, we have not attempted to foster any particular theories in this volume. Divergent opinions may appear in different chapters and represent the current thinking of the authors. We feel that only through exposure to differing opinions can questions be answered and controversies be clarified.

Consistent with this, although we have recently suggested a new fiber nomenclature (*Am. J. Clin. Nutr.* **28:**674, 1975), we have left it to each author to define ''fiber'' in his own terms.

We hope these differing opinions will make this book a more meaningful contribution to the knowledge of fiber in human nutrition.

Gene A. Spiller
Ronald J. Amen

ACKNOWLEDGMENTS
    We offer special appreciation to Myron A. Beigler, vice-president and director of The Institute of Agriscience and Nutrition, Syntex Research, for his encouragement, advice, and help during this endeavor. We wish to thank Dr. David Kritchevsky of the Wistar Institute, Dr. Hugh Trowell, formerly of the Department of Medicine, Markere University, Kampala, Uganda, and Ruth McClung Jones and Steven Sorensen for their technical advice. We also thank Peter Consigny and his staff for their aid in the graphic arts and Della Berntson, Penny Clark, and Alethe Echols for their secretarial and editorial assistance.

G.A.S.
R.J.A.

Contents

Chapter 2

The Chemistry of Dietary Fiber

David A.T. Southgate

Chapter 3

The Analysis of Dietary Fiber

David A.T. Southgate

Chapter 4
Physical Properties of Fiber: A Biological Evaluation
Martin A. Eastwood and W.D. Mitchell

Chapter 5
Microbial Activities Related to Mammalian
Digestion and Absorption of Food
Robert E. Hungate

Chapter 6
The Use and Function of Fiber in Diets
of Monogastric Animals
John A. Lang and George M. Briggs

Chapter 10
Gastrointestinal Diseases and Fiber Intake
with Special Reference to South African Populations
Alexander R.P. Walker

Chapter 11
The Effects of Dietary Fiber: Are They All Good?
Ian Macdonald

What Is Fiber?

John H. Cummings

I. Introduction

There is no definition of fiber that is concise and yet conveys fully the concept of fiber in human nutrition. What constitutes fiber depends on one's point of view. To the cereal chemist fiber is cellulose,[75] while to the botanist it is "a dispersed phase of microfibrils packed round with a complex continuous matrix"[104]; to the animal nutritionist it is "the insoluble matter indigestible by animal enzymes,"[153] and to the human nutritionist "unavailable carbohydrate and lignin."[138]

Outside the biological sciences the word fiber has a great variety of uses including the synthetic fibers of the textile industry, glass fiber, and even moral fiber. But in human medicine, particularly nutrition, fiber is now reserved for a group of substances of plant origin which are found largely but not entirely in the plant cell wall and which are thought to be neither digested nor absorbed in the upper gastrointestinal tract. Other parts of the diet may be legitimately regarded as fibrous, notably collagen fiber associated with meat, but in no way is this considered in the same context as fiber of plant origin.

The problem of a clear understanding of precisely what constitutes fiber has been magnified by the difficulties associated with the chemical analysis of so complex a substance. As chemical and nutritional techniques have evolved so has the concept of fiber. Early attempts during the last century to separate out the "indigestible" part of animal feeds led to the isolation of an acid- and alkali-resistant mixture called fiber or *crude fiber*, which was composed largely of cellulose. Cellulose was known to be a major part of the plant cell wall. Later other cell-wall constituents such as hemicellulose and lignin were found to be

JOHN H. CUMMINGS • Clinical Scientific Staff, Medical Research Council Gastroenterology Unit, Central Middlesex Hospital, London, England.

important in assessing the quality and digestibility of foodstuffs. In more recent years fiber has attracted the attention of human nutritionists, partly as a component of our diet that is indigestible and therefore has to be taken note of in calculating the energy value of diets, and partly because of the effect cereal fiber in particular exerts on bowel function.

Today the view of fiber is even broader. It will be quickly realized that the whole of the plant cell-wall structure may now be considered within the concept of fiber, together with other plant substances such as gums and mucilages that are not digested by the normal secretions of the human gut. Furthermore, not all fiber is in fact fibrous, nor is it totally indigestible. Its role in man has been extended to include a wide range of gastrointestinal, nutritional, and metabolic effects. The implication of fiber alongside other components of our diet in the prevention of diseases of major importance assures it of a continuing place in medical and biological thought.

II. Development of the Concept of Fiber in Human Nutrition

It was known in the Middle Ages and earlier that some part of our food from plants, particularly from cereals, was indigestible and had quite definite effects on bowel function,[25,42] but it was not until the beginning of the nineteenth century that an attempt was made to define this component. The initial impetus to this work came from the developing science of animal nutrition, whose followers used the terms "fiber" or "crude fiber" for the component. These words have since been hallowed by legal sanction and incorporated into the statute book, where they remain to this day.

A. Crude Fiber

The original method for measuring the crude fiber content of animal foodstuffs was developed in 1806 by Einhof.[157] The idea was to be able to estimate the indigestible fraction of the diet and thereby predict its nutritive value, a feature of obvious nutritional and economic importance. Later in the nineteenth century, the term Weende procedure was applied to the method after a town in Germany where research in this field was being undertaken. Today the method involves extraction of a forage sample with petroleum ether, after which it is dried, boiled with dilute sulfuric acid, filtered, washed with boiling water, boiled with dilute sodium hydroxide, filtered, washed again with boiling water, and then washed with 1% hydrochloric or sulfuric acid, alcohol, and ether. The residue is weighed and ashed, the crude fiber value being the final residue weight minus the ash weight.[3,48,66] The method was thought originally to measure the cellulose content of the diet and still today in some disciplines is equated with the measurement of cellulose.[75] The reason for the incorporation of the precise chemical methods into the feeding-stuffs regulations was to define the indigestible part of a feed that was being sold so that farmers might have some idea of the nutritional quality of the food.[38]

Despite its established place in the history of animal nutrition, deficiencies in this method have been recognized for over a hundred years.[27,38,59,92,126,171] The main problems with it are: First, it is not a good predictor of the nutritive value of a feed. Second, the determination is extremely method-dependent and insensitive at the low levels that are found in many foods. Third, it measures a variable proportion of the total plant cell-wall constituents. Cell-wall constituents are underestimated; for example, the crude-fiber content of wholemeal flour is 2%, while the total cell-wall constituents are approximately 11 or 12%.[38] With the development of more precise methods for measuring the components of the plant cell wall, it is now known that the crude-fiber method recovers approximately 50–80% of the cellulose, 10–50% of the lignin, and 20% of the hemicelluloses.[155,157] A more important deficiency in the method, however, is that it recovers a variable proportion of the total cell-wall constituents of a plant. In lettuce the crude-fiber fraction represents almost 80% of the total cell-wall constituents; in apple it is about 50%, but in bran only around 25%. The proportion of the total cell-wall constituents that are recovered depends to a large extent on the amount that the hemicellulose fraction contributes to the overall cell wall. The hemicelluloses are largely lost during the crude-fiber procedure, and therefore plants, such as wheat or banana,[137] that have a high proportion of hemicellulose in the cell wall give a low recovery of cell-wall constituents in the crude-fiber fraction.

Has the crude-fiber procedure any place in human nutrition? At the moment it is impossible to answer this, as the effects of fiber in man are still largely undocumented, but it seems unlikely that a method with so many deficiencies will find a place of any value. The animal nutritionists are giving serious thought to other methods of predicting forage digestibility so that the crude-fiber method may well be superseded by the methods of Van Soest,[153,158] while in human nutrition an analytical scheme such as that developed by Southgate[137] should prove a more valuable technique for investigating the role of fiber than the crude-fiber method.

B. Residue, Roughage, and Bulk

In the absence of a full understanding of the nature and properties of fiber a variety of terms were found to describe it, of which residue, roughage, and bulk came into common usage in medicine. These terms, however, are not dignified by any precise meaning nor by chemical definition and, in fact, are often used interchangeably. Furthermore, they imply effects on the gut by fiber that are either undocumented or in some cases untrue.

The term residue is used broadly to mean two different things. Firstly, it may refer to the indigestible part of food,[82,164,173] the residue of the diet being that part not digested by gut enzymes; in this context, residue is equivalent to crude fiber[82] or to the combined cellulose, hemicellulose, and lignin content.[164] Residue is also used to describe the increase in fecal output that occurs when certain "indigestible" foods are eaten.[21,173] Feces were at one time regarded purely as the remains of our diet,[21] and the expression fecal residue was used regardless of whether any actual dietary residue was present in the stool. In this context milk once became known as a high-residue food largely, perhaps, because of an inappropriate experiment on dogs.[67] The inadequacies of this term are self-evident. It is impossible to use it in any precise way when there is so little documentation of the effect of dietary constituents on stool volume and dry weight,[164] and the concept largely ignores the fact that feces contain a number of components other than dietary residue such as bacteria, intestinal secretions, desquamated epithelial cells, mucus, and water.

Roughage as a term to describe certain types of food in our diet came into being because of the one-time fashion for prescribing bland diets in the treatment of dyspeptic symptoms. Subjects with peptic ulceration and nonspecific gastrointestinal disturbances were often advised to avoid roughage in the diet in the belief that it might aggravate their problems by producing abdominal cramps, bloating, and an increase in fecal output. Examples of "rough" foods generally included brown bread, cereals, nuts, spiced foods, and fruit and vegetables, particularly if uncooked. The fashion for bland or low-roughage diets has now largely passed, although the implication that foods containing roughage might have an abrasive effect on the gut still lingers on. Apart, however, from the report by Williams and Olmsted[173] of passage of hard, blood-flecked stools in a subject on a high-fiber-containing diet, there is no evidence that high-fiber-containing food damages the gut in any way. In fact, the converse is now thought to be true, with high-fiber foods being advised for constipation,[23,35,74] diverticular disease,[108,135] and the irritable bowel syndrome.[107] Fiber or roughage is now considered to be either of therapeutic or preventive value in a wide range of colonic disorders.[124] It seems unlikely that fiber-containing foods are in general damaging to the gut, as their widespread consumption in some areas of Africa[86] is not associated with a high incidence of gastrointestinal diseases.[148,159] The concept of fiber-containing foods being roughage in any sense therefore is probably not a valid one.

Bulk is a term used to describe the capacity of certain foods to hold water and thereby increase the weight of stool. It is now mainly used to describe a particular group of laxative agents obtained from plants.[37,70,74] These bulk laxatives undoubtedly contain fiber, but the way in which they increase fecal bulk has not been documented. It does not necessarily follow that because a substance holds water *in vitro* it will do the same in the gut. In fact, the only formal study that has been done to answer the question of how fiber increases fecal weight[172,173] suggested that it was not so much the water-holding capacity of the fiber that was important as the fact that it was metabolized to short-chain fatty acids. This is an area where further research would seem useful.

The terms residue, roughage, and bulk clearly lack precision, and do not imply a concept of fiber which has been documented in man. Their use would seem to be of no particular value in the present context.

C. Unavailable Carbohydrate

In terms of human nutrition, the introduction of the idea of unavailable carbohydrate in the diet by McCance and Lawrence in 1929[88] signified a definite advance in our thinking about fiber. It followed from a formal attempt to measure the carbohydrate content of the diet and a realization that a significant part of plant carbohydrate was not digested by human enzymes. These workers recognized that this unavailable carbohydrate included the hemicelluloses, together with those substances measured by the crude-fiber procedure (mainly cellulose) and one or two less common plant substances such as inulin and algal polysaccharides. Despite the authors' awareness that a substantial proportion of unavailable carbohydrate was in fact digested during its passage through the gut and that the crude-fiber fraction contained a noncarbohydrate material called lignin, the term was introduced, and the concept it denotes has proved a focal point for further thinking about fiber and its role in the diet.[137,138]

D. Dietary Fiber

Because unavailable carbohydrate is neither truly unavailable nor entirely carbohydrate, Trowell in 1972[150] redefined fiber in the diet as "the skeletal remains of plant cells that are resistant to digestion by enzymes of man" and called this dietary fiber. This term is now quite widely used to describe fiber in the diet and shows how the idea of fiber has broadened over the years from the original limited view that it was mainly cellulose to the present-day concept of the plant cell wall. In 1974 Trowell modified his definition to broaden the scope even further and included under the term dietary fiber the structural polysaccharides of the cell wall, lignin, plant lipids, nitrogen, trace elements, and other unidentified substances.[151] While this definition is not ideal, since not all plant cell-wall structures are fibrous and the term dietary is perhaps too restrictive, nevertheless the concept is a valuable one.

Further attempts to clarify this nomenclature for fiber are being made, and doubtless as knowledge about fiber in human nutrition develops, a generally acceptable term will be found.[140]

E. Bran

With the resurgence of interest in plant fiber over the past 10 years, bran has become almost synonymous with fiber. Of two reasons for this interest, the first is the increasing incidence of a variety of common diseases of Western civilization that are thought to be related to changes in our consumption of cereal fiber rather than to other components of the diet,[17-20,149] although this belief is disputed. Secondly, bran is the most concentrated form of fiber and is readily and cheaply available to the general public from a variety of commercial sources. It has also been widely used in experimental situations as fiber and therefore merits some special consideration.

Bran is a by-product of the flour milling process; it ideally comprises all the outer layers of the wheat grain down to and including the aleurone layer.[75,91] These layers include the pericarp, which incorporates the epidermis, cuticle, and cross-cell and tube-cell layers; the seed coat, which incorporates the testa, tegmen, and pigment cells; the hyaline layer, and the aleurone layer. Together these outer layers form between 11 and 16% of whole wheat. In practice, bran usually contains more than these layers, as their separation during the milling process is not commercially practical. The aleurone layer is a very thin one lying close to the endosperm of the grain. Clearly, separation of this from endosperm during milling is not easy, and therefore variable amounts of endosperm are nearly always present in commercially available bran. Its presence may be one of the factors altering the physical properties of bran preparations.[80] The inclusion of some endosperm in the bran is also important in another context. Most of the proteins that make up gluten are contained in the endosperm and thus make bran unsuitable for patients with coeliac disease. Wheat germ may also be present or may be added to bran preparations. Wheat germ is about 2% of the whole wheat and comprises the embryo and scutellum; it is separated from bran and endosperm during the milling process. Some wheat germ may remain attached to the bran, the amount depending to some extent on the moisture content of the wheat and the method of milling.[177]

Bran is therefore not a substance with uniform composition, nor is it entirely composed of fiber. A typical breakdown of its composition would be: cellulose, 21%; pentosan, 20–26%; starch, 7.5–9%; sugar, 5%; protein, 11–15%; fat, 5–10%; ash 5–9%; and water, 14%.[75,91] The fatty acids of bran are 84% unsaturated (mostly linoleic and oleic acid) and 16% saturated. The chief minerals present are potassium, phosphorus, and magnesium. Bran also contains a variety of other substances including vitamins and, perhaps more important, phytic acid. The inclusion of the aleurone layer in bran is of some importance in nutrition, as this layer of the wheat grain contains about 61% of the total wheat grain minerals, 83% of the niacin, 61% of the pyridoxin, and most of the phytate and folic acid.[91] It is worth noting that not all the fiber of wheat is removed with the bran.

The endosperm, which forms the white floury part of wheat, also contains cell walls and therefore a proportion of the total wheat fiber.

III. The Plant Cell Wall

The diverse nature of fiber is well illustrated in a consideration of the development and maturation of the plant cell wall.

A. Primary Wall

The plant cell wall starts as a dense disk of cytoplasm first visible in telophase (i.e., toward the end of cell division), lying between the newly divided nuclei and penetrated by microtubules. Membrane-enclosed vesicles or droplets derived from the Golgi bodies then appear and coalesce to form the cell plate lying transversely across the cell. These vesicles, which form the cell plate, are thought to contain pectins. The membranes of the vesicles become the plasmalemma that lies on each side of the developing cell plate. On the surface of the plasmalemma is an enzyme system that is able to synthesize cellulose. The cell plate thus becomes coated on both sides with cellulose microfibrils, while other polysaccharides are added to its surface, building up the primary cell wall of each daughter cell. The central part of the cell plate eventually becomes the middle lamella, which is rich in pectins and forms a layer of cement or intercellular substance between the new cell walls. The primary cell wall, therefore, contains cellulose microfibrils, protein, hemicelluloses, pectins, and water.[24,104,118,122]

After cell division, the cell wall continues to grow by a process typified by extension growth. In this process cellulose microfibrils are laid down on the inside of the cell wall in a loose network, the fibers generally running transverse to the direction of extension. As the cell grows, however, the fibers become reorientated so that eventually the fibers on the outer surface of the wall run more in the direction of the length of the cell. Other polysaccharides are laid down simultaneously as a matrix for the cellulose.

The polysaccharide polymers of the cell wall are built up from individual sugars, like glucose, transported as nucleoside diphosphate sugars, commonly uridine diphosphate (UDP-G) or guanosine diphosphate (GDP-G). These sugar nucleotides are formed in the Golgi bodies and are transported to the cell wall in vesicles or via the endoplasmic reticulum or the microtubules.[24,105,122,134] Glucose is an important precursor in cell-wall synthesis since it is directly incorporated into cellulose; it is probably also a direct precursor of the 5-carbon sugars xylose and arabinose, which form a major part of the hemicellulose fraction.

These pentose sugars are produced from glucose by oxidation at C-6 to form uronic acid, with subsequent decarboxylation, the whole process occurring in the cytoplasm of the cell rather than at the cell wall.[118,134] Much of the information about cellulose synthesis has come from a study of its synthesis by the bacterium, *Acetobacter*.

B. Secondary Cell Wall

The secondary cell wall forms after the cell has reached maturity and is laid down inside the primary cell wall either as a continuous layer or as localized thickenings or bands; these bands may be seen particularly in the cells of the xylem. The composition of the cell wall changes substantially at this stage, with a much higher proportion of cellulose being laid down and with a wider variety of hemicelluloses and very little pectin.[104,118] A further important development at this stage in the cell wall is the start of the process of lignification. Lignin is an aromatic polymer, unlike the polysaccharides of the cell wall, which grows into the cellulose microfibrils and encrusts the cell wall matrix, lending strength and rigidity to the whole structure. Completely lignified plant cells are considered to be dead in that they have no nucleus or cytoplasm but form specialized tissues in the vascular parts of plants.

The path to lignin synthesis is in part a common one with that of a variety of plant substances including the plant hormone auxin. This pathway involves the synthesis of shikimic acid from phosphoenol-pyruvate and erythose-4-phosphate which then goes by phosphorylated intermediates to chorismic acid and ultimately to the aromatic amino acids phenylalanine and tyrosine. These are then converted by key enzymes in the process, phenylalanine and tyrosine ammonia lyase, to cinnamic and coumaric acids which give rise to coniferyl, sinapyl, and coumaryl alcohols. These are subsequently dehydrogenated to form the basic phenyl propane units, which are rapidly polymerized forming lignin.[104,134] Lignin synthesis probably requires the presence of cellulose in the cell wall and is an aerobic process. The extent to which cell walls are lignified lends them distinctive properties and is of great importance in animal nutrition (see Section IV).

C. Growth, Maturation, and Taxonomic Differences

The composition and properties of the plant cell wall change considerably as the plant grows and matures.[134,145,146] The growth rate of plant shoots can be related to their pliability or softness, the more rapidly growing shoots becoming more pliable. This pliability may be related to the gel-forming properties of pectin in the cell wall. These gelling properties depend on the structure of the pectin polymer as well as on the degree of esterification of its uronic acid groups and the proportion of salt formation that occurs.[122] Stronger gels are formed from more highly methylated pectins and in the presence of a high proportion of calcium salts. Pectins isolated from rapidly growing parts of the plant often have poorer gel-forming properties than those from the more mature parts such as the stems and leaves. The methyl ester content of pectin changes during growth and ripening,[73,134] and changes in cation binding can be shown during germination.[54] Other cell-wall substances may also play a part in this plasticity, particularly cell-wall proteins, which are present as large glycoproteins linked by sulfhydryl bridges.[24] The proportion of lignin to cellulose in the cell wall also changes with maturation and may be influenced by environmental conditions, such as temperature, in which the plant is grown.[33] The proportions of cell-wall constituents also vary among different plant species and show variations among different parts of the plant, such as the stem and leaf or flower.[155]

It can thus be seen that the plant cell wall is a complex matrix of substances, the proportions of which differ with the stage of maturity of the plant, the part of the plant, and the species. All these factors may influence the physiological role which plant fiber plays in human and animal nutrition. The problem is further complicated by the variety of structures that are associated with and incorporated into the basic cell wall (see Section V).

IV. Components of Fiber and Their Possible Physiological Significance

No generally agreed-upon classification of the components of fiber exists because of the difficulty of obtaining the various fractions in a pure and unaltered form for analytical and research purposes. Ideally, a classification would be best based on the molecular structure of these substances, but this is not always known, particularly with regard to the pectic substances, plant gums, and mucilages. It is also important to remember that much of the data relating to cell-wall substances has been derived from wood, cereal, and grasses, with little analytical work having been done on the commonly eaten fruits and vegetables. The plant gums, mucilages, and some storage polysaccharides, while not strictly part of the plant cell wall, are included here, as they have biochemical similarities to plant cell-wall structures and might be expected to have similar nutritional and physiological implications in man.

A. Cellulose

Cellulose is the best known, most widely distributed, and only truly fibrous component of the plant cell wall. The discovery of cellulose is attributed to the French chemist Payen early in the nineteenth century. While he undoubtedly made a major contribution in deducing its composition,[110] the word cellulose was already in use in the eighteenth century.

Its primary structure is that of an unbranched 1-4β-D-glucose polymer[31,65,160] containing about 3000 glucose units, although values up to 100,000 have been suggested,[134] with a molecular weight of 6×10^5. In its naturally occurring state, however, it may be of much higher molecular weight, since extraction procedures used to purify it are likely to lead to breakdown of the larger polymers. The whole molecule folds into a flat, ribbon-like structure,[119,122,160] which X-ray diffraction studies have shown to have a helical conformation with repeating units every 10.3 Å, each representing individual cellobiose units. The structure is stabilized by extensive hydrogen bonding, both inter- and intramolecular, which is the basis for the crystalline structure of cellulose that has been demonstrated.

Because of the linear, unbranched nature of the cellulose polymer, it is able to pack together quite closely in a three-dimensional lattice-work, forming microfibrils of cellulose. These form the basis of cellulose fibers, which are woven into the plant cell wall. Its density is 1.59, and the pore size of the overall cellulose structure is in the region of 5–20 Å.[119]

Several types of cellulose are sometimes described, namely cellulose I, II, III, and IV. Cellulose I is the naturally occurring cellulose as it is found in the plant cell wall; the other forms are derived celluloses, obtained usually after various extraction procedures.[119,122,160,174]

Of the polysaccharides that make up the plant cell wall, cellulose is one of the very few unbranched homopolymers that exist. Other linear polymers include lichenan, which replaces cellulose in the cell wall of Iceland moss; chitin, which is found in fungi and algae and in the animal kingdom; and alginic acid, abundant in algae.[65,76,113,160] The lack of side chains and the homogeneity of these polymers will clearly confer distinct physical properties on them in terms of their ability to pack together and form fibers and in their interaction with other substances. The extent to which other polymers interact with cellulose is not clear, particularly the relationship of lignin to cellulose,[26,160,174] although some polysaccharides may be adsorbed onto its surface.[174] The purest naturally occurring form of cellulose is the seed hair of the cotton plant, which contains between 92 and 96% on a dry-weight basis of cellulose,[160,174] but most plant cell walls contain only 15–40% on a dry-weight basis. Cellulose synthesis by a bacterium (*Acetobacter*) has been noted.[61]

The role of cellulose in human physiology must eventually be related to its structure and physical properties. However, this is not easy to assess because our diets rarely include either cellulose or any other polysaccharide in a purified form. One important property of cellulose is its ability to take up water and swell.[160] Cotton fiber is able to take up 0.4 g water/g fiber.[119] These water-imbibing properties of cellulose are a ready explanation for its ability to increase fecal weight in human subjects.[41] However, cellulose is digested widely in the animal kingdom and also in the human gut.[64,68,101,139,175] The reported values for the proportion that is digested in man vary enormously, but a mean figure of about 40% digested is reasonable. The factors affecting cellulose digestibility in man have not received much attention, but there is clearly wide individual variation[92] and possibly some relation to the subject's age.[68,139]

In animal nutrition there is a lot of interest in cellulose digestibility (see Chapter 6), and much is known about factors influencing its availability for digestion. These include the intrinsic properties of the cellulose microstructure; associated substances such as lignin and silica, which both impair cellulose digestibility; its moisture content; total intake and intake of other dietary constituents; and time available for digestion.[26,115,121,155] *In vitro* studies of cellulose digestibility have shown that the rate of digestion follows second-order kinetics, and suggestions have been made that there may be a truly unavailable fraction of cellulose in the plant cell wall.[155] Prepared celluloses digest more slowly but more completely. Cellulose is digested by bacterial enzymes, the bacteria themselves penetrating the lumen of the cellulose fibers and also attaching themselves to the outside.[10]

Inclusion of cellulose in the diet of both animals and man is associated with changes in cholesterol and bile-salt metabolism. Increased fecal bile-salt excretion has been demonstrated,[116,133,141] although not consistently.[41] With regard to blood cholesterol level, however, cellulose seems to be much less potent than other components of fiber in lowering this level,[77,117,120,132,165] unless there is already an abnormality of cholesterol metabolism or the subject is taking a high-cholesterol diet.[133,144]

Cellulose may also have the capacity to bind folic acid and some of its derivatives.[87]

B. Hemicelluloses

By contrast with the uniformity of cellulose, the hemicellulose component of the plant cell wall consists of a wide variety (at least 250 known) of polysaccharide polymers, which contain a mixture of pentose and hexose sugars and many of which are branched. Their classification, perhaps more than any other part of the plant cell wall, is likely to lead to confusion because of their varying behavior in different analytical procedures. In general, the hemicelluloses may be described as those cell-wall polysaccharides soluble in cold dilute alkali. A variety of terms have been used to describe them, including *pentosans*, which highlights the fact that they are made up largely of 5-carbon sugars, and *noncellulosic polysaccharides* implying their polydispersity. The hemicelluloses are frequently subdivided into groups depending on their biochemical characteristics. One such subdivision is the *polyuronide hemicelluloses*, so named because of their high uronic acid content. Others are the *A and B hemicelluloses*, their designation depending on their precipitation (after neutralizing the original extract) either by dilute acid (A) or by dilute acid and the subsequent addition of alcohol (B). Perhaps a more useful classification, in terms of their possible role in human nutrition, would be as the *acidic* and *neutral hemicelluloses,* depending on whether they have relatively small or large numbers of uronic acid residues. The term hemicellulose was proposed by Schulze[130] in 1891 to describe a group of substances that he found in the plant cell wall, which were "easily dissolved in hot dilute mineral acid."

As a broad generalization, the hemicelluloses consist of a "backbone" of xylose sugars with various substitutions and degrees of branching.[134,168,169] Most contain between two and four different sugars, the ones commonly present being xylose, arabinose, mannose, galactose, glucose, rhamnose, and galacturonic and glucuronic acids. Hemicellulose molecules are usually much smaller than those of cellulose, with between 150 and 200 sugar units making up the molecule. They are also more amorphous than cellulose molecules, although some xylans exhibit a crystalline structure. Together with pectin, the hemicelluloses form the matrix of the plant cell wall in which are enmeshed the cellulose fibers. Cell walls of annual plants contain between 15 and 30% of hemicellulose on a dry-weight basis, while wood contains 20–25%.[169] The presence in hemicelluloses of uronic acids may well be important in determining and explaining their properties. Uronic acids are present in about half the known plant polysaccharides, the commonest being D-glucuronic and D-galacturonic acids. These are derived by oxidation of the terminal CH_2OH to $COOH$ and, when present as glycosides, behave like simple hydrocarboxylic acids, forming metal salts, amides, and alkyl and methyl esters.[31,76,113]

The neutral hemicelluloses are typified by those isolated from cereals.[6,65,134,169] These have a backbone of 1-4β-D-xylose with short side chains of arabinose, often in the furnanose form, with occasional (about 5%) glucuronic acid residues. It has been shown that the ratio of xylose to arabinose present in these polymers varies with the specific genetic strain of the plant.[36] Relatively pure xylans have been isolated from maize cob.[134] Other neutral hemicelluloses contain a variety of sugar units, the common ones being galactomannans, glucomannans, and galactoglucomannans. An unusual "hemicellulose" which forms 60% of the endosperm of the palm seed (ivory nut) is an unbranched D-mannose polymer.[65,169] The acidic hemicelluloses show even more variety than the neutral ones and are in general smaller and more highly branched. They contain a high proportion of uronic acids, which usually occur as end groups or single-unit side chains. These are often attached to the D-xylose backbone at C-2 and C-3. Also present are aldobiuronic acids, with covalent links that are extremely resistant to hydrolysis.

Three properties of the hemicelluloses could prove important in human physiology: their water-holding capacity, their digestibility, and their capacity to bind ions. It is widely assumed that the hemicelluloses will take up water, and their intake has been shown to be proportional to the increase in fecal weight in children.[68] However, few experimental data are available for the pure substances because they are difficult to isolate from the cell wall without serious modification of their structure.[169] While the addition of hemicellulose to the diet undoubtedly increases fecal output, this has been attributed to the metabolites (short-chain fatty acids) rather than to water-binding capacity.[173] Hemicellulose is indeed metabolized as it passes through the human gut, to a greater extent, in fact, than cellulose. Values for the amount digested vary widely, but mean values for the percent digested taken from the work of Williams and Olmsted[173] are 56.0% (± 29 SD) of hemicellulose and 38.3% (± 22 SD) of cellulose; and from Southgate and Durnin[139] 87.2% of hemicellulose and 29.4% of cellulose. The factors influencing digestibility can only be surmised at present. In humans there seems to be some relationship with the subject's age,[68,139] older people digesting hemicellulose more completely, while in ruminants the source of the fiber is important,[78] particularly the amount of lignin present. Structural differences in the hemicelluloses seem to make much less difference.[12] Nonruminants (swine and rats) digest hemicellulose to a lesser extent than ruminants,[79] with increased intake of cell-wall material decreasing overall digestibility.

The ion-binding capacity of hemicellulose is less well documented. Weak bile-salt binding capacity for Mucilose® has been demonstrated,[13] although this substance is perhaps more appropriately classified under the gums and mucilages.

C. Pectins

Although present in smaller amounts than other cell-wall substances, nevertheless the pectins are common to all cell walls and are present also in the intercellular layers. They constitute between 1 and 4% of total cell-wall polysaccharides,[169] although they are more abundant in specialized tissues of certain plants. The rind of citrus fruit contains 30% pectin, apple, 15% and onion skins, 11–12%.[2,5] Between 0.5 and 1% of the total fresh weight of an apple is pectin.[76] The terms pectic substances, pectin, pectic acid, and pectinic acid are all used to describe various groups of pectins,[5,176] but these divisions have at present no valuable application in human nutrition. As a group they are biochemically less well defined than the other polysaccharides but in general are smaller, with molecular weights in the region of 60,000–90,000.[134,176] The "parent" molecule is a polymer of 1-4β-D-galacturonic acid. Most are heteropolysaccharides containing mainly D-galacturonic acid, D-galactose, L-arabinose, D-xylose, L-rhamnose, and L-fucose. Pectic substances containing only one type of residue do exist in mustard seed and sunflower heads but are unusual.[122] More commonly, the basic galacturonic acid polymer is modified to include between 10 and 25% of neutral sugar residues as either single-unit or longer side chains, while between 3 and 11% of the uronic acids have methyl substitutions; less commonly there is *o*-acetylation. Apple pectin usually has between 6 and 9% methyl substitutions and citrus fruits, 7–10%; in contrast strawberry pectin has 0.2%. Many fewer residues are acetylated, although an exception to this is beet, whose pectin is 5–10% acetylated.[76,122,134,176] Calcium and magnesium salts of uronic acids form with the primary cell-wall pectins in particular.[5]

Two properties of pectin have received some attention and may be important in nutrition: its ability to form gels and its ion-binding capacity. The gel-forming properties of pectin are well known to anybody who has tried to make jam; these properties depend primarily on the polygalacturonic acid backbone of the polymer rather than on its side-chain substituents. The methyl esters of the uronic acids are thought to be important, since gelling properties are impaired as the methyl esters are removed.[54,122] Nonesterfied uronic acids in the polymer keep the molecules apart and therefore impair gel formation, although this repulsion can be overcome with calcium ions. The sloughing of the outer layer of potatoes has been ascribed to the lower level of calcium and magnesium pectates in the outer parts of these tubers.[161]

The ion-binding properties of pectins are closely related to their free uronic acid content.[16,81] These acidic polysaccharides function as cation exchangers. It has been shown that the cation-exchange property of a particular pectin is uniform throughout all the tissues of the plant and characteristic for that particular species. This property is also independent of the conditions and place in which the plant is grown.[81] The cation-exchange properties of pectin could explain the antitoxic effects of this substance.[93] The extent to which pectins contribute to the cation-exchange capacity of whole cell-wall preparations remains to be assessed.[90]

The fate of ingested pectin in man is largely uninvestigated. *In vitro* pectin is metabolized by bacteria and may have some antibacterial effect.[167] In man it may be completely metabolized in the gut, probably in the colon, with less than 5% being recovered.[166] This almost complete disappearance of pectin in the gut is in keeping with its relatively poor ability to increase fecal weight in man when compared with other, undigested polysaccharides.[29]

One property of pectin that has received much attention is its ability to alter cholesterol metabolism. In a wide range of animal species receiving cholesterol-supplemented diets with pectin as 3–5% of total food intake, the increase in blood cholesterol from the dietary cholesterol supplementation was either impaired or cholesterol levels were reduced.[45,50,55,83,127,132,165] Fecal steroid and lipid excretion is usually[50,85] although not always[83] increased, and total liver cholesterol is reduced or cholesterol synthesis impaired.[50,83,102] An interesting observation is that in many of these experiments the growth rate and gain in body weight of the animals under investigation was impaired by the pectin supplements to the diet,[45,50,55,127] without necessarily a concomitant decrease in food intake.

The effect of pectin on non-cholesterol-supplemented diets is much less dramatic.[85] In swine, pectin supplementation may lead to a rise in blood lipids and increase in back fat.[47] In human studies with pectin, blood cholesterol levels have fallen[49,72,77,152] with pectin supplements of between 15 and 35 g a day; as a means of reducing blood cholesterol levels, pectin was second only to a reduction in dietary intake of cholesterol and fat.[4]

The manner in which pectin affects cholesterol metabolism is of some interest, although no satisfactory hypothesis has yet been established. The increase in fecal lipid excretion[85] is probably not sufficient to explain the entire effect, and it is more likely that pectin alters the rate and route by which cholesterol is absorbed.[52,69,83] As with the gel-forming properties, so the cholesterol-lowering effect has been related to the methyl content of the pectin polymer.[43]

D. Plant Gums, Mucilages, and Storage Polysaccharides

While the plant gums, mucilages, and storage polysaccharides are not strictly cell-wall components, they are related both biochemically and in their properties to the cell-wall constituents and thus may be justifiably included under the general heading of fiber.

Plant gums are sticky exudates formed at the site of injury to plants, which dry to produce hard protective nodules.[5,65,136] They may be obtained commercially by deliberate incision of the particular plant or tree and collection of the fluid that drains. Commercially they are used as emulsifiers, thickeners, and stabilizers by the food industry, while some are included in pharmaceutical preparations or used as bulk-forming laxatives.

Biochemically the plant gums present a complex group of highly branched uronic-acid-containing polymers, mainly of glucuronic and galacturonic acids, with neutral sugars such as xylose, arabinose, and mannose. A high proportion of calcium and magnesium salts are formed and a significant number of the residues are acetylated.[5] Gum arabic or gum acacia is one of the better known plant gums and is obtained from *Acacia senegal*. It is a 1-3 and 1-6-D-galacturonic acid polymer with side branches of L-arabinose and L-rhamnose. Its molecular weight has been estimated at 250,000, and it is water soluble.[5,65,142] Apart from its use in the food industry, it is sold commercially as an adhesive. Better known for their laxative properties are the plant gums obtained from *Sterculia* species, such as karaya gum obtained from an Indian plant. Like gum arabic, it is a galacturonic acid polymer but includes D-galactose and L-rhamnose residues and branches of xylose, fucose, and galactose.[5,142] Gum tragacanth is a similar highly branched acidic polysaccharide which falls into this group of substances and is used in some brands of salad cream.

Mucilages are found in an entirely different part of the plant. They are usually mixed in with the endosperm or storage polysaccharides of plant seeds. Their role is to retain water and so protect the seed against desiccation.[5] Structurally, they resemble the hemicelluloses but are not classed biochemically with them because of their occurrence in a distinct part of the plant.

By contrast with the plant gums, many of the mucilages are neutral polysaccharides, although some acidic ones exist. Of the neutral polysaccharide mucilages guar (guaran or guar gum), which is isolated from the ground endosperm of a leguminous vegetable cultivated in India for animal feeds *(Cyamopsis tetragonolobus)*, has been characterized. It is a 1-4β-D-galactomanan with single-unit side chains consisting of 1-6α-D-galactose. In small amounts it finds widespread use in the food and pharmaceutical industries, being included as a thickener and stabilizer in salad cream, ice cream, soup, tablets, and toothpaste; it is also used as a laxative. Structurally similar neutral polysaccharides are found in alfalfa and carob gum. Mucilages have also been isolated from the endosperm of cereals and are characteristically highly branched, with a backbone of 1-4β-D-xylose with L-arabinose side chains. Slippery elm bark, which contains a methylated D-galacturonic acid polymer, is an example of an acidic mucilage.[5,65] Although very little data exist concerning the exact composition of psyllium seed and ispaghula, it is possible that the active component of these widely used laxative substances are plant mucilages.[15] Both are obtained from seeds of the *Plantago* (plantain) family and contain mixtures of neutral and acidic carbohydrate polymers.[136]

All plants at some stage lay up reserves of food mainly as polysaccharides in either root or stem tubers or in fruits and seeds. Most often the storage polysaccharide is starch as in potato, which is readily digested in the human gut; however, some nonmetabolizable storage polysaccharides exist which, while distinct from plant cell-wall structures and gums and mucilages, may well be expected to have similar nutritional implications in man. An example of this is inulin, a storage polysaccharide found in dahlia tubers and the jerusalem artichoke. It is a β-D-2-1-fructan and is thought to be readily hydrolyzed in the gut.[88]

In the context of human physiology, the most notable characteristic of the mucilages and gums is their effect on cholesterol and bile-salt metabolism. In this respect they are very similar to the pectins in that they produce in animals a fall in blood cholesterol and an increase in fecal cholesterol excretion, and prevent the rise in liver cholesterol on cholesterol-supplemented diets.[44,45,85,127] Guar is particularly notable in this respect, several authors commenting that it is more potent in its cholesterol-lowering properties than pectin. Cholesterol-lowering properties of guar in man have been demonstrated.[46,72] A number of other mucilaginous substances have been reported to affect cholesterol and bile-salt metabolism in both man and animals. These include psyllium seeds and ispaghula.[8,9,45,51]

E. Lignin

Lignin is the odd man out in the plant cell wall. It is highly insoluble, being a major part of the residue left after treatment of the cell wall with 72% sulfuric acid. Unlike other cell-wall structures, it is not a carbohydrate and it is a small polymer having a molecular weight of between 1000 and 4500. The basic units of the polymer are joined by carbon-to-carbon bonds, unlike the glycoside and acetal links of the carbohydrates. Its role in the cell wall is to lend strength and support by permeating the other constituents.[111,129,134] In general the amount of lignin present in the cell wall is less than that of the polysaccharides, although there is enormous variation. Wood may contain up to 40–50% lignin in the cell wall, while wheat cell walls contain 23%, cabbage, 6%, and apples, 25%.[137] Most published values for lignin are probably too high, as isolation of pure lignin from cell wall preparations is not easy, the lignin fraction often containing nitrogen, cutin, waxes, etc.[157]

The word lignin (from Latin *lignum*, wood) has been in use for over 150 years, although serious study of its chemistry did not begin until the middle of the nineteenth century.[131] Most of the work analyzing the biochemical structure of lignin has been done on wood lignin. Three basic units go to make up lignin polymers, each of which is derived in turn from different alcohols. All are substituted phenylpropanes; 4-hydroxyphenylpropane is derived from coumaryl alcohol; guaiacyl propane (4-OH-3-methoxyphenylpropane) is from coniferyl alcohol, while sinapyl alcohol gives rise to the third basic unit 3,5-dimethyl-4-OH-phenylpropane. The links between these basic units are complex and have been extensively studied. Chain termination is usually by a coniferyl alcohol or aldehyde, and it is thought that these are responsible for the phloroglucinol staining properties that are characteristic of lignins.[95,111,134]

A considerable amount of work has been done on the effect that lignin has on the digestibility of cell-wall constituents in the ruminant. The impetus to this work, which has held the attention of animal nutritionists for well over a century, comes from the fact that lignin impairs the digestibility of other cell-wall polysaccharides and thereby reduces the potential energy available from many common forages used in ruminant nutrition.[34,115,121,154,155] Even the detailed structure of individual lignins may affect cell-wall digestibility.[30] In the nonruminant lignin is not thought to be so important a factor in limiting the digestion of these polysaccharides,[78] although Williams and Olmsted[173] suggested that the extent of lignification of the cereal and vegetable substances which they fed to their human volunteers was important in determining the amount metabolized.

Lignin is thought to have bile-salt-binding properties, which *in vitro* can be shown to be pH dependent; the affinity for bile salts is less than for the more polar bile salts, suggesting that adsorption occurs by hydrophobic interactions.[40] This bile-salt-binding property was used in the treatment of a patient with a type of diarrhea following small intestinal resection in which bile salts are thought to play a part.[39] *In vivo* studies in man are few and thus far have failed to demonstrate a clear effect of lignin on bile-salt metabolism or on vitamin A absorption.[7,56]

F. Algal Polysaccharides

The algal polysaccharides in the cell wall are usually considered separately because they are found in algae and seaweeds. Algal cell walls are distinctive in that cellulose is often replaced by xylans and mannans, which form fibrous linear polymers, or by cellulose analogs as in lichenan.[24,118] A further distinctive feature is the occurrence of a number of sulfated forms, of which agar and carrageenan are examples. From the structural standpoint, however, the cell walls of algae contain many polysaccharides that are common to the plant kingdom in general, such as pectins and hemicelluloses. Also derived from algae is alginic acid, a nonsulfated polysaccharide.

Alginic acid[76,114] is worthy of note for several reasons. It is a linear 1-4β-D-manuronic acid polymer with L-guluronic acid substitutions. Because of this largely unbranched structure it, like cellulose, forms microfibrils. It has a molecular weight of up to 200,000, and in its natural form exists as sodium, potassium, calcium, or magnesium salts (alginates), which represent as much as 40% of the dry matter of some seaweeds.

Alginic acid is widely used in the food, pharmaceutical, and paper industries and has also been the focus of some nutritional experimentation. Its role in altering cholesterol metabolism appears to be fairly undramatic,[102,132] but its effects on strontium absorption from the gut are notable. Sodium alginate significantly impairs strontium uptake from the gut and into isolated rat intestine. The way in which it does this is a good example of the importance of the detailed biochemical structure of these polysaccharide compounds in determining their physiological properties. Strontium and to a much lesser extent, calcium, are bound to alginates in direct proportion to the guluronic acid content of the polymer. Alginates with a high guluronic/manuronic acid ratio show a greater ability to prevent the absorption in man of radioactive strontium from milk, while the guluronic acid monomer itself has no such properties.[109,143,147]

Agar[5,114] is a sulfated galactan, the basic units of which are 1-3β-D-galactose alternating with (1-4)3,6-anhydro-L-galactose. D-Glucuronic acid monomers are also present with a significant proportion of half-ester sulfates.

Agar is obtained from seaweed by aqueous extraction and subsequently bleached with hypochlorite, sulfite or activated carbon. Its gel-forming properties and relative resistance to bacterial attack have led to its widespread use in microbiology. These properties have led also to its inclusion in laxative preparations,[15] and it has been shown to have a striking capacity to increase fecal bulk.[173]

Carrageenan is another sulfated galactan, which exists in two forms, k and λ. The basic units are (1-4)3,6-anhydro-D-galactose, many of which are sulfated, and 1-3, 1-4 galactose polymers, which are sulfated or even disulfated at the C-2 and C-3 positions. Considerable work has been done on the gel-forming properties of this substance and on its molecular structure, which is thought to be helical.[114,123] The ability of carrageenan to react with milk protein has led to its widespread use in preparations containing milk and chocolate. It has been shown to be a potent cholesterol-lowering agent in chickens,[45] but unlike almost any other plant polysaccharide, it has an adverse effect on the gut. Ulceration of the caecum of both rats and guinea pigs has been demonstrated on adding carrageenan to the diet.[1,162] Other sulfated polysaccharides and even sulfated non-polysaccharide polymers also seem to have this property.[94,163] It is thought that ulcer formation is due to the release of lysozyme by the mucosal macrophages after taking up carrageenan particles. This may be prevented by giving milk with the carrageenan, possibly because the milk protein and carrageenan form a large complex that cannot be taken up by the cells.[1]

V. Fiber-Associated Substances

It will be readily apparent, after consideration of the development of the plant cell wall and the properties of its individual constituents, that fiber is a complex matrix of substances into which are incorporated a variety of other things. While these might not be considered strictly fiber, nevertheless they are invariably present to some extent in fiber preparations and may be almost impossible to separate from it. The extent to which they are responsible for some of the supposed actions of fiber remains to be established.

A. Phytic Acid

Phytic acid (inositol hexaphosphate) is present throughout the plant kingdom in seeds in association with protein. Mostly it is present as salts complexed with the ions of calcium, magnesium, and potassium and is known as phytin. Its role is probably to act as a storage vehicle for these ions, because during germination phytase cleaves the phosphate residues from the inositol, thus freeing these ions for participation in a wide variety of synthetic reactions.[53,75] Since cereal grains are in fact seeds, most of the phytate we take in our diet comes from this source. In wheat phytic acid is present in the aleurone layer of grain, where up to 90% of the total phosphate is present as phytate.[91] Because the aleurone layer forms part of the bran removed during the milling process, phytate is therefore an important constituent of bran and wholemeal flour but not white flour. One hundred percent extraction flour contains eight times more phytate than does 72% extraction.[75]

Phytic acid has been a subject of considerable interest in nutritional circles for many years because of its capacity to bind calcium and magnesium in the diet. Its presence in wholemeal flour and bread has been related to the negative calcium balance that develops in subjects living largely on wholemeal bread.[89] This aspect has, perhaps, more than anything cast a shadow on the value of wholemeal bread in our diet, for in circumstances where calcium intake is only marginal, it is thought to lead to the development of osteomalacia[11] and rickets. For this reason, during the Second World War high-extraction flours were supplemented with calcium. Certainly adding pure phytic acid to the diet will induce negative calcium balance,[89,125] a property which has been used to good effect in the treatment of idiopathic hypercalcuria.[60] Phytate is also thought to impair iron[14,71,106] and zinc absorption.[32,125] Phytic acid is destroyed by heat and by the action of phytase during the rising and proving of dough,[100,112] although this may not necessarily impair the ion-binding capacity of these substances.[57,106] In view of the ion-binding properties of other cell-wall constituents, it may well be worth looking at the phytic acid story again.

B. Silica

The ash content of the plant cell wall, particularly of wheat, may be as high as 10%. Of this, the principal element present is often silicon. Plant silicon is sometimes referred to as opal, but in fact it is more commonly present as silica gel. It is concentrated usually in the aerial part of the plant (stem, leaf, flower), being deposited in the cell wall along with cellulose, the amount present varying according to the species, the soil content of silica, and the maturity of the plant. It is an important element in plant growth, particularly for species such as rice, and silica deficiency has been reported.[83,134,154] The significance of silica in nutritional terms relates to its capacity to impair the digestibility of cell-wall materials. In this context it is thought to be on a par with lignin in impairing digestibility.[121,154,156] A less important problem nutritionally, but perhaps of interest, is the trouble that the high silica content of some tropical species of wood poses when these timbers are sawed. The silica content of the cell wall of some trees in the tropics may be as high as 10%, making them difficult to cut and leading to the production of sparks during sawing.[154]

C. Cuticular Substances

Plants contain lipids in three major forms: first, as triglycerides, which are usually found in the seeds or fleshy parts of plants and which act as food storage reserves; second, as phospholipids and glycolipids, which are present in the cell membranes of the plant cell organelles such as mitochondria and chloroplasts, where they serve in a structural and metabolic role; and third, as lipids that are present in the cuticular substances which cover the surface of fruits, seeds, and leaves and form a waterproof coating.[63,99]

While the cuticular substances form a small proportion of the total plant lipids, they are very important to the plant, providing an extremely hydrophobic layer on its outer surface; as such they are associated with the plant cell wall.[24,63,134] Cuticular substances may be divided into two broad groups: waxes and cutins. Waxes, which may be extracted by simple solvents such as benzene, include a mixture of paraffins, aliphatic acids, and alcohols. The paraffins are unbranched, with chain lengths in the region of C-27 to C-31. The aliphatic acids, also of the long-chain variety (C-16 to C-34 mainly), are usually either saturated or oleic and linoleic acids. They are frequently present as monoesters. The bloom that is seen on many fruits and flowers has been attributed to the crystalline pattern produced by surface waxes. The wax content of apples increases with storage in some species and may show remarkable changes in composition. The linoleic acid content of the wax of Granny Smith apples shows an 11-fold increase during storage.[103]

The nonextractable component of the cuticular substances, i.e., that which needs saponification before extraction, is a group of complex polymers of which cutin is the most important. Cutin forms the ground substance of the cuticle and is a complex polymer of mono-, di-, tri-, and polyhydroxy fatty acids. These form complex molecules with esters forming between the hydroxyl and carboxyl groups. Two variations of the cuticular substances are found in specific places. Suberin is a cutin-like substance found in cork, and sporopellenin is a mixture of cuticular substances which form the highly resistant wall of pollen grains.[118] Cuticular substances are extremely resistant to digestion and in turn are thought to impair the digestibility of the other cell-wall constituents in a manner similar to lignin.[157] While they represent only a very small proportion of the fat content of the diet, their resistance to digestion means that they appear in the feces, where they may constitute a much larger proportion of the fecal fat and may account for a substantial part of the increase in fecal fat that is seen in subjects taking a high-cereal-fiber diet.[170]

D. Protein

All plant cell walls contain protein. The proliferating cell wall may contain up to 25% of the total cell nitrogen, although this proportion falls considerably in the mature plant. Proteins are incorporated in the cell wall, probably in a structural capacity, and contain a high proportion of the unusual amino acid, hydroxyproline. The cell walls of algae contain hydroxylysine. These proteins are probably present as glycoproteins and, because of their close association with the cell-wall structures, are extremely resistant to digestion.[24,118,134]

E. Other Substances

The plant cell wall contains an endless list of other substances whose role in human nutrition is not clear. These include the tannins,[24] a group of polyhydroxyphenolic compounds which are present in leaves and account for the bitter taste of tea and beer. They may have a protective role in the plant. Callose[24,28] is a cellulose-like polymer laid down during normal plant development in the sieve plates of the phloem. It is also formed when a plant is injured and may be subsequently reabsorbed. The cell wall also contains a variety of enzymes; vitamins and other aromatic constituents[96-98]; nonmetabolizable sugars such as raffinose, which may be important in gas production in the human gut,[62] and glycosides such as saponin.[22]

VI. Conclusion

It is clear that the concept of fiber has changed over the years and will continue to develop. From a simple beginning in which fiber was seen as the indigestible part of animal feeds and consisting mainly of cellulose, we are now faced with the task of investigating the physiological significance of a complex matrix of polymers and associated substances that form the plant cell wall. Other plant materials like the gums, mucilages, and some storage polysaccharides, because of their structural similarity with the cell-wall substances, should also be included under this heading. Fiber is not a "pure" substance, and therefore experimental work to define its effects is not straightforward. Any attempt to work with isolated or purified fractions of the cell wall must be interpreted in this light. It does not necessarily follow that because an isolated fiber constituent has a particular property, such as cholesterol-lowering by pectin, this will be retained when present in the whole cell wall. Nevertheless, the complexity of fiber makes it imperative to seek out close associations between its structure and its function in man, and to this end its components should be individually examined.

Much of our present knowledge of fiber comes from studies of wood, grasses, and animal feeds and from ruminant nutrition. What is badly needed is an assessment of the role in human physiology of fiber from the commonly eaten vegetables, fruits, and cereals. The "whole" substance as it occurs in the diet might be found important in disease prevention, while individual components or fractions of fiber could prove useful therapeutically in a variety of circumstances. Whatever conclusions are ultimately drawn about the role of fiber in human nutrition, its study will inevitably focus our attention on new aspects of human physiology, disease, and way of life.

ACKNOWLEDGMENTS

My thanks are due to Dr. Hugh Wiggins, Dr. David Southgate, and Dr. E. N. Rowlands for critical advice and help during the preparation of this chapter. I wish also to express my appreciation to Dr. Philip James, Celia Greenberg and Will Branch of the Dunn Nutrition Laboratory fiber group for many stimulating discussions about fiber.

References

1. ABRAHAM, R., R. J. FABIAN, L. GOLBERG and F. COULSTON. Role of lysozymes in carrageenan-induced cecal ulceration. *Gastroenterology* **67:**1169, 1974.
2. ALEXANDER, M. M., and G. A. SULEBELE. Pectic substances in onion and garlic skins. *J. Sci. Food Agric.* **24:**611, 1973.
3. ANALYTICAL METHODS COMMITTEE. Determination of the crude fibre in national flour. *Analyst* **68:**276, 1943.
4. ANDERSON, J. T., F. GRANDE and A. KEYS. Cholesterol-lowering diets. *J. Am. Diet. Assoc.* **62:**133, 1973.
5. ASPINALL, G. O. Pectins, plant gums and other plant polysaccharides. *In: The Carbohydrates,* W. Pigman and D. Horton (eds.), New York, Academic Press, 1970, Vol. IIb, Chap. 39, p. 515.
6. ASPINALL, G. O., and R. J. FERRIER. The constitution of barley husk hemicellulose. *J. Chem. Soc.* **840:**4188, 1957.
7. BARNARD, D. L., and K. W. HEATON. Bile acids and vitamin A absorption in man: The effects of two bile acid-binding agents, cholestyramine and lignin. *Gut* **14:**316, 1973.
8. BEHER, W. T., and K. K. CASAZZA. Effects of psyllium hydrocolloid on bile acid metabolism in normal and hypophysectomised rats. *Proc. Soc. Exp. Biol. Med.* **136:**253, 1971.
9. BEHER, W. J., B. M. SCHUMAN, M. A. BLOCK and K. K. CASAZZA. The effect of psyllium hydrocolloid and cholestyramine (Cuemid) on hepatic bile lipid ratios in man. *Gastroenterology* **60:**191(A), 1971.
10. BERG, B. B., B. VAN HOFSTEN and G. PETTERSSON. Electron-microscopic observations on the degradation of cellulose fibres by *Cellvibro fulvus* and *Sporocytophaga myxococcoides. J. Appl. Bacteriol.* **35:**215, 1972.

11. BERLYNE, G. M., J. B. ARI, E. NORD and R. SHAINKIN. Bedouin osteomalacia due to calcium deprivation caused by high phytic acid content of unleavened bread. *Am. J. Clin. Nutr.* **26:**910, 1973.

12. BEVERIDGE, R. J., and G. N. RICHARDS. Digestion of polysaccharide constituents of tropical pasture herbage in the bovine rumen. Part IV. The hydrolysis of hemicellulose from spear gram by cell-free enzyme systems from rumen fluid. *Carbohydr. Res.* **29:**79, 1973.

13. BIRKNER, H. J., and F. KERN. *In vitro* adsorption of bile salts to food residues, salicylazosulfapyridine and hemicellulose. *Gastroenterology* **67:**237, 1974.

14. BJORN-RASMUSSEN, E. Iron absorption from wheat bread. Influence of various amounts of bran. *Nutr. Metab.* **16:**101, 1974.

15. BLACOW, N. W., (ed.). *Martindale. Extra Pharmacopoeia.* London, The Pharmaceutical Press, 1972.

16. BRANCH, W. J., D. A. T. SOUTHGATE and W. P. T. JAMES. Binding of calcium by dietary fibre: Its relationship to unsubstituted uronic acids. *Proc. Nutr. Soc.* **34:**120 A, 1975.

17. BURKITT, D. P. Varicose veins, deep vein thrombosis and haemorrhoids: Epidemiology and suggested aetiology. *Br. Med. J.* **2:**556, 1972.

18. BURKITT, D. P. Cancer and other noninfective diseases of the bowel. Epidemiology and possible causative factors.*Rendiconti* **5:**33, 1973.

19. BURKITT, D. P., and P. A. JAMES. Low-residue diets and hiatus hernia. *Lancet* **2:**128, 1973.

20. BURKITT, D. P., A. R. P. WALKER and N. S. PAINTER. Effect of dietary fibre on stools and transit times, and its role in the causation of disease. *Lancet* **2:**1408, 1972.

21. CHEADLE, W. B. A clinical lecture on the pathology and treatment of chronic constipation in childhood, and its sequel, atony and dilation of the colon. *Lancet* **2:**1063, 1886.

22. CHEEKE, P. R. Alfalfa: A natural hypocholesterolemic agent. *Am. J. Clin. Nutr.* **26:**133, 1973.

23. CLEAVE, T. L. Natural bran in the treatment of constipation. *Br. Med. J.* **1:**461, 1941 (letter).

24. CLOWES, F. A. L., and B. E. JUNIPER. *Plant Cells.* Oxford, Blackwell, 1968.

25. COGAN, T. *The Haven of Health.* London, Thomas Creede, 1588.

26. COWLING, E. B., and W. BROWN. *Celluloses and their Applications. Advances in Chemistry Series.* No. 95, D. J. Hajny and E. T. Reese (eds.). Washington, D.C., American Chemical Society, 1969, p. 152.

27. CRAMPTON, E. W., and L. A. MAYNARD. The relation of cellulose and lignin content to the nutritive value of animal feeds. *J. Nutr.* **15:**383, 1938.

28. CRONSHAW, J. Phloem differentiation and development. *In: Dynamic Aspects of Plant Ultrastructure,* A. W. Robards (ed.). London, McGraw-Hill, 1974, p. 391.

29. CUMMINGS, J. H., D. J. A. JENKINS and H. S. WIGGINS. Personal communication.

30. CYMBALUK, N. F., A. J. GORDON and T. S. NEUDOERFFER. The effect of the chemical composition of maize plant lignin on the digestibility of maize stalk in the rumen of cattle. *Br. J. Nutr.* **29:**1, 1973.

31. DANISHEFSKY, I., R. L. WHISTLER and F. A. BETTELHEIM. Introduction to polysaccharide chemistry. *In: The Carbohydrates,* W. Pigman and D. Horton (eds.). New York, Academic Press, 1970, Vol. IIA, p. 375.

32. DAVIES, N. T., and R. NIGHTINGALE. Effect of phytate on zinc absorption and faecal zinc excretion and carcass retention of zinc, iron, copper and manganese. *Proc. Nutr. Soc.* **33:**8A, 1974.

33. DEINUM, B., A. J. H. VAN ES and P. J. VAN SOEST. Climate, nitrogen and grass. II. The influence of light intensity, temperature and nitrogen on *in vivo* digestibility of grass and the prediction of these effects from some chemical procedures. *Neth. J. Agric. Sci.* **16:**217, 1968.

34. DEKKER, R. F. H., and RICHARDS, G. N. Effect of delignification on the *in vitro* rumen digestion of polysaccharides of bagasse. *J. Sci. Food Agric.* **24:**375, 1973.

35. DIMOCK, E. M. The treatment of habitual constipation by the bran method. M. D. thesis. Cambridge, 1936.

36. DONNELLY, B. S., J. L. HELM and H. A. LEE. The carbohydrate composition of corncob hemicelluloses. *Cereal Chem.* **50:**548, 1973.
37. *Drug and Therapeutics Bulletin.* Bulk "laxatives" in medicine and surgery. **11:**77, 1973.
38. EASTWOOD, M. A., N. FISHER, C. T. GREENWOOD and J. B. HUTCHINSON. Perspectives on the bran hypothesis. *Lancet* **1:**1029, 1974.
39. EASTWOOD, M. A., and R. H. GIRDWOOD. Lignin. A bile-salt sequestering agent. *Lancet* **2:**1170, 1968.
40. EASTWOOD, M. A., and D. HAMILTON. Studies on the adsorption of bile-salts to nonabsorbed components of diet. *Biochem. Biophys. Acta* **152:**165, 1968.
41. EASTWOOD, M. A., J. R. KIRKPATRICK, W. D. MITCHELL, A. BONE and T. HAMILTON. Effects of dietary supplements of wheat bran and cellulose on faeces and bowel function. *Br. Med. J.* **4:**392, 1973.
42. ELYOT, T. The castel of helth, corrected and in some places augmented. *The Seconde Boke, Cap. 11 of Breade.* London, Th. Berthelet, 1541, p. 28.
43. ERSHOFF, B. H., and A. F. WELLS. Effects of methoxyl content on the anticholesterol activity of pectic substances in the rat. *Exp. Med. Surg.* **20:**272, 1962.
44. ERSHOFF, B. H., and A. F. WELLS. Effects of gum guar, locust bean gum and carrageenan on liver cholesterol of cholesterol-fed rats. *Proc. Soc. Exp. Biol. Med.* **110:**580, 1962.
45. FAHRENBACH, M. J., B. A. RICCARDI and W. C. GRANT. Hypocholesterolemic activity of mucilaginous polysaccharides in white leghorn cockerels. *Proc. Soc. Exp. Biol. Med.* **123:**321, 1966.
46. FAHRENBACH, M. J., B. A. RICCARDI, J. C. SAUNDERS, I. N. LOURIE and J. G. HEIDER. Comparative effects of guar gum and pectin on human serum cholesterol levels. *Circulation* **31/32** (Suppl. 2):1141, 1965.
47. FAUSCH, H. D., and T. A. ANDERSON. Influence of citrus pectin feeding on lipid metabolism and body composition of swine. *J. Nutr.* **85:**145, 1965.
48. *Fertilisers and Feedingstuffs Regulations.* London, Her Majesty's Stationery Office S. R. and O. No. 658, 1932.
49. FISHER, H., D. GRIMINGER, E. R. SOSTMAN and M. K. BRUSH. Dietary pectin and blood cholesterol. *J. Nutr.* **86:**113, 1965.
50. FISHER, H., W. G. SILLER and P. J. GRIMINGER. The retardation by pectin of cholesterol-induced atherosclerosis in the fowl. *J. Atheroscler. Res.* **6:**292, 1966.
51. FORMAN, D. T., J. E. GARVIN, J. E. FORESTNER and R. B. TAYLOR. Increased excretion of fecal bile acids by an oral hydrophilic colloid. *Proc. Soc. Exp. Biol. Med.* **127:**1060, 1968.
52. GOFF, D. V., L. M. CASTELL, J. I. JENSEN, E. R. LEEDS, C. NEWTON and D. J. A. JENKINS. The effects of dietary fibre on fat absorption in man. *5th Int. Symp. Drugs Affecting Lipid Metab.*, Milan, Italy, 1974, p. 55.
53. GOODWIN, T. W., and E. I. MERCER. *Introduction to Plant Biochemistry.* Oxford, Pergamon Press, 1972.
54. GOULD, S. E. B., D. A. REES, N. G. RICHARDSON and I. W. STEELE. Pectic polysaccharides in the growth of plant cells: Molecular structural factors and their role in the germination of white mustard. *Nature* **208:**876, 1965.
55. GRIMINGER, P., and H. FISHER. Anti-hypercholesterolemic action of scleroglucan and pectin in chickens. *Proc. Soc. Exp. Biol. Med.* **122:**551, 1966.
56. HEATON, K. W., S. T. HEATON and R. E. BARRY. An *in-vivo* comparison of two bile salt binding agents, cholestyramine and lignin. *Scand. J. Gastroenterol.* **6:**281, 1971.
57. HEGSTED, D. M., C. A. FINCH and T. D. KINNEY. The influence of diet on iron absorption. II. The interrelation of iron and phosphorus. *J. Exp. Med.* **90:**147, 1949.
58. HELLENDOORN, E. W. Physiological importance of indigestible carbohydrates in human nutrition. *Voeding* **34:**619, 1973.
59. HENNENBURG, W., and F. STOHMANN. Beitr. zur Begründung einer rationellen Fulterung de Widerkäuer I and II. [as cited by Mangold (Ref. 92)].

60. HENNEMAN, P. M., P. H. BENEDICT, A. P. FORBES and H. R. DUDLEY. Idiopathic hypercalcuria. *N. Engl. J. Med.* **259:**802, 1958.
61. HESTRIN, S., and M. SCHRAMM. Synthesis of cellulose by acetobacter xylinum. 2. Preparation of freeze-dried cells capable of polymerising glucose to cellulose. *Biochem. J.* **58:**345, 1954.
62. HICKEY, C. A., E. L. MURPHY and D. H. CALLOWAY. Intestinal gas-production following ingestion of commercial wheat cereals and milling fractions. *Cereal Chem.* **49:**276, 1972.
63. HITCHCOCK, C., and B. W. NICHOLS. *Plant Lipid Biochemistry.* London, Academic Press, 1971.
64. HOPPERT, C. A., and A. J. CLARK. Digestibility and effect on laxation of crude fibre and cellulose in certain common foods. *J. Am. Diet. Assoc.* **21:**157, 1945.
65. HORTON, D., and M. L. WOLFRAM. Polysaccharides. *In: Comprehensive Biochemistry,* M. Florkin and E. H. Stotz (eds.). London, Elsevier, 1963, Vol. 5, Carbohydrates, p. 185.
66. HORWITZ, W. Crude fibre. *In: Official Methods of Analysis of the Association of Official Analytical Chemists, U. S. A.,* 11th ed. Washington, D.C., Association of Official Analytical Chemists, 1970, p. 129.
67. HOSOI, K., W. C. ALVAREZ and F. C. MAN. Intestinal absorption. A search for a low residue diet. *Arch. Intern. Med.* **41:**112, 1928.
68. HUMMEL, F. C., M. L. SHEPHERD and I. G. MACEY. Disappearance of cellulose and hemicellulose from the digestive tracts of children. *J. Nutr.* **25:**59, 1943.
69. HYUN, S. A., G. V. VAHOUNEY and C. R. TTREADWELL. Effect of hypocholesterolaemic agents on intestinal cholesterol absorption. *Proc. Soc. Exp. Biol. Med.* **112:**496, 1963.
70. IRESON, J. D., and G. B. LESLIE. An *in-vitro* investigation of colloidal bulk-forming laxatives. *Pharm. J.* **205:**540, 1970.
71. JENKINS, D. J. A., M. J. HILL and J. H. CUMMINGS. Effect of wheat fibre on blood lipids, fecal steroid excretion and serum iron. *Am. J. Clin. Nutr.* **28:**1408, 1976.
72. JENKINS, D. J. A., C. NEWTON, A. R. LEEDS, and J. H. CUMMINGS. Effect of pectin, guar gum and wheat fibre on serum cholesterol. *Lancet* **1:**1116, 1975.
73. JERMYN, M. A., and F. A. ISHERWOOD. Changes in the cell wall of the pear during ripening. *Biochem. J.* **64:**123, 1956.
74. JONES, F. A., and E. W. GODDING. *Management of Conspipation.* London, Blackwell, 1972.
75. KENT-JONES, D. W., and A. J. AMOS. *Modern Cereal Chemistry.* 6th ed. London, Food Trade Press, 1967.
76. KERTESZ, Z. I. Polyuronides. *In: Comprehensive Biochemistry,* M. Florkin and E. Stotz (eds.). London, Elsevier, 1963, Vol. 5, Carbohydrates, p. 233.
77. KEYS, A., F. GRANDE and J. T. ANDERSON. Fiber and pectin in the diet and serum cholesterol concentration in man. *Proc. Soc. Exp. Biol. Med.* **106:**555, 1961.
78. KEYS, J. E., P. J. VAN SOEST and E. P. YOUNG. Comparative study of the digestibility of forage cellulose and hemicellulose in ruminants and nonruminants. *J. Anim. Sci.* **29:**11, 1969.
79. KEYS, J. E., P. J. VAN SOEST and E. P. YOUNG. Effect of increasing dietary cell wall content on the digestibility of hemicellulose and cellulose in swine and rats. *J. Anim. Sci.* **31:**1172, 1970.
80. KIRWAN, W. O., A. N. SMITH, A. A. McCONNELL, W. D. MITCHELL and M. A. EASTWOOD. Action of different bran preparations on colonic function. *Br. Med. J.* **4:**187, 1974.
81. KNIGHT, A. H., W. M. CROOKE and H. SHEPHERD. Chemical composition of pollen with particular reference to cation exchange capacity and uronic acid content. *J. Sci. Food Agric.* **23:**263, 1972.
82. KRAMER, P. L. The meaning of high and low residue diets. *Gastroenterology* **47:**649, 1964.
83. LEVEILLE, G. A., and H. E. SAUBERLICH. Mechanism of the cholesterol-depressing effect of pectin in the cholesterol-fed rat. *J. Nutr.* **88:**209, 1966.
84. LEWIN, J., and B. E. F. REIMANN. Silicon and plant growth. *In: Annual Review of Plant Physiology,* L. Machlis (ed.). Palo Alto, California: Annual Reviews Inc., 1969, Vol. 20, p. 289.

85. LIN, T. M., K. S. KIM, E. KARVINEN and A. C. IVY. Effect of dietary pectin, "proto-pectin" and gum arabic on cholesterol excretion in rats. *Am. J. Physiol.* **188**:66, 1967.
86. LUBBE, A. M., and C. M. MAREE. Dietary survey in the Mount Ayliff district: A preliminary report. *S. Afr. Med. J.* **47**:304, 1973.
87. LUTHER, L., R. SANTINI, C. BREWSTER, E. PEREZ-SANTIAGO and C. E. BUTTERWORTH. Folate binding by insoluble components of American and Puerto Rican diets. *Ala. J. Med. Sci.* **3**:389, 1965.
88. McCANCE, R. E., and R. D. LAWRENCE. The carbohydrate content of foods. Med. Res. Counc. Spec. Rep. Ser. **135**, 1929.
89. McCANCE, R. E., and E. M. WIDDOWSON. Mineral metabolism of healthy adults on white and brown bread dietaries. *J. Physiol.* **101**:44, 1942–1943.
90. MacCONNELL, A. A., M. A. EASTWOOD and W. D. MITCHELL. Physical characteristics of vegetable foodstuffs that could influence bowel function. *J. Sci. Food Agric.* **25**:1457, 1974.
91. MacMASTERS, M. M., J. J. C. HINTON and D. BRADBURY. Microscopic structure and composition of the wheat kernel. *In: Wheat Chemistry and Technology,* Y. Pomeranz (ed.). St. Paul, Minnesota, American Association of Cereal Chemists, 1971, p. 51.
92. MANGOLD, D. E. The digestion and utilisation of crude fibre. *Nutr. Abstr. Rev.* **3**:647, 1934.
93. MANVILLE, I. A., E. M. BRADWAY and A. S. McMINIS. Pectin as a detoxication mechanism. *Am. J. Dig. Dis. Nutr.* **3**:570, 1936.
94. MARCUS, R., and J. WATT. Ulcerative disease of the colon in laboratory animals induced by pepsin inhibitors. *Gastroenterology* **67**:473, 1974.
95. MARTON, J. Lignin structure and reactions. *Advances in Chemistry Series 59.* Washington, American Chemical Society, 1966.
96. MASON, J. B., N. GIBSON and E. KODICEK. The chemical nature of the bound nicotinic acid of wheat bran: Studies of nicotinic acid-containing macromolecules. *Br. J. Nutr.* **30**:297, 1973.
97. MASON, J. B., and E. KODICEK. The chemical nature of the bound nicotinic acid of wheat bran: Studies of partial hydrolysis products. *Cereal Chem.* **50**:637, 1973.
98. MASON, J. B., and E. KODICEK. The identification of *o*-aminophenol and *o*-aminophenyl glucose of wheat bran. *Cereal Chem.* **50**:646, 1973.
99. MECHAM, D. K. Lipids. *In: Wheat Chemistry and Technology,* Y. Pomeranz (ed.). St. Paul, Minnesota, American Association of Cereal Chemists, 1971, p. 343.
100. MELLANBY, E. Phytic acid and phytase in cereals. *Nature,* **154**:394, 1944.
101. MILTON-THOMPSON, G. J., and B. LEWIS. The breakdown of dietary cellulose in man. *Gut* **12**:853, 1971.
102. MOKADY, S. Effect of dietary pectin and algin on the biosynthesis of hepatic lipids in growing rats. *Nutr. Metab.* **16**:203, 1974.
103. MORICE, I. M., and F. B. SHORLAND. Composition of the surface waxes of apple fruits and changes during storage. *J. Sci. Food Agric.* **24**:1331, 1973.
104. NORTHCOTE, D. H. Differentiation in higher plants. *Oxford Biology Readers No. 44.* London, Oxford University Press, 1974.
105. NORTHCOTE, D. H., and J. D. PICKETT-HEAPS. A function of the Golgi apparatus in polysaccharide synthesis and transport in the root-cap cells of wheat. *Biochem. J.* **98**:159, 1966.
106. *Nutrition Reviews.* Effect of phytate on iron absorption. **25**:218, 1967.
107. PAINTER, N. S. Irritable or irritated bowel. *Br. Med. J.* **2**:46, 1972.
108. PAINTER, N. S., A. Z. ALMEIDA and K. W. COLEBOURNE. Unprocessed bran in the treatment of diverticular disease of the colon. *Br. Med. J.* **2**:137, 1972.
109. PATRICK, G. Inhibition of strontium and calcium uptake by rat duodenal slices: Comparison of polyuronides and related substances. *Nature* **216**:815, 1967.
110. PAYEN, A. Mémoire sur la composition du tissu propre des plantes et du ligneux. *Compt. Rend.* **7**:1052, 1838.
111. PEARL, I. A. *The Chemistry of Lignin.* London, Arnold, 1967.

112. PEERS, F. G. The phytase of wheat. *Biochem J.* **53:**102, 1953.
113. PERCIVAL, E. Aldonic, uronic, oxoaldonic and ascorbic acids. *In: Comprehensive Biochemistry,* M. Florkin and E. H. Stotz (eds.). London, Elsevier, 1963, Vol. 5, Carbohydrates, p. 67.
114. PERCIVAL, E. Algal polysaccharides. *In: The Carbohydrates,* W. Pigman and D. Horton (eds.). New York, Academic Press, 1970, Vol. IIb, Chap. 40, p. 537.
115. PIGDEN, W. J., and D. P. HEANEY. Lignocellulose in ruminant nutrition. *In: Cellulases and their Applications, Advances in Chemistry Series 95,* G. J. Hajny and E. T. Reese (eds.). Washington, American Chemical Society, 1969, p. 245.
116. PORTMAN, O. W., and D. MURPHY. Excretion of bile acids and β-hydroxysterols by rats. *Arch. Biochem. Biophys.* **76:**367, 1958.
117. PRATHER, E. S. Effect of cellulose on serum lipids in young women. *J. Am. Diet. Assoc.* **45:**230, 1964.
118. PRESTON, R. D. Plant cell walls. *In: Dynamic Aspects of Plant Ultrastructure,* A. W. Robards (ed.). Maidenhead, England, McGraw-Hill, 1974, Chapt. 7, p. 256.
119. RÅNBY, B. Recent progress on the structure and morphology of cellulose. *In: Cellulases and their Applications, Advances in Chemistry Series 95,* G. J. Hajny and E. T. Reese (eds.). Washington, American Chemical Society, 1969, p. 139.
120. RANHOTRA, G. S. Effect of cellulose and wheat mill-fractions on plasma and liver cholesterol levels in cholesterol-fed rats. *Cereal Chem.* **50:**358, 1973.
121. RAYMOND, W. F. The nutritive value of forage crops. *In: Advances in Agronomy,* N. C. Brady (ed.). New York, Academic Press, 1969, Vol. 21, p. 1.
122. REES, D. A. *The Shapes of Molecules: Carbohydrate Polymers.* Contemporary Science Paperbacks. Edinburgh, Oliver and Boyd, 1967.
123. REES, D. A. Shapely polysaccharides. *Biochem. J.* **126:**257, 1972.
124. REILLY, R. W., and KIRSNER, J. (ed.). *Fiber Deficiency and Colonic Disorders.* New York, Plenum, 1975.
125. REINHOLD, J. G., K. NASR, A. LAHIMGARZADEH and H. HEDAYATI. Effect of purified phytate and phytate-rich bread upon metabolism of zinc, calcium, phosphorous and nitrogen in man. *Lancet* **1:**283, 1973.
126. REMY, E. Experimentelle studien zur biochemie und biologie der ronfaser. *Biochem. Z.* **236:**1, 1931.
127. RICCARDI, B. A., and M. J. FAHRENBACH. Effect of guar gum and pectin N. F. on serum and liver lipids of cholesterol-fed rats. *Proc. Soc. Exp. Biol. Med.* **124:**749, 1967.
128. ROBERTSON, J. Change in the fibre content of the British diet. *Nature* **238:**290, 1972.
129. SCHUBERT, W. J. *Lignin Biochemistry.* New York, Academic Press, 1965.
130 SCHULZE, E. Zur Kenntnis der chemischen Zusammensetzung der pflanzlschen zellmembranen. *Ber. Dtsch. Chem. Ges.* **24:**2277, 1891.
131. SCHULZE, F. Beitrag zur kenntniss des Lignins. *Chem. Cent.* **21:**321, 1857.
132. SHUHACHI, K., Y. OKAZAKI and A. YOSHIDA. Hypocholesterolaemic effect of polysaccharides and polysaccharide-like food stuffs in cholesterol-fed rats. *J. Nutr.* **97:**382, 1969.
133. SHURPALEKAR, K. S., T. R. DORAISWAMY, O. E. SUNDARAVALLI and M. NARAYANA RAO. Effect of inclusion of cellulose in an "atherogenic" diet on the blood lipids of children. *Nature* **232:**554, 1971.
134. SIEGAL, S. M. Biochemistry of the plant cell wall. *In: Comprehensive Biochemistry,* M. Florkin and E. H. Stotz (eds.). New York, Elsevier, 1968, Vol. 26A, p. 1.
135. SMITH, A. N., W. O. KIRWAN and S. SHARIFF. Motility effects of operations performed for diverticular disease. *Proc. R. Soc. Med.* **67:**1041, 1974.
136. SMITH, F., and R. MONTGOMERY. *The Chemistry of Plant Gums and Mucilages and some Related Polysaccharides.* New York, Rheinhold, 1959.
137. SOUTHGATE, D. A. T. Determination of carbohydrate in food. II. Unavailable carbohydrate. *J. Sci. Food Agric.* **20:**331, 1969.

138. SOUTHGATE, D. A. T. Fibre and the other unavailable carbohydrates and their effects on the energy value of the diet. *Proc. Nutr. Soc.* **32**:131, 1973.

139. SOUTHGATE, D. A. T., and J. V. G. A. DURNIN. Calorie conversion factors. An experimental reassessment of the factors used in the calculation of the energy value of human diets. *Br. J. Nutr.* **24**:517, 1970.

140. SPILLER, G. A., and R. J. AMEN. Plant fibers in nutrition: Need for better nomenclature. *Am. J. Clin. Nutr.* **28**:675, 1975.

141. STANLEY, M. M., D. PAUL, D. GACKE and J. MURPHY. Effects of cholestyramine, Metamucil®, and cellulose on fecal bile salt excretion in man. *Gastroenterology* **65**:889, 1973.

142. STECHER, P. G. (ed.). *The Merck Index. An Encyclopedia of Chemicals and Drugs.* Rahway, New Jersey, Merck and Co. Inc., 1968.

143. SUTTON, A. Reduction of strontium absorption in man by the addition of alginate to the diet. *Nature* **216**:1005, 1967.

144. THIFFAULT, C., M. BÉLANGER and M. POULIOT. Traitement de l'hyperlipoprotéinèmie essentielle de type II par un nouvel agent thérapeutique la celluline. *Can. Med. Assoc. J.* **103**:165, 1970.

145. THORNBER, J. P., and D. H. NORTHCOTE. Changes in the chemical composition of a cambial cell during its differentiation into xylem and phloem tissue in trees. I. Main components. *Biochem J.* **81**:449, 1961.

146. THORNBER, J. P., and D. H. NORTHCOTE. Changes in the chemical composition of a cambial cell during its differentiation into xylem and phloem tissue in trees. II. Carbohydrate constituents of each component. *Biochem J.* **81**:455, 1961.

147. TRIFFITT, J. T. Binding of calcium and strontium by alginates. *Nature* **217**:457, 1968.

148. TROWELL, H. C. *Noninfective Disease in Africa.* London, Arnold, 1960.

149. TROWELL, H. C. Ischemic heart disease and dietary fibre. *Am. J. Clin. Nutr.* **25**:926, 1972.

150. TROWELL, H. C. Crude fibre, dietary fibre and atherosclerosis. *Atherosclerosis* **16**:138, 1972.

151. TROWELL, H. C. Definitions of fibre. *Lancet* **1**:503, 1974.

152. TRUSWELL, A. S., and R. M. KAY. Absence of effect of bran on blood lipids. *Lancet* **1**:922, 1975.

153. VAN SOEST, P. J. Nonnutritive residue. A system of analysis for the replacement of crude fibre. *J. Assoc. Off. Agric. Chem.* **49**:546, 1966.

154. VAN SOEST, P. J. Composition, maturity and the nutritive value of forages. *In: Cellulases and their Applications, Advances in Chemistry Series 95.* G. J. Hajny and E. T. Reese (eds.). Washington, American Chemical Society, 1969, p. 262.

155. VAN SOEST, P. J. The uniformity and nutritive availability of cellulose. *Fed. Proc.* **32**:1804, 1973.

156. VAN SOEST, P. J., and L. H. P. JONES. Effect of silica in forages upon digestibility. *J. Dairy Sci.* **51**:1644, 1968.

157. VAN SOEST, P. J., and McQUEEN R. W. The chemistry and estimation of fibre. *Proc. Nutr. Soc.* **32**:123, 1973.

158. VAN SOEST, P. J., and R. H. WINE. Use of detergents in the analysis of fibrous feeds. IV. Determination of plant cell wall constituents. *J. Assoc. Off. Agric. Chem.* **50**:50, 1967.

159. WALKER, A. R. P., B. F. WALKER, B. D. RICHARDSON and A. WOOLFORD. Appendicitis, fibre intake and bowel behaviour in ethnic groups in South Africa. *Postgrad. Med. J.* **49**:243, 1973.

160. WARD, K., and P. A. SEIB. Cellulose, lichenan, and chitin. *In: The Carbohydrates,* W. Pigman and D. Horton (eds.). New York, Academic Press, 1970, Vol. IIA, p. 413.

161. WARREN, D. S., and J. S. WOODMAN. Distribution of cell wall components in potato tubers: A new turbidimetric procedure for the estimation of total polyuronide (pectic substances) and its degree of esterification. *J. Sci. Food Agric.* **24**:769, 1973.

162. WATT, J., and R. MARCUS. Ulceration of the colon in guinea-pigs fed carrageenan. *Proc. Nutr. Soc.* **29:**4A, 1970.
163. WATT, J., and R. MARCUS. Effect of lignosulphate on the colon of guinea-pigs. *Proc. Nutr. Soc.* **33:**65A, 1974.
164. WEINSTEIN, L., R. E. OLSEN, T. B. VAN ITALLIE, E. CASO, D. JOHNSON, and F. J. INGEL-FINGER. Diet as related to gastrointestinal function. *JAMA* **176:**935, 1961.
165. WELLS, A. F., and B. H. ERSHOFF. Beneficial effects of pectin in prevention of hypercholesterolemia and increase in liver cholesterol in cholesterol-fed rats. *J. Nutr.* **74:**87, 1961.
166. WERCH, S. C., and A. C. IVY. On the fate of ingested pectin. *Am. J. Dig. Dis.* **8:**101, 1941.
167. WERCH, S. C., R. W. JUNG, H. PLENK, A. A. DAY and A. C. IVY. Pectin and galacturonic acid and the intestinal pathogens. *Am. J. Dis. Child.* **63:**839, 1942.
168. WHISTLER, R. L., and B. D. E. GAILLARD. Composition of xylans from several annual plants. *Arch. Biochem. Biophys.* **93:**332, 1961.
169. WHISTLER, R. L., and E. L. RICHARDS. Hemicelluloses. *In: The Carbohydrates,* W. Pigman and D. Horton (eds.). New York, Academic Press, 1970, Vol. IIA, p. 447.
170. WIGGINS, H. S. Personal communication.
171. WILLIAMS, R. D., and OLMSTED, W. H. A biochemical method for determining indigestible residue (crude fibre) in faeces. Lignin, cellulose and nonwater soluble hemicelluloses. *J. Biol. Chem.* **108:**653, 1935.
172. WILLIAMS, R. D., and W. H. OLMSTED. The manner in which food controls the bulk of faeces. *Ann. Intern. Med.* **10:**717, 1936.
173. WILLIAMS, R. D., and W. H. OLMSTED. The effect of cellulose, hemicellulose and lignin on the weight of stool. A contribution to the study of laxation in man. *J. Nutr.* **11:**433, 1936.
174. WOOD, T. M. Cellulose and cellulolysis. *In: World Review of Nutrition and Dietetics,* G. H. Bourne (ed.). Basel and Munchen, Karger, 1970, p. 227.
175. WOODMAN, H. E. The role of cellulose in nutrition. *Biol. Rev.* **5:**273, 1930.
176. WORTH, H. G. J. The chemistry and biochemistry of pectic substances. *Chem. Rev.* **67:**465, 1967.
177. ZIEGLER, E., and E. N. GREER. Principles of milling. *In: Wheat Chemistry and Technology,* Y. Pomeranz (ed.). St. Paul, Minnesota, American Association of Cereal Chemists, 1971, p. 115.

2

The Chemistry of Dietary Fiber

David A. T. Southgate

I. Introduction

A. Definition of Dietary Fiber

Before one can attempt to review the chemistry of dietary fiber, it is essential to establish a definition of the term itself. Dietary fiber was suggested as a term to apply to all the constituents derived from the plant cell walls in the diet which are not digested by the endogenous secretions of the human digestive tract.[138] The main reason for the choice of this term was to establish a distinction between *dietary fiber* and *fiber* or *crude fiber*. The term crude fiber has been used for many years to denote an estimate of the indigestible matter in animal feeds.[146] This distinction is necessary because crude fiber is an empirically determined fraction of a food which gives an approximate estimate of the cellulose and lignin in the food and does not include many of the cell-wall polysaccharides that are undigested by man.[134,142]

Dietary fiber as defined above includes all of the so-called unavailable carbohydrates and is equivalent to the original use of this term by McCance and Lawrence,[93] who included the noncarbohydrate lignin in the unavailable carbohydrate fraction.

The human diet also includes a number of plant polysaccharides which are structurally related to those found in the plant cell wall and which are similarly undigested in the human tract, and it seems appropriate to consider these as also forming part of dietary fiber.

DAVID A. T. SOUTHGATE • Dunn Nutritional Laboratory, Medical Research Council, Cambridge, England.

The plant cell wall contains small amounts of other constituents, either as integral parts of its structure, such as protein and inorganic material, or in very close association with the structural elements of the wall, such as cutin, suberin, waxes, and other lipids. Trowell has suggested that dietary fiber should include all these minor constituents. This extended definition of the term is not wholly acceptable; however, a brief account of these constituents has been included because they may, in certain circumstances, influence the properties of the dietary fiber in a food or diet.

B. Components of Dietary Fiber

Dietary fiber as defined in the previous section is the sum of the various indigestible constituents of plant origin in the diet. As such it does not have a defined composition but will vary from foodstuff to foodstuff and therefore from diet to diet, depending on the foods making up the diet. This variable composition renders a detailed description of the chemistry of dietary fiber *per se* rather difficult, and this account will primarily be concerned with describing the chemistry of the various constituents which together make up the total dietary fiber in the diet.

The components which make up dietary fiber are listed in Table I. They can be considered to fall into two broad classes, those derived from the plant cell wall and those indigestible plant polysaccharides that are not directly associated with the structure of the plant cell wall. The first category will be regarded by many as

TABLE I. Components of Dietary Fiber

Principal sources in the diet	Description	Classical nomenclature
Structural materials of the plant cell wall	Structural polysaccharides	Pectic substances Hemicelluloses Cellulose
	Noncarbohydrate constituents	Lignin Minor components
Nonstructural materials either found naturally or used as food additives	Polysaccharides from a variety of sources	Pectic substances Gums Mucilages Algal polysaccharides Chemically modified polysaccharides

forming true dietary fiber. It includes the structural polysaccharides of the plant cell wall, the so-called unavailable carbohydrates, and the noncarbohydrate lignin. In the table these are listed under the classical (trivial) terminology for the components of the plant cell wall. These terms are derived from the classical schemes for the fractionation of the plant cell wall, which is discussed below (p. 85) and have some limitations. The fractionation of the plant cell wall has, however, provided the starting materials for most of the detailed studies of the chemical structures of these polysaccharides, and for this reason it is necessary to retain these older terms in the subsequent sections describing the chemistry of these components of dietary fiber.

The second category comprises a range of plant polysaccharides that are frequently present in the diet. These polysaccharides are structurally similar to many components of the plant cell wall. Furthermore, they are not hydrolyzed by the intestinal secretions, and they frequently produce effects in the large intestine similar to those attributed to what may be thought of as dietary fiber in the strict sense (that is, the structural materials of the plant cell wall). They are found in many foods and are increasingly being used as food additives.[62,77,157]

C. Fractionation of the Plant Cell Wall

The nomenclature used to describe the components of the plant cell wall is based on the classical schemes of fractionation,[109] and the definition of these terms is essentially more dependent on the method of isolation than on the basis of chemical structure as such. It is therefore necessary, at this stage, to consider briefly the procedures that are used to fractionate the plant cell wall into its components.

The fractionation procedures will also be discussed in great detail from the viewpoint of their use as analytical methods in Chapter 3, Section III-A.

A typical fractionation scheme is set out in Figure 1, which is based on the review by Siegel.[130]

The plant tissue is usually dried and then extracted with organic solvents to remove lipids and pigments. In most studies it is then customary to remove lignin by treating the sample with chlorine, either as the gas or from hypochlorite, followed by extraction of the chlorinated aromatic material.

The residue is then extracted with dilute alkali under nitrogen, and a large proportion of the matrix polysaccharides dissolve at this stage.[151] The polysaccharides are recovered from the extract by neutralization when the hemicellulose A fraction precipitates, together with some pectic substances. Further addition of alcohol results in the precipitation of the hemicellulose B fraction with some pectic material. The pectic substances are removed by extraction of the hemicel-

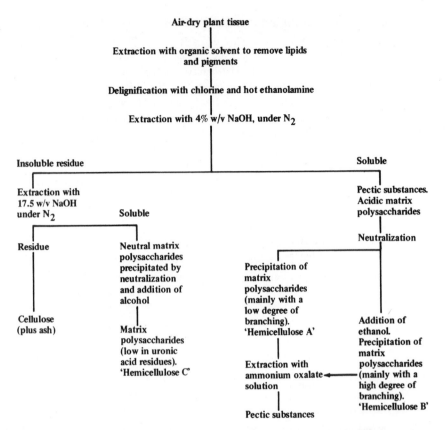

Figure 1. Basic scheme for the fractionation of plant cell walls (Southgate, 1975, after Siegel, 1968).

lulose fractions with ammonium oxalate solution, from which they are recovered by the addition of acetone or alcohol.

The residue insoluble in dilute alkali is reextracted with strong alkali under nitrogen. The residue from this extraction is virtually only cellulose contaminated with some inorganic material.[150] The remainder of the matrix polysaccharides dissolve in the alkali, and these are recovered by neutralization and the addition of alcohol to give the hemicellulose C fraction.

Lignin can be isolated by a number of different procedures; possibly the most common is to prepare the residue insoluble in 72% w/w H_2SO_4, that is, Klason lignin.

There are many variants of this scheme and most workers have found that some minor modifications are required for each type of plant tissue. Table II summarizes the basic features of the relation between the fractionation of the cell-wall components and the terminology used to define the components.

TABLE II. Terminology and Procedures Used to Isolate
the Components of Plant Cell Wall

Pectic substances	Extraction with hot water with the addition of a chelating agent such as ammonium oxalate or EDTA
Hemicelluloses (A and B)	Extraction with dilute alkali under nitrogen (A) Precipitated on neutralization (B) Precipitated on neutralization and addition of alcohol
Hemicellulose (C)	Extraction with strong alkali under nitrogen
Cellulose	Residue insoluble in 17.5% NaOH (α cellulose)
Lignin	Residue insoluble in 72% w/w H_2SO_4 (Klason lignin)

II. The Chemistry of the Individual Components

A. Pectic Substances

The pectic substances form a group of complex polysaccharides that are found in the middle lamellae, primary cell walls, and intercellular material of most plants. Although pectic substances are characteristically soluble in hot water, those that make up an integral part of the wall are less readily soluble in water and appear to be present in the wall as calcium salts. Aqueous solutions of pectic substances gel on cooling, a property which makes them of great commercial importance. There is a considerable body of literature relating to the technological uses of these substances.[17,131] The rinds of citrus fruits and sugar beet pulp are rich sources and may contain up to 25–30%; apple is a particularly important commercial source of pectin and may contain up to 15% on a dry-weight basis.

For technological purposes pectic substances are defined as complex plant polysaccharides of which D-galacturonic acid is the principal component; they are further classified into three groups.

Pectins are readily soluble in water and have a high degree of methylation of the carboxylic acid group. *Pectinic acids* are pectins in which only a portion of the acidic groupings are methylated, and *pectic acids* refers to the polysaccharides which are devoid of methyl groups.

Pectic substances are either extracted with hot water or with water containing a chelating agent such as ethylene diaminetetraacetic acid (EDTA) or ammonium oxalate. The agent presumably chelates the calcium in the insoluble calcium pectates and releases the pectic substances into solution. The recovery from aqueous solution is usually by ethanol precipitation or by preparation of the insoluble calcium salt and decomposition of this material.

1. Structure

A number of recent reviews are available tracing the development of modern views on the structure of pectic substances; this section is a brief summary of the present position.[7,8,70]

Hydrolysis of the pectic substances usually gives a mixture of sugars in addition to galacturonic acid. The sugars usually found are D-galactose, L-arabinose, D-xylose, L-rhamnose, and L-fucose; 2-O-methyl-D-xylose and 2-O-methylfucose are often present.

The early structural studies appeared to show that D-galacturonic acid was a major constituent, and it was for a long while uncertain if the neutral sugars were components of the polysaccharide or derived from material contaminating the galacturonan. In this respect the classical study by Bishop[21,158] of the pectic substance from sunflower heads was possibly misleading, as this material appears to be atypical in being virtually a pure polyuronide.[2,8]

A considerable number of pectic substances have now been examined, and in most cases the fraction has been shown to contain a number of different types of polysaccharide.

Three types of homopolysaccharide have been described: galacturonans, galactans, and L-arabinans, in addition to a range of acidic heteropolysaccharides, of which substituted galacturonans are most common. Neutral arabinogalactans have been isolated from some sources. It is probable that these are actually degraded products of the complex native heteropolysaccharides present in the cell wall.[8] This is particularly true for many of the detailed early structural studies where the extremely labile nature of many glycosidic linkages, particularly those involving arabinofuranoside residues, was not appreciated.

Although pectic substances are apparently universal in the plant kingdom, detailed studies are available for only a relatively few preparations.[1,9,10,12,18,19,71,132] However, as more structural studies accumulate, the general features of pectic substances are becoming clearer, and these few examples will illustrate some of these points.

2. Homopolysaccharides

L-Arabinans (Figure 2a) have been isolated from the pectic substances of mustard seed[75] and subjected to detailed structural studies. These show that the polysaccharide is a $(1\rightarrow5)\alpha$-linked arabinofuranoside chain carrying $(1\rightarrow3)$ linked arabinofuranoside residues at intervals.

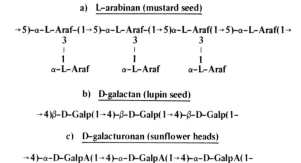

a) L-arabinan (mustard seed)

→5)-α-L-Araf-(1→5)-α-L-Araf-(1→5)α-L-Araf(1→5)-α-L-Araf(1→
   3   3   3
   | | |
   1   1   1
α-L-Araf   α-L-Araf   α-L-Araf

b) D-galactan (lupin seed)

→4)β-D-Galp(1→4)-β-D-Galp(1→4)-β-D-Galp(1–

c) D-galacturonan (sunflower heads)

→4)-α-D-GalpA(1→4)-α-D-GalpA(1→4)-α-D-GalpA(1–

Figure 2. Partial structures of homopolymeric pectic substances. (In this and in subsequent figures a simplified nomenclature has been used for reasons of clarity.)

Sugar beet also contains an arabinan that appears to be a heteropolysaccharide with the arabinose linked in the same way as in the pure L-arabinan.[76]

A D-*galactan* (Figure 2b) has been isolated from the seed of the lupin and is a linear (1→4)α galactopyranosyl polysaccharide.[72]

D-*Galacturonans* (Figure 2c) have been regarded as the typical pectic substances; however, one of the few examples of a pure polyglycuronan is found in the pectic material isolated from sunflower heads examined by Bishop.[21,158] This was a linear (1→4)α-D-galactopyranosyluronan.

3. Heteropolysaccharides

The heteropolysacchards more commonly form the major part of the pectic material. There appear to be two main types of polysaccharide.

L-*Arabinogalactans* (Figure 3a) have been isolated from *Centrosoma* sp. and soya bean,[13] and found to have a linear (1→4)β-D-galactopyranose chain with side chains containing arabinofuranoside groups linked 1→5 and 1→3 onto the galactan main chain.

Substituted galacturonans: The majority of pectic materials appear to contain neutral sugars as an integral component of the predominantly galacturonan chain. In general these complex heteropolysaccharides also contain other sugars in side chains.[14] Although detailed structural work is not available for a large range of pectic substances, Aspinall[8] has suggested the main features of this complex heteropolysaccharide (Figure 3b). The main backbone appears to be a (1→4)α-D-galactopyranosyluronic acid chain with 1–2% of rhamnopyranoside residues at intervals. The side chains are of several kinds: long galactan or arabinan portions terminated by a xylopranose residue, e.g., fucosyl xylopyranose or galactosyl xylopyranose.

a) L-arabino–D-galactan (soya bean cotyledons)

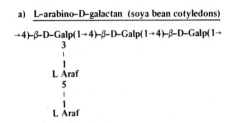

b) Hypothetical structure incorporating structural features likely to be present
   in the native pectic substances

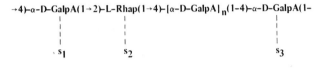

Substituent sidechains

s₁   (D-Galp)ₙ(1→   s₃   β-D-Xylp(1-

     or   α-L-Fuc(1→2)-D-Xylp(1→

s₂   (L-Araf)ₙ(1→   or

                                                β-D-Galp(1→2)-D-Xylp(1→

Figure 3. Partial structures of heteropolymeric pectic substances.

This structure would incorporate many of the features required in the hypothetical native pectic substance in the plant tissues prior to extraction. If the galactan side chains also carried arabinofuranoside residues, then the majority of the structural characteristics would be present.

4. Properties

Gel Formation.   Pectic substances are, in the main, acidic polysaccharides which dissolve in hot water to form a gel on cooling. This gel melts as the temperature is raised. Because of its technological importance, this property has attracted a considerable amount of work.

The proportion of methylation of the pectic substances has an influence on gel formation, a higher degree of methylation giving the best gelling characteristics.[17] The formation of pectic gels also depends on the acidity and solute content of the gelling solution. Sugar is the solute employed in making jam, but other substances are equally efficient in promoting gel formation.

The structural features of the pectin molecule involved in gel formation have been examined by Rees.[119,120,122] Gel formation is believed to involve intermolecular association of the galacturonan chains.

Salt Formation. Pectic substances form salts, and calcium and magnesium pectates are often found in cell walls. They are relatively insoluble, and precipitation as the calcium salt forms the basis of a method for measuring pectin. Salt formation appears to involve the hydroxyl groups and is not solely dependent on the carboxyl groups of the uronic acids.[36,42,85,86,101,102]

Demethylation. The methoxyl groups can be removed by acid, alkaline, or enzymatic hydrolysis; deesterification can be complete, producing the pectic acid form, or it may be partial.[17] Some degree of depolymerization may accompany the deesterification unless the conditions are controlled. Alkaline treatment can result in the formation of unsaturated sugar compounds by a process of β-elimination.[8]

B. Hemicelluloses

The term hemicellulose was used originally in 1891 by Schulze to describe the polysaccharides that could be extracted from the plant cell wall with dilute alkali.[6] Subsequently it has become clear that these polysaccharides are related neither biosynthetically nor structurally to cellulose. The term, however, has been retained in modern carbohydrate chemistry as a trivial name for the complex mixture of polysaccharides that can be extracted from most plant cell walls with dilute alkali. The carbohydrates so extracted have undoubtedly undergone some degree of degradation,[90] and Albersheim[2] feels that the fractions obtained in the classical alkali fractionation are in the main artifacts, products of a progressive degradation of the native polysaccharides.

Most of the structural studies on these components have been carried out on material extracted from woody tissues,[137] although increasing attention has been given recently to the hemicelluloses from grasses.[27,43,54,153] However, there has been little work on the hemicelluloses from plant tissues that characteristically form part of the human diet.

1. Structure

The material extracted by dilute alkali is usually a complex mixture of polysaccharides, most of which are heteroglycans containing between two and four (or exceptionally five or six) different sugar residues.[155] Classically the mixture was fractionated into hemicellulose A, which precipitated from the extract on acidification, and hemicellulose B, precipitated on the addition of ethanol to the neutralized solution. Hemicelluloses A usually contain arabinoxylans with some uronic acid residues, whereas those in the B classification tend to be of lower molecular weight and comprise a greater range of types of polysaccharides.

The older classical fraction procedures, of which O'Dwyers's work is typical,[112] have been superseded in recent studies by procedures involving fractionation of complexes formed with iodine and quaternary ammonium compounds or by gel-filtration techniques.[5,22,51,57,135] Some major types of structure have been isolated from hemicellulose fractions from a variety of sources, mainly woody tissues.[137]

D-*Xylans*.[152] A group of polymers having a main chain of $(1-4)\beta$-D-xylopyranosyl residues is found in all land plants and in most plant organs. The pure homoglycan (Figure 4a) is rare, and most preparations contain side chains of which the commonest is 4-*O*-methyl-α-D-glucopyranosyluronic acid; arabinose is also a common side chain and wheat-flour xylan is an arabinoxylan, with the L-arabinose linked as nonreducing furanose end groups.[15,110] The xylan chain appears to contain open regions, with up to five xylose residues without substitution; arabinose/xylose ratios are variable, which suggests that some polysaccharides contain large areas of highly branched sequences while others have frequent open regions.

The average number of xylopyranosyl residues is of the order of 75. The basic chain is cellulosic in character and differs in the absence of projecting primary alcohol groups and the presence of branches. Comparison of the xylans from different plant families shows some systematic differences; for example, the Leguminosae xylans contain less arabinose and more uronic acid residues than xylans from the Gramineae.[58]

4-*O*-Methylglucurono-D-*xylans* (Figure 4b). Most hemicellulose preparations give 4-*O*-methylglucuronic acid on hydrolysis, and 4-*O*-methylglucorono-

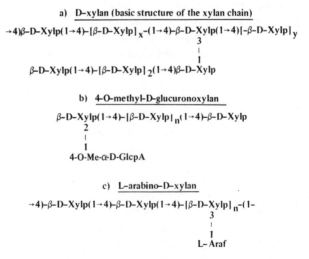

Figure 4. Partial structures of hemicelluloses—xylans.

D-xylans are important components of the hemicellulose fraction of the walls of many woody plants and appear to be widely distributed among the leaf and stem tissues of other plants.[33]

Most detailed studies apply to these substituted xylans from woody tissues. In the native tissue there is some degree of O-acetylation, but these groups are frequently lost during the alkali fractionation. In studies based on holocellulose preparations that have been delignified, some modifications of structure involving depolymerization and oxidation are often found, and some 10–20% of the total cell-wall polysaccharides may dissolve during delignification.[137]

In woody tissues yields of between 5 and 27% are reported, with a median of around 14. The polysaccharides have a negative rotation and a degree of polymerization of between 200 and 150. There is apparently one side chain per 9–12 xylose residues, although some woods yield a more substituted xylan. The distribution of 4-O-methylglucuronoside chains is not well established, but they are probably randomly distributed and not arranged in blocks. It is not possible to define the condition of the carboxyl groups *in situ*.[137]

The molecules are essentially linear, and the main xylan chain has a threefold screw axis (Figure 5) compared with twofold in cellulose.[39,91,121]

The xylan chain is fairly readily hydrolyzed by dilute acid; the disaccharide xylobiose is, for example, hydrolyzed seven times faster than cellobiose in 0.5 M H_2SO_4 and, as judged by tritium exchange, the xylan molecule is more accessible than cellulose.

The undegraded 4-O-methylglucuronoxylans are not soluble in water but dissolve in aqueous methyl sulfoxide. Studies in this solvent and methanol show that the ionized acids are more readily solvated with water and the xylan chain more readily with methyl sulfoxide.[137] Arabino (4-O-methylglucurono) xylans have been found in some gymnosperms containing L-arabinofuranosyl side chains linked 1→3.

Arabinoxylans (Figure 4c). Arabinose-substituted xylans are uncommon in woody tissues, but arabinoxylans are widely distributed in the cell walls of many cereals,[11,34,35,65,97,98,108] and the husks of many grains have provided the starting material for the arabinoxylans in most structural studies.

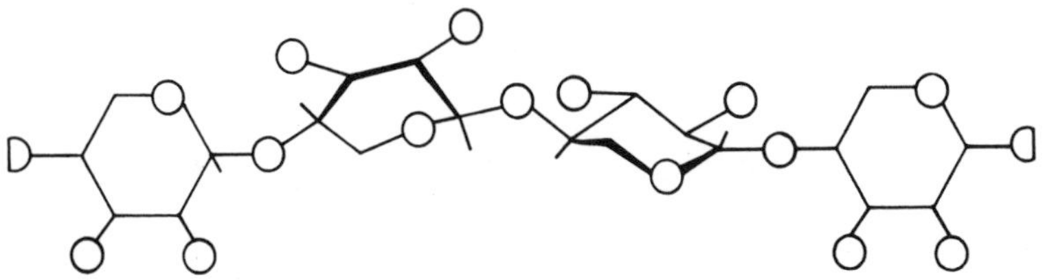

Figure 5. Conformation of the xylan chain (after Marchessault and Liang, 1962).

The main chain is composed of $(1\rightarrow4)\beta$-D-xylopyranosyl units with an occasional branching in some preparations.[115,154] The side chains are usually single L-arabinofuranosyl groups linked $1\rightarrow3$, although 2-O-D-xylopyranosyl-L-arabinofuranosyl and glucuronosyl substituents are often found. The side chains seem to be linked randomly. An arabinoxylan from barley husks[11] contained one arabinosyl residue per six xylan residues, whereas the corresponding material from rye flour[52] appears to have one arabinosyl unit for every two xylose residues.

Mannans (Figure 6a). The hemicellulose fraction from many cell walls contains mannose[29]; the classic example of a polysaccharide of this type is ivory nut mannan. This is a linear molecule with $(1-4)\beta$-linked D-mannopyranose residues. These polymers seem to be storage polysaccharides, and substituted mannans are found in many seeds.[63,83,88,123]

Glucomannans (Figure 6b). These occur in relatively small amounts (2%) in hardwood hemicelluloses.[99] They have negative rotations and a degree of polymerization of around 70. The glucomannans appear to be linear polymers with both mannose and glucose in the chain; the ratio of mannose/glucose is between 1:1 and 2.4:1. Hardwood glucomannans appear to contain no galactose and are relatively insoluble, but the glucomannans from gymnosperms have galactose side chains and a higher mannose/glucose ratio (3:1) and are more properly called galactoglucomannans (Figure 6c). Many of these form insoluble complexes with $Ba(OH)_2$.

The presence of side chains tends to make these polysaccharides more soluble in water, possibly because the side chains prevent the formation of intermolecular hydrogen bonding.

Arabinogalactans. Substituted galactans form part of the hemicellulose complex in many tissues,[28,103,107,156] and although most of the detailed structural work relates to arabinogalactans from woody tissues, polymers of this type are very widely distributed. One of these polysaccharides that has received considerable study is the water-soluble arabinogalactan of larch.

a) <u>D-mannan</u> (basic structure, as in the ivory nut)

$\rightarrow$4)-β-D-Manp(1$\rightarrow$4)[-β-D-Manp]$_n$(1$\rightarrow$4)-β-D-Manp (1$\rightarrow$

b) <u>Glucomannan</u>

$\rightarrow$4)-β-D-Manp(1$\rightarrow$4)-β-D-Glcp(1$\rightarrow$4)-β-D-Manp(1–

c) <u>Galactoglucomannan</u>

$\rightarrow$4)-β-D-Manp(1$\rightarrow$4)-β-D-Glcp(-1$\rightarrow$4)[β-D-Manp]$_3$(1$\rightarrow$4)β-D-Glcp(1$\rightarrow$4)-β-D-Man
6
|
1
Galp

Figure 6. Partial structures of hemicelluloses—mannans.

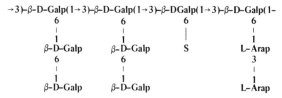

Sidechains β-D-galactopyranosyl or less frequently
L-arabinofuranosyl or D-glucopyranosyl

Figure 7. Partial structures of hemicelluloses—arabinogalactan (larch).

These preparations often contain xylose, rhamnose, and uronic acid, which appear to be parts of the molecule. The rotations are low and positive (+11°), and the degree of polymerization is around 380. A typical summary formula is given by Timell (Figure 7).[137] There appear to be two components with molecular weights of 100,000 and 16,000.

The arabinogalactans from other coniferous woods contain side chains of five types: galactopyranosyl chains linked 1↑6, Galp-Araf, Galp-Araf-Arap, $(Galp)_x$-Glcp A and Galp-Araf-Xylp.

An arabinorhamnogalactan has been isolated from maple syrup: This has a negative rotation of −41° and appears to be a β1-3 galactan carrying single or double arabinose side chains and single rhamnose side chains.

In the preparation from wheat flour[53] the ratio of galactose to arabinose was 1.4:1, with some evidence of β-D(1→6) and (1→3) linkage of the galactose and with the L-arabinosyl group linked to C3. This polysaccharide was also associated with protein. This highly branched arabinogalactan was apparently covalently bound to a protein, and the linkages appear to involve the hydroxyproline residues in the protein.

Glucans. The noncellulosic polysaccharides of most cell walls contain variable amounts of predominantly water-soluble glucans,[30,31,113] in which the linkages are β1-3 or 1-4. The proportion of 1-3 to 1-4 linkages varies a little with the source of the glucan and is usually between 4:3 and 7:3. These polysaccharides form part of the so-called cereal gums,[16,64] and although these may be regarded as falling outside the hemicellulose classification, they do form part of the cell wall.

2. Hemicelluloses —Summary

The hemicelluloses seem, therefore, to comprise a complex series of heteroglycans based on three types of homopolymeric backbone chain, xylans, mannans, and galactans, and one type of mixed chain, the glucomannan. These main backbones carry side chains containing arabinose, galactose, 4-*O*-methylglucuronic acid, and less frequently both xylose and arabinose. In the case of the galactans the side chains may be several residues long, and these molecules are typically highly branched. The other polymers usually carry side chains containing one or two residues and, rarely, three.

The chemical properties of these polysaccharides have not been studied in great depth. In general the linear backbone chains tend to be less soluble in water than the substituted molecules.[80]

The glucuronoxylans are soluble in dilute alkali from which they are precipitated by acidification with acetic acid (so-called hemicellulose A). The hemicellulose-B fraction, precipitated with ethanol after removal of hemicellulose A, consists of branched heteroglycans (including the galactans) and the linear arabinoxylans that form complexes with iodine in $CaCl_2$.[57]

The galactomannans seem to fall into two classes depending on their solubility in alkali. In some plants appreciable amounts of galactose and mannose remain associated with the cellulose after extraction with strong alkali. In general the alkali-soluble galactomannans do not form a complex with iodine, whereas those insoluble in strong alkali precipitate with iodine and are presumably linear with fewer side chains.[59]

C. Plant Gums and Mucilages

The plant gums and mucilages constitute a very heterogeneous group of complex polysaccharides, which are not part of the cell wall structure but which are in general indigestible. This discussion will include the common exudate gums of a number of other *Acacia* spp.; mesquite gum also appears to be of this polysaccharides possess complex, very highly branched structures and usually contain uronic acid residues in addition to those of two or more neutral sugars.

1. Gums

The ideal system of classification of gums on purely structural grounds has yet to be achieved, and side-chain variations appear to be extremely common within a plant genus. A system of classification has been suggested by Aspinall based on the polysaccharide cores.[7,8]

Galactan Group. This includes gum arabic (Figure 8) and the exudate gums of a number of other *Acacia* spp. mesquite gum also appears to be of this type. The core polysaccharide is a β-galactan linked $\beta(1\rightarrow3)$ with side chains of $\beta(1\rightarrow6)$ galactopyranosyl chains terminated with glucuronyl residues, which themselves carry a range of side chains. These are typically L-arabinofuranosyl $(1\rightarrow3)$, L-rhamnopyranosyl $(1\rightarrow3)$, α-D-galactopyranosyl $(1\rightarrow3)$ L-arabinofuranosyl, with, less commonly, β-L-arabinopyranosyl $(1\rightarrow3)$ L-arabinofuranosyl chains. There is some variation in side chains which appears to be species specific.

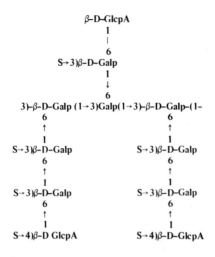

$$\beta\text{-D-GlcpA}$$
$$1$$
$$|$$
$$6$$
$$\text{S}\rightarrow 3)\beta\text{-D-Galp}$$
$$1$$
$$\downarrow$$
$$6$$

3)-β-D-Galp (1→3)Galp(1→3)-β-D-Galp-(1–

6		6
↑		↑
1		1

S→3)β-D-Galp   S→3)β-D-Galp

6		6
↑		↑
1		1

S→3)β-D-Galp   S→3)β-D-Galp

6		6
↑		↑
1		1

S→4)β-D GlcpA   S→4)β-D-GlcpA

S ≡ L-Araf (1–; L-Rhap (1–; α-D Galp (1→3)–L-Araf-(1–

and rarely β-L-Arap (1→3) L-Araf(1–

Figure 8. Partial structures of gums—galactan group (gum arabic).

These gums undergo autohydrolysis in solution in very weak acid, leading to a loss of arabinose from the molecule and giving a degraded gum virtually free of arabinose.

Glucuronomannan Group.   Three plant gums have been shown to contain a glucuronomannan core: damson and cherry gums and gum ghatti. In these the core chain consists of alternate glucopyranosyluronic acid and mannopyranosyl residues (Figure 9). The side chains are of galactosyl residues linked $\beta(1\rightarrow 6)$ attached to the core chain by an arabinopyranosyl group. The galactosyl side chains also carry short arabinosyl branches.

Galacturonorhamnan Group. The gums in this group contain interior chains of D-galacturonic acid and L-rhamnosyl groups. The proportion of these residues varies and in some *Stercularia* gums and some *Khaya* gums D-galactose is also present.

Aspinall [7,8] divides the gums in this group into three subgroups, depending on the proportion of galacturonic acid in the interior chain. The major component of gum tragacanth (Figure 10a), tragacanthic acid for example, has interior chains that are virtually those of a galacturonan. The links are 1-4 and some branching may be present. The side chains contain xylose, xylose and fucose, or xylose and galactose and in some respects resemble the side chains observed in some pectic substances. The rhamnose content of this gum appears to be very small.

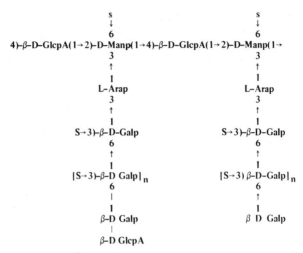

Figure 9. Partial structures of gums—glucuronomannan group (gum ghatti).

The *Khaya* gums show some specific variations, and although the structures are not fully established, rhamnosyl residues are combined with galacturonic acid residues in the core in an alternating sequence.[118] The side chains are of galactose or galactose and 4-*O*-methylglucuronic acid residues (Figure 10b). In the *Stercularia* gums (Figure 10c) the interior chain contains alternating galacturonic acid and rhamnosyl residues and also galacturonic acid residues linked to D-galactose; there are some differences in the sites of attachment of the side chains.

These gums can be considered to have an interior core of the pectic type to which are attached more complex side chains than are usual in the pectic substances.

Xylan Group.   Only a few xylan gums have been described: sapote gum (*Sapota activas*) and the polysaccharides from the corms and seed cases of *Watsonia*. These are structurally related to, but more complex than, the xylans found as cell-wall components. The partial structure of the *Watsonia* gum is given in (Figure 11a).

Xyloglucan.   The gel-forming polysaccharide from the seed of tamarind (*Tamarindus indica*) is the best studied of a group of polysaccharides isolated from plant seeds which give a blue color with iodine and which have been called amyloids, although they are structurally unrelated to amylose.[149] The interior chain resembles cellulose II (Figure 11b). Several other polysaccharides of this type with a variety of substituent side chains have been reported, and the xyloglucans may be of relatively widespread occurrence.

Aspinall's classification in terms of basal chain types provides a convenient framework for the consideration of these complex structures and emphasizes the relationships with the other plant polysaccharides. Biogenesis does not appear to involve the addition of more residues to the outer chains of preexisting polymers such as cell-wall components.[19] The relationship between molecular structure and physical structure in these highly branched polysaccharides has not been explored in detail. It is quite possible that the physical properties of these molecules, which are important as far as their biological functions are concerned, are due to the secondary and tertiary structure of the outer chains, in which case an alternative classification could be desirable.

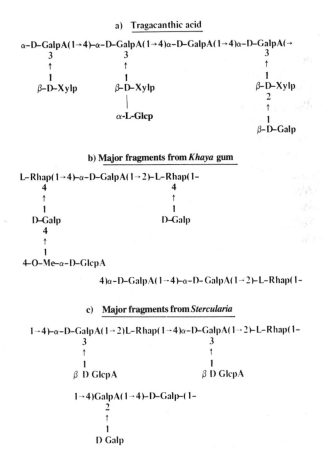

Figure 10. Partial structures of gums—galacturonorhamnan group.

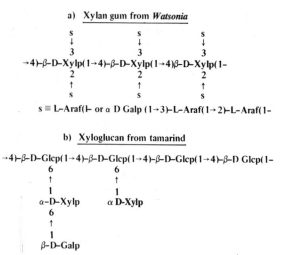

Figure 11. Partial structures of gums—xylan and xyloglucan groups.

2. Mucilages

Many seeds contain polysaccharide components that are extractable with water and possess water-binding properties. In the seed they seem to act as a reservoir for the retention of water and so protect the seed against excessive desiccation. They are customarily divided into neutral and acidic components.[139,140]

Neutral. The neutral components include galactomannans, glucomannans, arabinoxylans, and in one case xyloarabinans, although the latter are often very intimately associated with acidic polymers. The galactomannans are components of many Leguminosae seeds such as alfalfa and guar.[141] Of these, guar gum has been studied in some depth. Structurally it appears to be closely related to the hemicellulose galactomannans; β-D-mannopyranosyl chains with 1→4 linkages carrying 1→6 galactopyranosyl side chains occur in guar gum, with approximately one in two mannose residues carrying a galactosyl side chain.

The glucomannans found in seeds and tubers have glucose and mannose in the proportions of 1:1 or 1:3.

Arabinoxylans are found in the extracts of many cereals, and the mucilages of *Plantago* spp. and linseed contain complex branched polysaccharides of this type.[74] A xyloarabinan has been found in the mucilage of cress seeds.[139]

Acidic. Acidic components of many mucilages have been observed but, in general, structural studies on these polysaccharides have been limited in scope. Most contain interior chains of D-galacturonic acid and L-rhamnose (Figure 12). The mucilage of linseed has two acid polysaccharides in addition to an arabinoxylan. These acidic mucilages and the mucilage of slippery elm bark

→4)-α-D-GalpA(1→2)-L-Rhap(1→4)α-D-GalpA(1→2)L-Rhap(1-
                                3
                                ↑
                                s

S ≡ D-GalpA(1→4)D-Galp(1-

D-Galp(1→3) D-Galp

3-O-methyl ester groups are often found on the
sidechains.

Figure 12. Fragments of slippery elm mucilage.

contain some L-galactose, and this sugar may be widely distributed among polysaccharides of this type.[20,61,71]

D. Algal Polysaccharides

A range of different types of polysaccharides can be extracted from the tissues of algae. In the main the marine algal preparations have attracted the most attention. Some of these polysaccharides have valuable properties for the food technologist, and many foods contain small amounts of algal polysaccharides. They are not hydrolyzed by the mammalian digestive enzymes and are therefore part of dietary fiber as defined earlier.

1. Reserve Polysaccharides

Polysaccharides of the starch type are found in a number of species, and structural studies have confirmed their similarity to starches from higher plants.[114] Polysaccharides with both the amylose and amylopectin structures have been isolated. In general, the molecular weights of the algal starches are lower, and the starch granules are less highly organized than those of tubers or cereals.

Glucans with β-linked glucose residues are widely distributed, and *laminarin* is a typical example with an essentially linear β-D$(1{\rightarrow}3)$ structure.[49] A variable proportion (around 40%) of the chains is terminated by D-mannitol and the remainder by a reduced D-glucosyl residue. These polysaccharides are in general water soluble with a low negative rotation and have molecular weights of between 2600 and 4000.

2. Structural Polysaccharides

Cellulose occurs in many species of algae and has essentially the same structure as the cellulose of land plants.

Glucans with linear $\beta(1\rightarrow4)$ and $\beta(1\rightarrow3)$ glucosyl units are found in some species,[49] and D-mannans and D-xylans have also been described. Structural studies on many of these have confirmed that they are usually linear β-D$(1\rightarrow4)$ mannans and β-D$(1\rightarrow4)$- and $(1\rightarrow3)$-linked xylans, respectively.

Alginic acid is an important algal polysaccharide in the context of dietary fiber, as alginates are frequently used as food additives.

The sodium salt has a high negative rotation $[\alpha]_D = -120{\sim}150°$ and is a glycuronan containing variable proportions of D-mannuronic and L-guluronic acids. It occurs in all Phaeophycae.

In the plant it is present as a mixture of the Na, K, Ca, and Mg salts and as such is insoluble. Extraction involves treatment with acid and extraction with Na_2CO_3, from which it can be isolated by precipitation with alcohol, or as the acid or Ca salt.

Attempts to fractionate alginic acid into a D-mannuronan and L-guluronan have been unsuccessful, and the definitive structure of alginates has not been elucidated. The high negative rotation suggests a considerable proportion of β-D$(1\rightarrow4)$ D-mannuronic acid residues, and the L-guluronic acid is probably $1\rightarrow4$ linked. X-ray analysis of alginate fibers suggests an essentially linear, unbranched molecule. There appear to be blocks of mannuronic acid residues and blocks of guluronic acid residues interspersed with chains containing both residues. These mixed regions are more susceptible to hydrolysis, and it is possible to obtain mannuronan and guluronan fragments from hydrolysates.[114]

Alginic acid is insoluble in water but readily soluble in aqueous solutions of alkali metal hydroxides and carbonates, and these alginate salts yield viscous solutions in water. The amounts of polyvalent ions required to precipitate alginates varies in the order Ba < Pb < Ca < Sr < Cd < Ca < Zn < Ni < Co < Mn < Fe < Mg, and some other properties such as gel formation are affected in a similar or reverse order.

Alginates rich in L-guluronic acid have higher pK values. Higher calcium affinity in Na/Ca ion-exchange reaction is observed for alginates richer in mannuronic acid, although the physicochemical significance of these differences is not understood.

Alginates are used widely as the Na, NH_4, or K salts or propane-(1-2) diol esters for improving the texture of many types of foods (e.g., cream, fruit drinks).[95,157]

The other major class of algal polysaccharides that is important in the context of dietary fiber is the sulfated D- and L-galactans, which are extracted from algae with boiling water. Three major groups, agar, carrageenan, and porphyran, have been recognized, and both agar and carrageenan find wide use in food industry.[62]

Agar[44,145] consists of two components: agarose and agaropectin. Agarose is composed of D-galactose and 3,6-anhydro-L-galactose chains of alternating $(1\rightarrow3)$-linked β-D-galactopyranosyl and $(1\rightarrow4)$ 3,6 anhydro-α-Lgalactopyranose residues. The reducing end is galactose, with an anhydro residue at the nonreducing end. Side chains containing 6-O-methyl-D-galactose residues are linked to E-4 of the anhydro residues.

Agaropectin is probably a mixture of polysaccharides containing D-galactose, 3,6-anhydro-L-galactose, half-ester sulfate, and D-glucuronic acid.

κ *Carrageenan* is precipitable with KCl and contains 3,6-anhydro-D-galactose instead of the L form as in agarose. The proportion of D-galactose, 3,6-anhydro-D-galactose, and ester sulfate is approximately 6:5:7, and the molecule consists of alternating (1→3)D-galactose and (1→4)-linked 3,6-anhydro-D-galactose residues carrying sulfate groups mainly at C-4 in the case of D-galactose and C-2 of the anhydrogalactose.

λ *Carrageenan* is a much more highly sulfated polysaccharide with a similar basic structure.

Molecular weights of agar range from 5000 to 110,000; for κ carrageenan from 260,000 to 320,000; and for λ carrageenan, 330,000 to 790,000. Strong heating is necessary to bring these polysaccharides into solution, especially those with high anhydrogalactose contents which tend to set to hard gels.

Gel formation depends on the proportion of 3,6-anhydro residues. When viscous solutions are required in food processing, preparations are used that are less hydrophilic, with less anhydro and relatively more ester sulfate in the molecule.

E. Cellulose

Cellulose is the major structural polysaccharide in the cell wall and is often said to be the earth's most abundant organic substance. Because of its important role in the paper industry and in the preparation of synthetic fibers, it has been the subject of extensive chemical and physical investigation. As a consequence, the volume of the scientific and technical literature relating to the chemistry of cellulose is very great. This account of the chemistry of cellulose, therefore, is intended as a summary of these more extensive reviews,[82,147] concentrating on the features of cellulose chemistry that appear relevant to the topic of dietary fiber. The plant foods in the human diet have undergone only mild treatment during preparation and cooking, and most of the cellulose consumed in the diet is, therefore, more closely related to the celluloses in the plant cell wall than to isolated industrial celluloses.

1. Structure

The cellulose in most plants appears to be a single homopolysaccharide in which the D-glucose residues are linked $\beta 1 \to 4$ in long chains (Figure 13). Traces of other sugars are found in many cellulose preparations, and it is not

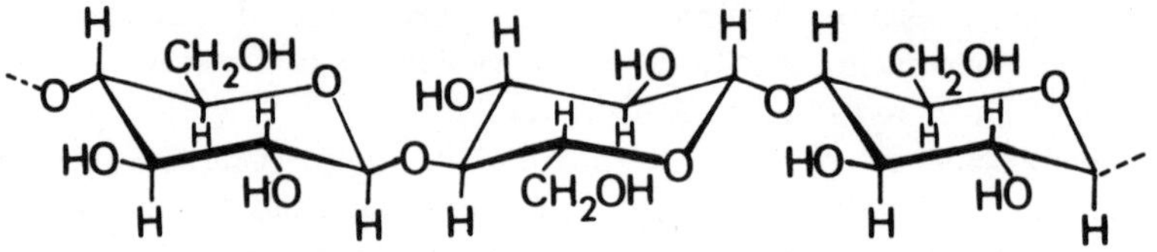

Figure 13. Conformation of the cellulose chain.

clear whether these sugars are derived from contaminating hemicelluloses or are incorporated in the glucan molecule.

In many cellulose fibers the cellulose has a predominantly crystalline form; this is thought to be the form in which cellulose occurs in most cell walls. The fibers are extremely strong, and this is believed to be due to the length of the chains involved and to the hydrogen-bonding capacity of the three hydroxyl groups. The X-ray diffraction patterns from these crystalline regions of cellulose fibers[78] have given rise to a number of models for the detailed arrangement of the cellulose molecules within the microfibrils that make up the fibers themselves. Models such as those of Meyer and Misch (Figure 14) show the close packing arrangement of the cellulose chains and the high degree of hydrogen bonding in the crystalline regions of the cellulose.[8,100]

The disordered noncrystalline regions seen in all cellulose fibrils may represent areas where other sugar or uronic acid residues are incorporated in the molecule. These regions are more accessible to many reagents because the hydroxyl groups are not capable of forming hydrogen bonds as strongly as in the crystalline regions.[129] The X-ray diffraction pattern does not change when cellulose fibrils are wetted, suggesting that water uptake is limited to these disordered regions.

The degree of polymerization in native cellulose may be greater than 10,000 or 15,000; all isolation procedures undoubtedly involve some degree of degradation.

2. Properties

Within the context of dietary fiber, the most important properties of cellulose concern, first, the susceptibility of the molecule to hydrolysis and, second, its capacity to absorb water.

Enzymolysis. Cellulases are widely distributed amongst the fungi and bacteria and may be involved in the germination of the seeds of higher plants. Many members of the microflora of the rumen secrete cellulases, but no cellulase activity has been demonstrated in mammalian digestive secretions.[66] Fungal cellulases seem to lead to a breakdown of the primary wall and conversion into water-soluble products; bacterial action involves a surface pitting of the wall.[147]

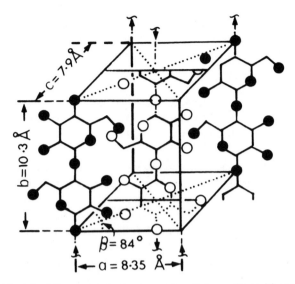

Figure 14. Model of the crystal structure of cellulose (from Meyer and Misch, 1927) showing close packing of the chains.

The initial changes in the cellulose fibers on treatment with cellulases are not associated with great changes in the degree of polymerization, and the initial sites of attack are apparently limited to the noncrystalline regions. Substitution of cellulose tends to increase susceptibility to enzymolysis, possibly because the substituents reduce the degree of intermolecular hydrogen-binding in the fiber. The changes in the susceptibility of the molecule following physical or chemical treatment emphasize the difficulty in relating the rate of enzymolysis of cellulose preparations to the behavior of native celluloses.[66]

The majority of cellulase preparations give a random cleavage of glycosidic bonds producing cellulodextrins and D-glucose. Cellobiose is also occasionally seen where the preparations are devoid of cellobiase activity.

Acid Hydrolysis. Aqueous acids hydrolyze the glycosidic linkages between C-1 and oxygen. Hydrolysis appears to be random along the length of the chain with some evidence of an increased rate at the ends of chains. It is probably here that the carboxylic acids introduced during purification have an inductive effect that influences the rate of hydrolysis. Complete hydrolysis to D-glucose is possible under homogeneous conditions and can be achieved by an initial treatment with strong sulfuric acid followed by a dilute acid hydrolysis.

The heterogeneous hydrolysis of solid cellulose follows complex kinetics. The initial rapid rate of hydrolysis followed by a slower rate can be interpreted in terms of an initial attack on the accessible disordered regions or the fibrillar surfaces followed by a slower hydrolysis of the less accessible regions.[82]

There is a rapid loss in fiber strength possibly associated with a rapid decrease in the average molecular chain length leading to microcrystalline forms.

Oxidation and other changes associated with isolation involve the formation of carbonyl and carboxylic acid groups; chlorite is capable of oxidizing terminal groups.

Sorption of Water. Cellulose, although insoluble, has a great affinity for water and will absorb water from the air. At relative humidities of between 90 and 100%, the water absorbed may be equal to 15–24% of the weight of the cellulose.

Sorption of water is accompanied by swelling and, as this does not affect the X-ray diffraction pattern, the absorption is probably limited to the noncrystalline regions. Aqueous sodium hydroxide causes changes in the X-ray diffraction pattern, and some separation of the chains in the lattice occurs.

The numerous chemical studies of cellulose illustrate a feature that is of great importance when considering the behavior of cellulose in the dietary fiber complex in the gastrointestinal tract and when studying the behavior of purified celluloses in man.

In native cellulose the susceptibility to hydrolysis, both enzymatic and acid, is dependent at least initially on the proportion of disordered noncrystalline regions. Any treatment that brings about changes in the crystalline regions of the cellulose will increase the reactivity of the cellulose.

F. Lignin

The other major component of the plant cell wall is the complex aromatic polymer lignin. This component is very resistant. to chemical degradation and appears to be more resistant to enzymatic digestion than any other naturally occurring polymer. There are probably many different types of lignin polymer, and on structural grounds this component should be described as "the lignins." It is, however, convenient to use the term lignin in a collective sense to mean the lignin fraction.

Much of the information concerning the structure of lignin has been derived from studies on lignin isolated from wood.[55,126] Lignin itself is extremely inert, and in the majority of cases vigorous chemical procedures are required to free the lignin from contaminating polysaccharides. The extent of modifications introduced into the lignin molecule by these procedures is difficult to assess, but it is important to recognize the fact that all structural studies are based on lignin preparations that have probably undergone some degree of chemical modification during their isolation.

The current concepts of the structure of the lignin molecule are based primarily on the analysis of the degradation products of lignin produced by various oxidation procedures. In recent years these have been augmented by studies of the biosynthetic routes leading to lignin formation.[48]

Studies on the lignin in nonwoody tissues are relatively few in number, although considerable attention has been given to the lignin in grasses and other plants used as forages.[25,67,68] These studies seem to indicate that many of the structural features seen in the lignin from woody tissues are also present in lignin from other sources.[24]

Molecular weights have been measured for preparations from spruce lignin; these were of the order of 8000, corresponding to more than 40 phenyl propanoid units. Carboxyl and aliphatic and phenolic hydroxyl groups were present in the molecule.

It is not possible at the present time to give a structure for lignin; in fact, the reactions now known to be involved in the biosynthesis of lignin are such that one would expect them to produce a random type of polymer.[56] Neish[106] gives the structural elements of guaiacyl lignin (Figure 15), which incorporates the major features of the lignin molecule.

The variety of types of linkage and the various functional groups distributed through the molecule is well illustrated in this composite figure. The actual spatial configuration of a molecule of this type of structure is difficult to establish, but one may conjecture a complex three-dimensional system which, in the cell wall, is infiltrated around the cellulose fibrils. The aromatic character of the lignin molecule may also be imagined to create hydrophobic regions in the cell-wall structure.

G. Minor Components of the Plant Cell Wall

In addition to the polysaccharides and lignin, the typical plant cell wall contains small amounts of other substances.[2,26,116] It has been suggested that the term dietary fiber should include all the materials in the plant cell wall. While not subscribing to this all-embracing definition, one must be aware of the nature of these substances which may be associated with the components of dietary fiber, as they may have some effect on the chemistry of the complex.

1. Protein

Small amounts of protein are usually associated with isolated cell walls. Until recently it was not clear whether these proteins were contaminants derived from the cytoplasm or were part of the wall structure itself. It is now clear that there is a fairly well-defined class of plant proteins, sometimes called extensins, which form an integral part of the cell wall. These proteins are characteristically

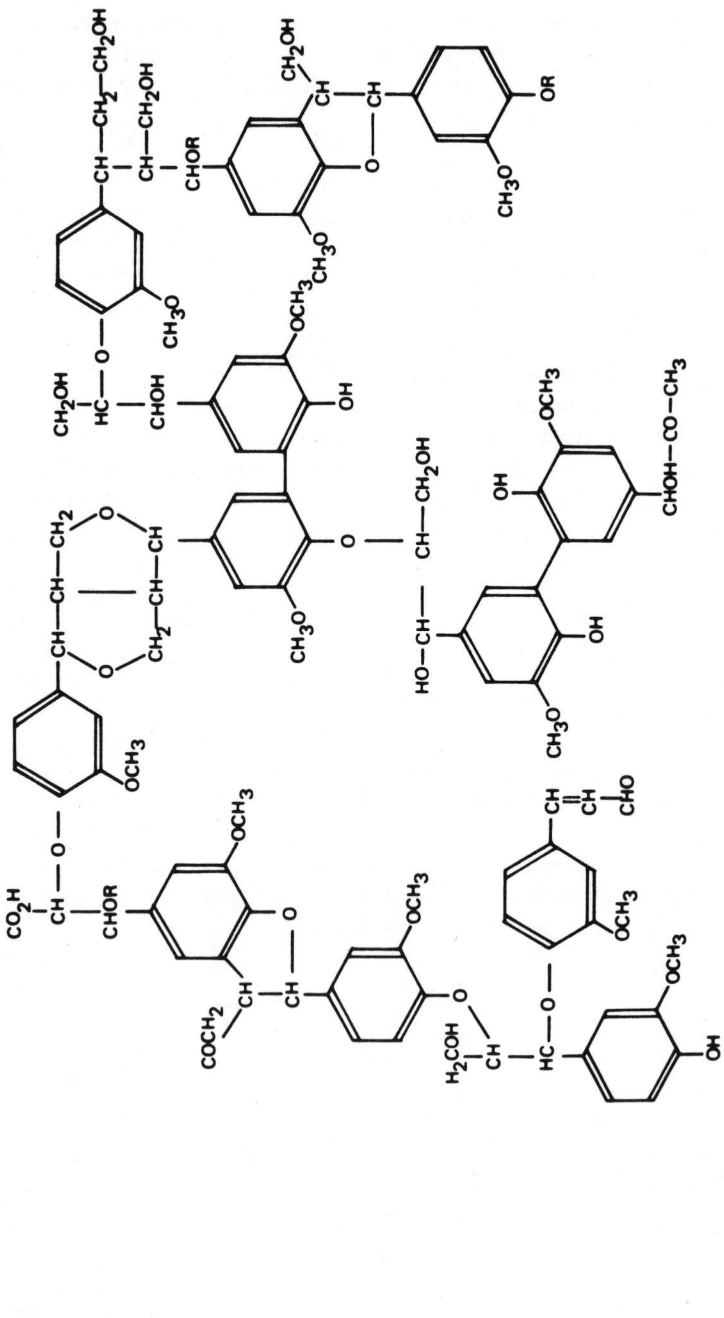

Figure 15. Structural features of the lignin molecule (after Neish, 1965).

rich in hydroxyproline and have been viewed by some authors as playing an analogous role to the collagen of connective tissues in animals.[87]

The hydroxyl group of the hydroxyproline forms bonds with the hemicellulosic polysaccharides. It has been postulated that the orientation of polysaccharide chains by the protein moiety of the wall may be important in determining the structure of the wall during elongation of the growing cell.

2. Inorganic Constituents

Many plant cells contain appreciable amounts of inorganic constituents as insoluble inclusions,[50,105] and some of these are deposited within the cell wall itself. Depositions of calcium and magnesium carbonates and calcium oxalate are common, and in many plants they are formed along vascular bundles in leaves. Silicon is found in all plants grown in soil, and in some plants considerable amounts of silicates are associated with cell walls. The Equisitales are very remarkable in this respect but many Gramineae have extensive deposits in both stem and leaf structures. The extent to which silica-rich plant tissues are eaten by man is uncertain, but in animals the proportion of silica in the cell wall of the forage is known to be a factor in determining the rate of degradation in the rumen.[40,142]

3. Cutin and Suberin

All plant organs that are above the ground have a thin external cuticular layer. This is made up of inert extracellular secretions and, although it is primarily on the surface of the epidermal cells, in some cases it appears to be partially encrusted into the structure of the cell wall.

The major component of this cuticular layer is cutin; this is a very hydrophobic substance which contains a number of hydroxy monocarboxylic acids with 16 or 18 carbon atoms in linear chain with two or three hydroxyl groups. These may spontaneously form polyesters with one another in the native cutin.[92]

Suberin appears to be a related type of substance.[125]

4. Waxes

A thin layer of wax is characteristically found over the stem, leaf, flower, and fruit of many plants. These waxes contain esters of higher fatty acids and higher aliphatic alcohols. Some alkanes are often present with odd carbon numbers ranging from C_{21} to C_{37}. The chain lengths of the acids range from C_{14} to C_{34} and of the alcohols from C_{22} to C_{32}.[47]

The amounts of these waxes consumed with the dietary fiber will obviously depend very much on the type of plant organ being consumed.

III. Arrangement and Distribution of the Components of Dietary Fiber in Foods

A. General Features of the Arrangement within the Cell Wall

Most dietary fiber is eaten in the form of plant cell wall structures and not as substances distributed randomly in the diet. This implies that any account of the chemistry of dietary fiber in relation to nutrition must also take into account the effects of the chemical and physical interrelations between the various components in the plant cell wall.

1. Ultrastructure of the Plant Cell Wall

Most of the detailed studies of the plant cell wall have been made on plants that are not normally considered part of the human diet. There has been, as one might expect, a considerable concentration on the development of the wall in woody tissues. Studies in these areas and studies of young cells growing in tissue culture,[14] taken in conjunction with a number of elegant studies on the cell walls of some algae, suggest, however, that some general features of ultrastructure are common to many plants. Therefore it is relevant to review briefly these findings in relation to the chemistry of dietary fiber.

Development of the Cell Wall.[2,105] The first visual evidence of wall formation is commonly the appearance of the cell plate between the daughter nuclei. This gradually extends to join onto the existing walls and becomes the middle lamella, on which is formed the cell wall proper. This structure appears, on histochemical evidence, to be rich in pectic substances. In general, the intercellular material in most plants appears to belong to this classification. Isolated cell preparations can be prepared by treatment with hot solutions of EDTA, which characteristically dissolve pectic material that has been deposited as calcium pectate.

The primary wall, which is deposited on the middle lamella, consists of cellulose fibrils that appear to be deposited as a random network. The matrix surrounding these fibrils contains noncellulosic polysaccharides with both pectic and hemicellulosic characteristics.[148]

The secondary wall, which is usually considerably thicker than the primary wall, is characteristically composed of a number of reasonably distinct layers. Usually it is possible to recognize three layers. In the layers of the secondary wall, the cellulose fibrils are laid down in a regular parallel arrangement. The angle that these fibrils present to the axis of the cell differs in the distinct layers.[105]

The matrix surrounding the fibrils in the secondary wall is composed principally of noncellulosic polysaccharides of the hemicellulosic type. Only a few intensive studies have been made of secondary walls, and these showed variations in the proportions of the different hemicellulosic polysaccharides in the different layers.

Lignification usually starts in the middle lamella and often at the point where several cell boundaries come together. The process then continues towards the cell lumen. Lignification of a wall is almost invariably associated with an apparent increase in the thickness of the wall. This swelling on lignification appears to indicate that the lignin is deposited within the original cellulosic structure of the wall and that lignification does not involve replacement of the noncellulosic matrix.[38]

Chemical Interactions between the Components of the Plant Cell Wall. The actual nature of interactions between the components has yet to be established with any certainty. Undoubtedly there may be a considerable amount of hydrogen bonding between the hydroxyl groups of the component polysaccharides.

Linkages between the side chains of some hemicellulosic polysaccharides and the many hydroxyproline residues that appear to be characteristic of the cell-wall proteins have been demonstrated. These linkages have been suggested as a method for controlling the orientation of the polysaccharide fibers in the wall.[87,117]

The intermolecular associations involved in the formation of polysaccharide gels are also indicative of the type of binding one would expect in the matrix.[119,120]

Many of the hemicellulosic polysaccharides have some of the conformational characteristics of the cellulose chains. It is highly probable that some regions of these backbone chains may be hydrogen bonded to some extent with the cellulose fibrils.

The nature of linkages between the lignin and the polysaccharides of the wall is similarly unclear.[89,104] What is often called the lignocellulose complex can be isolated from tissues by mild hydrolytic procedures, and its resistance to all but the most vigorous chemical treatment is indicative of some degree of covalent bonding. Some xylans also appear to be very closely associated with this lignocellulose complex.

B. Distribution of Dietary Fiber and Its Components in Foodstuffs

The human diet includes examples of all types of plant organs which, in turn, contain a number of different types of tissue, each with its own characteristic cellular structure. Although there have been only a few detailed quantitative studies of the composition of the cell wall in different tissues, those that are available show that the composition of the wall varies from tissue to tissue.[32,73,124,128,136,147]

This variation means that the proportions of the different components of dietary fiber consumed will be greatly influenced by the types and amounts of the various plant organs that are included in the diet. In this section an attempt has been made to bring together some of the salient features of plant anatomy and cell-wall structure in the foods eaten by man and to use these features to indicate in broad terms the effects of variations in the diet on the chemical composition of the dietary fiber it provides. Present knowledge does not permit a discussion of the detailed quantitative aspects of this topic at the present time, but a large body of information derived from anatomical studies[38,45] can be used to indicate in qualitative terms the differences between dietary fiber from the different types of plant organs.

1. Stem Structures

The stem of most food plants contains large amounts of thin-walled parenchymatous tissues. The walls of these cells are relatively richer in pectic substances than more mature walls. In the main, the supporting tissues in the stems consumed in the human diet are the collenchyma with thicker, unlignified walls. In many foods the xylem elements in the vascular bundles are relatively lightly lignified with only a few xylem vessels in each bundle giving the characteristic lignin stain with phloroglucinol in acid. The epidermal tissues of the stem have thicker walls, and the outer surface carries a cuticular layer. In more mature stems, the deposition of cutin and suberin occurs; in stem structures that are formed beneath the soil, such as the potato, a considerable amount of suberinization occurs. Analagously the stems of plants grown in dry climates tend to have a more heavily cutinized epidermis, and the cell walls of the epidermis are thicker and may be lignified.

As stem structures mature, the proportions of the vascular material increases until a continuous cylinder composed of lignified xylem elements is formed, leading eventually to the formation of the structure seen in woody tissues. Most foods, however, are consumed in a relatively immature condition, and in these the vascular bundles are still discrete and lightly lignified.

2. Leaf Structures

The petiole carries the vascular bundles emerging from the stem, the arrangement of which varies with the species of plant. In general these bundles are lightly lignified and rich in thick-walled collenchyma cells, which provide the stem with its main structural strength. The size of the bundles and the number of lignified elements per bundle decrease as the veins divide toward the periphery of the leaf.

In the leaf blade itself there are upper and lower epidermal layers, the walls of which are thickened. The outer surfaces carry a cuticular layer, which in many leaves is protected by a distinctive waxy coating. The cells of the mesophyll between the two epidermal layers are thin walled and form the photosynthetic tissues.

In the leaves of monocotyledonous plants, the arrangement of the vascular bundles is somewhat different, and they are characteristically associated with the heavily lignified sclerenchyma cells. In grasses the proportions of lignified elements is often higher than in the leafy tissues of dicotyledonous plants.

3. Root Structure

In the developed countries, the roots of plants are usually eaten before any pronounced secondary thickening has occurred. In the very young root, the xylem elements develop alongside the phloem and are surrounded by thin-walled cells; as the root matures, the xylem forms a continous cylinder. The immature xylem contains relatively few lignified elements, but very mature roots may contain a very heavily lignified cylinder of xylem. The epidermal tissues of roots show depositions of suberin early in their development, and even in immature roots areas of suberinization can be seen where lateral root development is taking place internally.

4. Fruits

In addition to the conventional fruits, in the realm of the plant anatomist, many so-called vegetables (e.g., cucumber, squash, tomato) are, strictly speaking, fruits. It is customary to divide fruits into two classes, fleshy and hard, the latter including the majority of nuts eaten by man. All fruits have the same basic vascular elements as the floral parts from which they are derived. The vascular supply is usually strengthened as the fruit matures, although in most fruits the xylem elements in the fruit are lightly lignified.

The epidermal tissues may have thin-walled cells, as in many berries, but in other fruits the cells are thick-walled and have heavy cuticles. In some fruits, for example, some varieties of apples and pears, a peridermal layer develops which is suberinized.

The major portion of fleshy fruits consists of succulent parenchyma; in dry fruits lignified sclerenchyma and nonsucculent parenchyma are present. The parenchyma of the pericarp in fleshy fruits may be homogeneous, as in the apple,[84] peach, and cherry, or it may contain succulent parenchyma cells which are more or less turgid with fluid and sclereids. These sclereids frequently form in clusters and give a gritty texture to the flesh, as in the pear.[81]

The typical structure of citrus fruits is formed by thin-walled multicellular emergences derived from the inner surfaces of the carpels. The macroscopically distinct vascular elements in these fruits often contain only a few lignified elements.

In many fruits several distinct layers can be seen that frequently reflect differences in cell type and possibly differences in the cell-wall structure and composition.

Drupes typically have a fleshy outer layer enclosing a hard or stony layer. These hard layers consist of sclereids and fibers; in some fruits such as the apple this layer is described as cartilaginous and is also heavily lignified. The lignification of these hard tissues is usually quite complete, and the stony layer of these fruits is very resistant to chemical and biological degradation.

Dry fruits are basically similar in structure to fleshy fruits in the immature stages. The changes occurring during development involve the formation of sclerenchyma in the pericarp, which is often both lignified and suberinized.

The placenta forms a large part of many fruits such as the tomato or watermelon. The placenta usually consists of soft thin-walled parenchyma cells which may be quite large. The vascular elements in the placenta are usually only lightly lignified.

5. Seeds

The seed coat is often very closely applied to the embryo and may be fused with the inner layers of the pericarp. The seed coat of most species has a protective function and is frequently impervious to water by virtue of a thick cuticle. The walls of the cells in the seed coat are often thickened and frequently lignified.

In other species the outer seed coat develops as a fleshy layer and in some, mucilaginous cells are present that absorb water and swell, rupturing the seed coat and cuticle. The vascular elements in the seed coat are usually limited in number unless the seed is large and the seed coat of complex structure.

The embryo usually contains undifferentiated vascular elements; in seeds with cotyledons that expand on germination, the vascular elements may be at the procambium stage. In these, lignification is not usually detectable, and most of the tissues are composed of parenchymatous cells with thin walls. In the cereal grains and many seeds, the endosperm cells are packed with starch granules. Where other reserve polysaccharides are present, these are usually deposited in the walls of the endosperm, which are correspondingly thickened.[3,4]

IV. Conclusions

This review of the chemistry of dietary fiber shows very clearly that many of the details concerning the chemical structures and properties of the individual components have yet to be established. For this reason, it is difficult to draw any firm conclusions about particular features of polysaccharide or lignin structure and chemical or physicochemical properties that may be important in determining the role that dietary fiber plays in the physiology of the large intestine.[37,46,94]

It is, however, possible to draw two conclusions. The first concerns the way in which the components of fiber in the human diet should be classified (and therefore the terminology which should be used to describe them), and the second concerns a general view of the nature of the dietary fiber complex.

A. Classification of Dietary Fiber

The account of the chemistry of the components of dietary fiber has, of necessity, been presented using the classical terminology for the different fractions of the plant cell wall. The structural studies on these isolated components show that there are many structural features that are shared by polysaccharides falling into the different groupings: pectic substances, hemicelluloses, and gums and mucilages.

Furthermore, modern views on the chemistry of the plant cell wall suggest that some of these isolated components are artifacts,[2] resulting from the action of the reagents used in fractionation on the native polysaccharides in the plant tissue. This appears to be true for many of the components of the pectic substances[8] and of the hemicellulose fractions following alkali extraction.

In view of these two considerations, it is suggested that the terminology used in the classification of the components of dietary fiber should be modified and to a certain extent simplified.

The primary groupings would be under three headings.

Cellulosic. This group is limited to the linear $(1\rightarrow4)\beta$-glucans, usually with a high degree of polymerization.

Noncellulosic Polysaccharides. This group includes all the matrix polysaccharides from the cell wall, together with all the other indigestible polysaccharides other than cellulose. The group includes a large number of different structural types of polysaccharide and ideally should be subdivided on a structural basis.

The analytical problems associated with the fractionation of this complex group are immense[134,143] and until studies are available that show the types of polysaccharide structure that are related to the observed physiological properties of dietary fiber, some simplified compromise will be necessary.

The use of the term pectic substances should be restricted to polysaccharides that are predominantly galacturonans. The term hemicellulose, however, is more difficult to defend, and it should correctly be restricted to alkali-extracted substances and not used as a collective term for the matrix polysaccharides as a whole.

Lignin. This category includes all the inert aromatic polymeric matter in the cell wall. Analytically, it is frequently associated with nitrogenous matter and may be contaminated with cutin and suberin. The fraction includes a range of polymeric forms, and a more precise definition, which is less dependent on the method used for analysis, may be difficult to establish.

The main features of this classification are summarized in Table III, which also includes some subsidiary structural features that may be important in determining the physiological role of dietary fiber.

B. The Nature of the Dietary Fiber Complex

Although there still remains a considerable amount of work to be done, it is possible to present a general overview of the dietary fiber complex in the human diet.

The exact nature of the complex will depend on the foods present in the diet, but there are some features that will be common to most diets. Dietary fiber is mainly provided by the cell-wall structures from the plant foods in the diet, with

TABLE III. A Structural Classification of the Components of Dietary Fiber

Primary grouping	Structural types	Subsidiary structural features of possible importance in determining properties
Cellulose	Long-chain 1,4-β-D-glucans	Degree of polymerization Distribution of noncrystalline regions in fibrils Presence and distribution of other sugars

Noncellulosic polysaccharides	Galacturonans
	Galacturonorhamnans
	Arabinans
	Xylans
	  glucurono-
	  arabino-
	Mannans
	  galacto-
	  gluco-
	
	Galactans
	  arabino-
	Glucans
	  $\beta 1 \rightarrow 3$: 1,4
	  xylo-
		Degree of polymerization
		Extent of methyl esterification of
		  uronic acids
		Presence of acetyl groups
		Number and distribution of side chains
		
		Type and extent of branching
		Intermolecular bonding - covalent
		Intermolecular bonding - hydrogen

Lignin	Unknown but different
	  structural types are
	  probable
		Degree of polymerization
		Distribution of functional groups
		Bonding with polysaccharides

small amounts of related polysaccharides contributed by many manufactured or processed foods. These latter polysaccharides are dispersed throughout the foods and are usually consumed in water-soluble forms.

The cell-wall structures, however, are contained in discrete particles which are distributed very unevenly through the diet. The lignified portions of these structures are particularly unevenly distributed.

In both lignified and unlignified tissues there are a range of noncellulosic polysaccharides. These heteropolysaccharides are based on a limited number of basic polysaccharide chains which carry a variety of side chains. The structures of many of these basic chains possess some similarities in conformation to cellulose[127] but usually have a lower degree of polymerization and lack the capacity to form highly ordered fibers. The tertiary structure of these molecules may be such that the side chains are more important as determinants of the properties of the polysaccharide than the basic core chain itself. In the arabinoxylans, for example, the screw symmetry of the xylan chain means that the external character of the molecule will be very much that of the arabinofuranose residues in the side chains, which would effectively shield the xylan core.

In the heteropolysaccharides with uronic acid in the main chain (the pectic substances), the degree of methyl esterification of the carboxyl groups is a major determinant of the physicochemical properties of the molecule.

In the cell wall there is a considerable amount of intermolecular bonding involving both covalent and hydrogen bonding.[23,60] The exact nature and extent of these linkages is uncertain as yet, but in the plant and therefore in the diet, these intermolecular associations and linkages will probably be important in determining the properties of the complex as a whole.

This is particularly true for the cellulose and lignin; for both these components there is evidence for some complex bonding, and the presence of lignin in a wall structure limits the enzymatic degradation of the cellulose.[40,41] The aromatic character of the lignin may result in hydrophobic regions in the wall where enzymolysis would be restricted.

The dietary fiber eaten by man is a highly complex mixture of many components, and the properties of the individual components may well be extensively modified by the interrelationships within the cell wall structures. For this reason the physiological properties of dietary fiber may not depend solely on the chemical properties of its individual components but may, in part at least, depend on the fact that these components are eaten as cell-wall structures.

ACKNOWLEDGMENTS

During the preparation of this chapter and of Chapter 3, I have had the benefit of discussions with many associates and especially my coworkers, Ms. Celia Greenberg and Mr. W. J. Branch. I would like to acknowledge their help and support.

References

1. ABDEL-FATTAH, A. F., and M. EDREES. The pectic substances of pigmented onion skins. Part II. Some structural features of the pectin and pectic acid. *Carbohydr. Res.* **28:**114, 1973.
2. ALBERSHEIM, P. Biogenesis of the cell wall. *In: Plant Biochemistry,* J. Bonner and J. E. Varner (eds.). New York and London, Academic Press, 1965, p. 298.
3. AMIN, EL. S., and M. M. EL-SAYED. The polysaccharides of mango seeds (*Mangifera indica* Var Bullocks Heart). *Carbohydr. Res.* **27:**39, 1973.
4. AMIN, EL. S., and A. M. PALEOLOGOV. A study of the polysaccharides of the kernel and endocarp of the fruit of Doum palm (*Hyphaene thebaica*). *Carbohydr. Res.* **27:**447, 1973.
5. ANDERSON, E., and L. SANDS. A discussion of methods of value in research on plant polyuronides. *In: Advances in Carbohydrate Chemistry,* W. W. Pigman and M. L. Wolfrom (eds.). New York, Academic Press, 1945, Vol. 1, p. 329.
6. ASPINALL, G. O. Structural chemistry of the hemicelluloses. *In: Advances in Carbohydrate Chemistry,* M. L. Wolfrom (ed.). New York and London, Academic Press, 1959, Vol. 14, p. 429.
7. ASPINALL, G. O. Gums and mucilages. *In: Advances in Carbohydrate Chemistry and Biochemistry,* M. L. Wolfrom and R. S. Tipson (eds.). New York and London, Academic Press, 1969, Vol. 24, p. 333.
8. ASPINALL, G. O. Pectins, plant gums and other plant polysaccharides. *In: The Carbohydrates, Chemistry and Biochemistry,* 2d ed., W. W. Pigman and D. Horton (eds.). New York and London, Academic Press, 1970, Vol. IIB, p. 515.
9. ASPINALL, G. O., J. W. T. CRAIG and J. L. WHYTE. Lemon peel pectin. Part 1. Fractionation and partial hydrolysis of water-soluble pectin. *Carbohydr. Res.* **7:**442, 1968.
10. ASPINALL, G. O., and R. S. FANSHAWE. Pectic substances from lucerne (*Medicago sativa*). Part 1. Pectic acid. *J. Chem. Soc.* **1969:**4215, 1969.
11. ASPINALL, G. O., and R. J. FERRIER. The constitution of barley husk hemicellulose. *J. Chem. Soc.* **1957:**4188, 1957.
12. ASPINALL, G. O., B. GESTETNER, J. A. MOLLOY and M. UDDIN. Pectic substances from lucerne (*Medicago sativa*). Part II. Acidic oligosaccharides from partial hydrolysis of leaf and stem pectic acids. *J. Chem. Soc., Ser. C* **1968:**2554, 1968.
13. ASPINALL, G. O., K. HUNT and I. M. MORRISON. Polysaccharides of soybean. Part V. Acidic polysaccharides from the hulls. *J. Chem. Soc. Ser. C* **1967:**1080, 1967.
14. ASPINALL, G. O., and KUO-SHI JIANG. Rapeseed hull pectin. *Carbohydr. Res.* **38:**247, 1974.
15. ASPINALL, G. O., and K. M. ROSS. The degradation of two periodate-oxidised arabinoxylans. *J. Chem. Soc.* **1963:**1681, 1963.
16. ASPINALL, G. O., and R. G. J. TELFER. Cereal gums. Part 1. The methylation of barley glucosan. *J. Chem. Soc.* **1954:**3519, 1954.
17. BAKER, G. L. High-polymer pectins and their de-esterification. *Adv. Food Res.* **1:**395, 1948.
18. BARRETT, A. J., and D. H. NORTHCOTE. Apple fruit pectic substances. *Biochem. J.* **94:**617, 1965.
19. BEAVAN, G. H., and J. K. N. JONES. Pectic substances. Part V. The molecular structure of strawberry and apple pectic acids. *J. Chem. Soc.* **1947:**1218, 1947.
20. BEVERIDGE, R. J., J. F. STODDART, W. A. SZAREK and J. K. N. JONES. Some structural features of the mucilage from the bark of *Ulmus fulva* (slippery elm mucilage). *Carbohydr. Res.* **9:**429, 1969.
21. BISHOP, C. T. Carbohydrates of sunflower heads. *Can. J. Chem.* **33:**1521, 1955.
22. BLAKE, J. D., and G. N. RICHARDS. An examination of some methods for fractionation of plant hemicelluloses. *Carbohydr. Res.* **17:**253, 1971.
23. BLAKE, J. D., and G. N. RICHARDS. Evidence for molecular aggregation in hemicelluloses. *Carbohydr. Res.* **18:**11, 1971.

24. BLAND, D. E., A. F. LOGAN and M. MENSHUN. The lignin of *Sphagnum*. *Photochemistry*. **7:**1373, 1968.

25. BONDI, A., and H. MEYER. Lignins in young plants. *Biochem. J.* **43:**248, 1948.

26. BONNER, J., and J. E. VARNER (eds.). *Plant Biochemistry*. New York and London, Academic Press, 1965.

27. BUCHALA, A. J., C. G. FRASER and K. C. B. WILKIE. Quantitative studies on the polysaccharides in the nonendospermic tissues of the oat plant in relation to growth. *Phytochemistry* **10:**1285, 1971.

28. BUCHALA, A. J., C. G. FRASER and K. C. B. WILKIE. An acidic galactoarabinoxylan from the stem of *Avena sativa*. *Phytochemistry* **11:**2803, 1972.

29. BUCHALA, A. J., and H. MEIER. A galactomannan from the leaf and skin tissues of red clover. *Carbohydr. Res.* **31:**87, 1973.

30. BUCHALA, A. J., and H. MEIER. A hemicellulosic β D-glucan from maize stem. *Carbohydr. Res.* **26:**421, 1973.

31. BUCHALA, A. J., and K. C. B. WILKIE. The ratio of $\beta(1{\to}3)$ to $\beta(1{\to}4)$ glucosidic linkages in nonendospermic hemicellulosic β glucans from oat plant (*Avena sativa*) tissues at different stages of maturity. *Phytochemistry* **10:**2287, 1971.

32. BUCHALA, A. J., and K. C. B. WILKIE. Total hemicellulose from wheat at different stages of growth. *Phytochemistry* **12:**499, 1973.

33. BUCHALA, A. J., and K. C. B. WILKIE. Uronic acid residues in the total hemicelluloses of oats. *Phytochemistry* **12:**655, 1973.

34. COLE, E. W. Isolation and chromatographic fractionation of hemicelluloses from wheat flour. *Cereal Chem.* **44:**411, 1967.

35. COLE, E. W. Some physicochemical properties of a wheat flour hemicellulose in solution. *Cereal Chem.* **46:**382, 1969.

36. COOK, W. J., and C. E. BUGG. Calcium interactions with D-glucans, crystal structure of $\alpha\alpha$ trehalose-calcium bromide monohydrate. *Carbohydr. Res.* **31:**265, 1973.

37. CUMMINGS, J. H. Dietary fiber, progress report. *Gut* **14:**69, 1973.

38. CUTTER, E. G. *Plant Anatomy — Experiment and Interpretation. Part 2. Organs*. London, Arnold, 1971.

39. DEA, I. C. M., D. A. REES, R. J. BEVERIDGE and G. N. RICHARDS. Aggregation with change of conformation in solutions of hemicellulose xylans. *Carbohydr. Res.* **29:**363, 1973.

40. DEKKER, R. F. H., and G. N. RICHARDS. Digestion of polysaccharide constituents of tropical pasture herbage in the bovine rumen. *Carbohydr. Res.* **22:**1, 1973.

41. DEKKER, R. F. H., and G. N. RICHARDS. Effect of delignification on the *in vitro* rumen digestion of polysaccharides of bagasse. *J. Sci. Food Agric.* **24:**375, 1973.

42. DELUCAS, L., C. BUGG, A. TERZIS and P. RIVEST. Calcium binding to D-glucuronate residues: Crystal structure of a hydrated calcium bromide salt of D-glucuronic acid. *Carbohydr. Res.* **41:**19, 1975.

43. DONNELLY, B. J., J. L. HELM and H. A. LEE. The carbohydrate composition of corncob hemicellulose. *Cereal Chem.* **50:**548, 1973.

44. DUCKWORTH, M., and W. YAPHE. The structure of agar. Part 1. Fractionation of a complex mixture of polysaccharides. *Carbohydr. Res.* **16:**189, 1971.

45. EAMES, E. J., and L. H. McDANIELS. *An Introduction to Plant Anatomy*, 2d ed. New York, McGraw-Hill, 1951.

46. EASTWOOD, M. A. Vegetable fibre: its physical properties. *Proc. Nutr. Soc.* **32:**137, 1973.

47. EGLINGTON, G., A. G. GONZALES, R. J. HAMILTON and R. A. RAPHAEL. Hydrocarbon constituents of the wax coatings of plant leaves: A taxonomic survey. *Phytochemistry* **1:**89, 1962.

48. EL-BASYOUNI, S. Z., A. C. NEISH and G. H. N. TOWERS. The phenolic acids in wheat. III. Insoluble derivatives of phenolic cinnamic acids as natural intermediates in lignin biosynthesis. *Phytochemistry* **3:**627, 1964.

49. ELYAKOVA, L. A., and T. N. ZVYAGINTSEVA. A study of the laminarins of some far eastern brown seaweeds. *Carbohydr. Res.* **34:**241, 1974.
50. EPSTEIN, E. Mineral metabolism. *In: Plant Biochemistry*, J. Bonner and J. E. Varner (eds.). New York, Academic Press, 1965, p. 460.
51. ERSKIN, A. J., and J. K. N. JONES. Fractionation of polysaccharides. *Can. J. Chem.* **34:**821, 1956.
52. EWALD, C. M., and A. S. PERLIN. The arrangement of branching in an arabinoxylan from wheat flour. *Can. J. Chem.* **37:**1254, 1959.
53. FINCHER, G. B., W. H. SAWYER and B. A. STONE. Chemical and physical properties of an arabinogalactan-peptide from wheat. *Biochem. J.* **139:**535, 1974.
54. FRASER, C. G., and K. C. B. WILKIE. A hemicellulose glucan from oat leaf. *Phytochemistry* **10:**199, 1971.
55. FREUDENBERG, K. *In: The Formation of Wood in Forest Trees*, M. H. Zimmermann (ed.). New York, Academic Press, 1964, p. 203.
56. FREUDENBERG, K. Lignin, its constitution and formation from p-hydroxycinnamyl alcohols. *Science* **148:**595, 1965.
57. GAILLARD, B. D. E. Separation of linear from branched polysaccharides by precipitation as iodine complex. *Nature* **191:**1295, 1961.
58. GAILLARD, B. D. E. Comparison of the hemicelluloses from plants belonging to two different plant families. *Phytochemistry* **4:**631, 1965.
59. GAILLARD, B. D. E., and R. W. BAILEY. The distribution of galactose and mannose in the cell wall polysaccharides of red clover (*Trifolium pratense*) leaves and stems. *Phytochemistry* **7:**2037, 1968.
60. GAILLARD, B. D. E., and G. N. RICHARDS. Presence of soluble lignin-carbohydrate complexes in the bovine rumen. *Carbohydr. Res.* **42:**135, 1975.
61. GILL, R. E., E. L. HIRST and J. K. N. JONES. Constitution of the mucilage from the bark of *Ulmus fulva* (slippery elm mucilage). Part II. The sugars formed in the hydrolysis of the methylated mucilage. *J. Chem. Soc.* **1946:**1025, 1946.
62. GLICKSMAN, M. Utilization of natural polysaccharide gums in the food industry. *Adv. Food Res.* **11:**109, 1962.
63. GOLDBERG, R. Etude des polysaccharides de réserve de deux graines de Liliacees: *Asparagus officinale* et *Endymion nutansdumer*. *Phytochemistry* **8:**1783, 1969.
64. GREENBERG, D. C. Variation of gum content in barley. Ph.D. dissertation, University of Cambridge, 1974.
65. GREMLI, H., and B. O. JULIANO. Studies on alkali-soluble rice-bran hemicellulose, *Carbohydr. Res.* **12:**273, 1970.
66. HALLIWELL, G. The enzymic decomposition of cellulose. *Nutr. Abs. Rev.* **29:**747, 1959.
67. HIGUCHI, T., Y. ITO and R. A. KAWAMURA. p-Hydroxyphenyl propane component of grass lignin and role of tyrosine ammonia lyase in its formation. *Phytochemistry* **6:**875, 1967.
68. HIGUCHI, T., Y. ITO, M. SHIMADA and R. A. KAWAMURA. Chemical properties of milled wood lignin of grasses. *Phytochemistry* **6:**1551, 1967.
69. HIRST, E. L., L. HOUGH and J. K. N. JONES. Constitution of the mucilage from the bark of *Ulmus fulva* (slippery elm mucilage). Part III. The isolation of 3-monomethyl D galactose from the products of hydrolysis. *J. Chem. Soc.* **1951:**323, 1951.
70. HIRST, E. L. and J. K. N. JONES. The chemistry of pectic materials. *In: Advances in Carbohydrate Chemistry*, W. W. Pigman and M. L. Wolfrom (eds.). New York, Academic Press, 1946, Vol. 2, p. 235.
71. HIRST, E. L., and J. K. N. JONES. Pectic substances. Part VI. The structure of the araban from *Arachis hypogea*. *J. Chem. Soc.* **1947:**1221, 1947.
72. HIRST, E. L., J. K. N. JONES and W. O. WALDER. Pectic substances. Part VII. The constitution of the galactan from *Lupinus albus*. *J. Chem. Soc.* **1947:**1225, 1947.

73. Hirst, E. L., D. J. Mackenzie and C. B. Wylam. Analytical studies on the carbohydrates of grasses and clovers. IX. Changes in carbohydrate composition during the growth of lucerne. *J. Sci. Food Agric.* **10:**19, 1959.

74. Hirst, E. L., E. G. V. Percival and C. B. Wylam. Studies on seed mucilages. Part VI. The seed mucilage of *Plantago arenaria*. *J. Chem. Soc.* **1954:**189, 1954.

75. Hirst, E. L., D. A. Rees and N. G. Richardson. Seed polysaccharides and their role in germination. *Biochem. J.* **95:**453, 1965.

76. Hough, L., and D. B. Powell. Methylation and periodate oxidation studies of the alkali-stable polysaccharide of sugar-beet pectin. *J. Chem. Soc.* **1960:**16, 1960.

77. Howling, D. Modified starches for the food industry. *Food Technol. Aust.* **26:**464, 1974.

78. Jeffrey, G. A., and R. D. Rosenstein. Crystal-structure analysis in carbohydrate chemistry. *Adv. Carbohydr. Chem.* **19:**7, 1964.

79. Jelaca, S. L., and I. Hlynka. Water-binding capacity of wheat flour crude pentosans and the relation to mixing characteristics of dough. *Cereal Chem.* **48:**211, 1971.

80. Jelaca, S. L., and I. Hlynka. Effect of wheat-flour pentosans on dough gluten and bread. *Cereal Chem.* **49:**489, 1972.

81. Jermyn, M. A., and F. A. Isherwood. Changes in the cell wall of the pear during ripening. *Biochem. J.* **64:**123, 1956.

82. Jones, D. M. Structure and some reactions of cellulose. *Adv. Carbohydr. Chem.* **19:**219, 1964.

83. Khanna, S. N., and P. C. Gupta. The structure of a galactomannan from the seeds of *Ipomoea muricata*. *Phytochemistry* **6:**605, 1967.

84. Knee, M. Polysaccharide changes in cell walls of ripening apples. *Phytochemistry* **12:**1543, 1973.

85. Kohn, R. The activity of calcium ions in aqueous solutions of the lower calcium oliogogalacturonates. *Carbohydr. Res.* **20:**351, 1971.

86. Koreed, A., T. Harada, K. Ogawa, S. Sato and N. Kasai. Study of the ultrastructure of gel-forming $(1-3)\beta$-D-glucoan (curdlan type polysaccharide) by electron microscopy. *Carbohydr. Res.* **33:**396, 1974.

87. Lamport, D. T. A. Cell wall metabolism. *Annu. Rev. Plant Physiol.* **21:**235, 1970.

88. Leschzner, C., and A. S. Cerezo. The structure of a galactomannan from the seed of *Gleditsia tricanthos*. *Carbohydr. Res.* **15:**291, 1970.

89. Lindgren, B. O. The lignin-carbohydrate linkage. *Acta Chem. Scand.* **12:**447, 1958.

90. Lintas, C., and B. L. d'Appolonia. Note on the effect of purification treatment on water-soluble pentosans. *Cereal Chem.* **49:**731, 1972.

91. Marchessault, R. H., and C. Y. Liang. The infrared spectra of crystalline polysaccharides. VIII. Xylans. *J. Polym. Sci.* **59:**357, 1962.

92. Matic, M. The chemistry of plant cuticles: A study of cutin from *Agave americana* L. *Biochem. J.* **63:**168, 1956.

93. McCance, R. A., and R. D. Lawrence. The carbohydrate content of foods. *Med. Res. Counc. Spec. Rep. Ser.* **135** (HMSO), 1929.

94. McConnell, A. A., N. A. Eastwood and W. D. Mitchell. Physical characteristics of vegetable foodstuffs that could influence bowel function. *J. Sci. Food Agric.* **25:**1457, 1974.

95. McDowell, R. H. New developments in the chemistry of alginates and their use in foods. *Chem. Ind.* **9:**391, 1975.

96. Medcalf, D. G., and P. W. Cheung. Composition and structure of glucofructans from durum wheat flour. *Cereal Chem.* **48:**1, 1971.

97. Medcalf, D. G., B. L. d'Appolonia and K. A. Gilles. Comparison of chemical composition and properties between hard red spring and durum wheat endosperm pentosans. *Cereal Chem.* **45:**539, 1968.

98. MEDCALF, D. G., and K. A. GILLES. Structural characteristics of a pentosan from the water-insoluble portion of durum wheat endosperm. *Cereal Chem.* **45:**550, 1968.
99. MEIER, H. Studies on hemicellulose from pine (*Pinus silvestris L.*) *Acta Chem. Scand.* **12:**1911, 1958.
100. MEYER, K. H., and L. MISCH. Positions des atomes dans le nouveau modele spatial de la cellulose. *Helv. Chim. Acta* **20:**232, 1937.
101. MOLLOY, L. F., and E. L. RICHARDS. Complexing of calcium and magnesium by the organic constituents of Yorkshire fog (*Holcus lanatus*). I. The organic acids, lignin and cell wall polysaccharides of Yorkshire fog. *J. Sci Food Agric.* **22:**393, 1972.
102. MOLLOY, L. F., and E. L. RICHARDS. Complexing of calcium and magnesium by the organic constituents of Yorkshire fog (*Holcus lanatus*). II. Complexing of Ca^{2+} and Mg^{2+} by cell wall fractions and organic acids. *J. Sci. Food Agric.* **22:**397, 1972.
103. MONRO, J. A., R. W. BAILEY and D. PENNY. Arabinogalactan of hypocotyl cells walls. *Phytochemistry* **11:**1597, 1972.
104. MORRISON, I. M. Some investigations of the lignin-carbohydrate complexes of *Lolum perenne*. *Biochem. J.* **139:**197, 1974.
105. MUHLETHALER, K. Plant cell walls. *In: The Cell,* J. Brachet and A. E. Minsky (eds.). New York, Academic Press, 1961, Vol. 2, p. 85.
106. NEISH, A. C. Coumarins, phenyl propanes and lignin. *In: Plant Biochemistry,* J. Bonner and J. E. Varner (eds.). New York, Academic Press, 1965, p. 581.
107. NEUKOM, H., and H. MARKWALDER. Isolation and characterisation of an arabinogalactan from wheat flour. *Carbohydr. Res.* **39:**387, 1975.
108. NEUKOM, H., L. PROVIDOLI, H. GREMLI and P. A. HUI. Recent investigations on wheat flour pentosans. *Cereal Chem.* **44:**238, 1967.
109. NORMAN, A. G. *The Biochemistry of Cellulose, the Polyuronides etc.* London, H. Milford, Oxford Univ. Press. 1937.
110. NORRIS, F. W., and I. A. PREECE. IX. Studies on hemicelluloses. 1. The hemicelluloses of wheat bran. *Biochem. J.* **24:**59, 1930.
111. NORTHCOTE, D. H. The synthesis and metabolic control of polysaccharides and lignin during the differentiation of plant cells. *In: Essays in Biochemistry.* P. N. Campbell and G. D. Greville (eds.) London and New York, Academic Press, 1969, Vol. 5, p. 89.
112. O'DWYER, M. H. The hemicelluloses of beech wood. *Biochem. J.* **20:**656, 1926.
113. PEAT, S., W. J. WHELAN and J. G. ROBERTS. The structure of lichenin. *J. Chem. Soc.* **1957:**3916, 1957.
114. PERCIVAL, E. Algal polysaccharides. *In: The Carbohydrates,* W. W. Pigman and D. Horton (eds.). New York and London, Academic Press, 1970, Vol. 11B, p. 537.
115. PERLIN, A. S. Isolation and composition of the soluble pentosans of wheat flours. *Cereals Chem.* **28:**370, 1951.
116. PRIDHAM, J. B. Phenol-carbohydrate derivatives in higher plants. *In: Advances in Carbohydrate Chemistry,* M. L. Wolfrom and R. S. Tipson (eds.). New York and London, Academic Press, 1965, Vol. 20, p. 371.
117. PUSZTAI, A., R. BOBIE and I. DUNCAN. Fractionation and characterisation of water-soluble polysaccharide-protein complexes containing hydroxyproline from the leaves of *Vicia faba. J. Sci. Food Agric.* **22:**514, 1971.
118. RAYMOND, W. R., and C. W. NAGEL. Microbial degration of gum karaya. *Carbohydr. Res.* **30:**293, 1973.
119. REES, D. A. Structure, conformation and mechanism in the formation of polysaccharide gels and networks. *In: Advances in Carbohydrate Chemistry,* M. L. Wolfrom and R. S. Tipson (eds.). New York and London, Academic Press, 1969, Vol. 24, p. 267.
120. REES, D. A. Shapely polysaccharides. The Eighth Colworth Medal Lecture. *Biochem. J.* **126:**257, 1972.

121. REES, D. A., and R. J. SKERRETT. Conformational analysis of celliobiose, cellulose and xylan. *Carbohydr. Res.* **7**:334, 1968.
122. REES, D. A., and A. W. WRIGHT. Polysaccharide conformation. Part VII. Model building computations for α 1-4 galacturonans and the kinking function of L rhamnose residues in pectic substances. *J. Chem. Soc., Ser. B* **1971**:1366, 1971.
123. REID, J. S. G., and H. MEIER. The formation of reserve galactomannan in the seeds of *Trigonella foenum-gracum. Phytochemistry* **2**:513, 1970.
124. REID, J. S. G., and K. C. B. WILKIE. Polysaccharides of the oat plant in relationship to plant growth. *Phytochemistry* **8**:2045, 1969.
125. ROELOFSEN, P. A. *In: Encyclopedia of Plant Anatomy,* W. Zimmerman and P. G. Ozenda. (eds.). Vol. 3, Part 4. The plant cell wall. Berlin, Borntraeger, 1959.
126. SARKANEN, K. V., and C. H. LUDWIG. *Lignins: Occurrence, Formation, Structure and Reactions.* New York and London, Wiley, 1971.
127. SATHYANARAYANA, B. K., and V. S. R. RAO. Conformational studies on β-D-(1→3) linked xylan. *Carbohydr. Res.* **15**:137, 1970.
128. SELVENDRAN, R. R. Analysis of cell wall material from plant tissues: Extraction and purification. *Phytochemistry* **14**:1011, 1975.
129. SHAFIZADEH, F., and G. D. McGINNIS. Morphology and biogenesis of cellulose and plant cell-walls. In: *Advances in Carbohydrate Chemistry,* M. L. Wolfrom and R. S. Tipson (eds.). New York and London, Academic Press, 1971, Vol. 26, p. 297.
130. SIEGEL, S. M. The biochemistry of the plant cell wall. *In: Comprehensive Biochemistry,* M. Florkin and E. H. Stotz (eds.). Amsterdam, Elsevier, 1968, Vol. 26A, p. 1.
131. SMITH, F., and R. MONTGOMERY. The chemistry of plant and mucilages. New York, Reinhold, 1959.
132. SOLOV'EVA, T. F., L. V. ARSENYUK and S. OVODOV. Some structural features of *Panax ginseng* C. A. Mey pectin. *Carbohydr. Res.* **10**:13, 1969.
133. SOUTHGATE, D. A. T. Determination of carbohydrates in foods. II. Unavailable carbohydrate. *J. Sci. Food Agric.* **20**:331, 1969.
134. SOUTHGATE, D. A. T. Problems in the analysis of polysaccharides in foodstuffs. *J. Assoc. Pub. Anal* **12**:114, 1974.
135. STEPANENKO, B. N., and L. B. UZDENIKOVA. Precipitation of neutral polysaccharides and separation of their mixtures by use of various quarternary salts. *Carbohydr. Res.* **25**:526, 1972.
136. THORNBER, J. P., and D. H. NORTHCOTE. Changes in the chemical composition of cambial cell during its differentiation into xylem and phloem tissue in trees. *Biochem. J.* **81**:449, 1961.
137. TIMELL, T. E. Wood hemicelluloses. *In: Advances in Carbohydrate Chemistry,* M. L. Wolfrom and R. S. Tipson (eds.). New York and London, Adademic Press, 1964, Part I, Vol. 19, p. 247; 1965, Part II. Vol. 20, p. 410.
138. TROWELL, H. Ischemic heart disease and dietary fiber. *Am. J. Clin. Nutr.* **25**:926, 1972.
139. TYLER, J. M. The seed mucilage of *Lepidium sativum* (cress). Part 1. Identification of the components and some neutral oligosaccharides derived from the mucilage, together with the examination of a pentosan fraction. *J. Chem. Soc.* **1965**:5288, 1965.
140. TYLER, J. M. The seed mucilage of *Lepidium sativum* (cress). Part II. Products of hydrolysis of the methylated mucilage and the methylated degraded mucilage.*J. Chem. Soc.* **1965**:5300,1965.
141. UNRAU, A. M., and Y. M. CHOY. Identification of linkages of a galactomannan isolated from seed of *Caesalpinia pulcherima. Carbohydr. Res.* **14**:151, 1970.
142. VAN SOEST, P. J. The uniformity and nutritive availability of cellulose. *Fed. Proc.* **32**:1804, 1973.
143. VAN SOEST, P. J., and R. W. McQUEEN. The chemistry and estimation of fibre. *Proc. Nutr. Soc.* **32**:123, 1973.
144. VAN SOEST, P. J., and R. H. WINE. Method for determination of lignin, cellulose and silica. *J. Anim. Sci.* **26**:940, 1967.

145. VATTUONE, M. A., E. A. DEFLORES and A. R. SAMPIENTRO. Isolation of neoagarabiose and neoagaratetarose from agarose digested by *Pseuodomonas elongate*. *Carbohydr. Res.* **39**:164, 1975.
146. WAITE, R., and A. R. N. GORROD. The structural carbohydrates of grasses. *J. Sci. Food Agric*, **10**:308, 1959.
147. WARD, K., JR., and P. A. SEIB. Cellulose, lichenan and chitin. *In: The Carbohydrates*. W. W. Pigman and D. Horton (eds.). New York, Academic Press, 1970, Vol. IIA, p. 413.
148. WARDROP, A. B. Cell wall organisation in higher plants. I. The primary wall. *Bot. Rev.* **28**:241, 1962.
149. WHITE, E. V., and P. S. RAO. Constitution of the polysaccharide from tamarind seed. *J. Am. Chem. Soc.* **75**:2617, 1953.
150. WHISTLER, R. L., J. BACHARACH and D. R. BOWMAN. Preparations and properties of corncob holocellulose. *Arch. Biochem.* **18**:25, 1948.
151. WHISTLER, R. L., and M. .S. FEATHER. Hemicellulose. Extraction from annual plants with alkaline solutions. *In: Methods in Carbohydrate Chemistry*, R. L. Whistler (ed.). New York and London, Academic Press, 1965, Vol. V. p. 144.
152. WHISTLER, R. L., and B. D. E. GAILLARD. Comparison of xylans from several annual plants. *Arch. Biochem. Biophys.* **93**:332, 1961.
153. WHISTLER, R. L., and G. E. LAUTERBACH. Isolation of two further polysaccharides from corncob. *Arch Biochem. Biophys.* **77**:62, 1958.
154. WHISTLER, R. L., and D. I. McGILLVRAY. 2-0-αD-Xylopyranosyl-L-arabinose from hemicellulose of corncob. *J. Am. Chem. Soc.* **77**:1864, 1955.
155. WHISTLER, R. L., and E. L. RICHARDS. Hemicelluloses. *In:The Carbohydrates*, W. W. Pigman and D. Horton (eds.). London and New York: Academic Press, 1970 Vol. IIA, p. 447.
156. WOOD, P. J., and I. R. SIDDIQUI. Isolation and structural studies of a water-soluble galactan from potato (*Solanum tuberosum*) tubers. *Carbohydr. Res.* **22**:212, 1972.
157. WYLIE, A. Alginates as food additives. *R. Soc. Health J.* **93**:309, 1973.
158. ZITKO, V., and C. T. BISHOP. Structure of a galacturonan from sunflower pectic acid. *Can. J. Chem.* **44**:1275, 1966.

The Analysis of Dietary Fiber

David A. T. Southgate

I. Introduction and Historical Review

The measurement of the complex mixture of polysaccharides and lignin making up total dietary fiber presents many problems to the analyst. The complete analysis of the range of substances present in a mixed diet, for example, would require a very complex and time-consuming scheme; at the present time it is not possible to suggest any one method that could be used, without modification, for any sample of a food or a mixed diet.

This chapter will be concerned with reviewing and describing the procedures that have been used in the measurement of the constituents of dietary fiber. The existing analytical procedures can be thought of as falling into two categories according to the "analytical school" from which they were derived, and it is useful at this stage to consider some aspects of the historical development of the analytical methods under review.

In early nutritional studies with the ruminant, it was quickly appreciated that a significant fraction of many forages was not digested in the mammalian tract and that the amount of this fraction in a ration had a proportionately depressing effect on the availability of other constituents in the diet. The workers at the Weende Research Station[56] devised a procedure for the measurement of this indigestible fraction, which, because it was associated with the vegetable fibers in the diet, became known as the crude fiber fraction, in agricultural circles quickly shortened to the fiber fraction.

The method was described by Henneberg and Stohmann in 1860[38] although it appears to have been in use for some time before then. Careful reading of the

DAVID A. T. SOUTHGATE • Dunn Nutritional Laboratory, Medical Research Council, Cambridge, England.

reports from the station suggests that even before the method was described in the literature,[86] some deficiencies in its value had been observed.

It must be remembered, however, that at that time the chemistry of natural products was in its infancy, and the chemical nature of many food constituents, let alone their role in nutrition, was poorly understood.

The crude-fiber method as originally proposed involved treatment of the food with hot acid and hot alkali and was in all essentials, although not in technical specifications, very similar to the crude-fiber procedures in wide use today.[3] This method became firmly established and eventually incorporated in many statutory provisions. As an index of the indigestible matter in animal feeds, it had some practical value and this, to many, justified its retention. Mangold,[56] when reviewing studies of the digestibility of crude fiber, states very clearly that crude fiber is an empirical fraction, defined solely by the analytical method used to measure it.

The empirical nature of the method imposed severe constraints on the food analyst, and very strict adherence to the provisions of the analytical protocol was necessary if consistent results were to be obtained. The procedure was also time consuming, and many modifications were introduced by individual workers in attempts to reduce the time involved in an analysis and to improve reproducibility. Many of these modifications were of limited use and did not have much influence on standard methodology.

As the knowledge of the chemical composition of foods improved and the limitations of the crude-fiber method became more apparent, several groups of workers became concerned with developing less empirical procedures. The studies of these workers can be be seen to have a more significant influence on methodology, because they became concerned with measuring all the indigestible components in the diet derived from the structural matter of the plant cell wall.

Since the late 1930s, therefore, one has been able to distinguish two main streams in food analysis as applied to what we now call dietary fiber. The "fiber" and the "cell-wall constituent" analytical schools resulted from this division.

The former was mainly concerned with animal nutrition and led to improved procedures for the analysis of the indigestible matter in animal feeds.[32]

The latter school primarily comprised those who were concerned with human nutrition and whose interest, therefore, extended to all the indigestible components of the plant cell wall in the human diet.[75] Originally the procedures of this school were scarcely less empirical than the original fiber procedure. In recent years, however, the development of a third major area of interest in studies of the chemistry of the plant cell wall[73] has provided a greater insight into the chemistry of the structural components of the plant cell wall. These studies should, in turn, encourage the development of methods for the analysis of these dietary constituents of greater refinement than was possible hitherto.

The work in these three fields of study provides the basis for the analytical measurement of dietary fiber. As yet no single procedure is entirely satisfactory, if one considers the complexity of the dietary fiber in a normal mixed diet eaten in the United States and in the United Kingdom. One of the aims of this chapter is that, in attempting to draw together the body of existing analytical experience in these three areas, the type of analytical approach that appears to offer the most promise in evaluating the physiological role of dietary fiber will emerge.

II. Measurement of Fiber

In this section the methods of the "fiber school" are described and reviewed. However, before describing the methods, it is useful to consider the various stages in the evolution of these methods.

These methods have been dominated by the needs of agricultural analysts for a reasonably rapid and reproducible procedure which provides an accurate measure of the indigestible matter in foods eaten by animals, the ruminant in particular. In approximate terms, a measure of the cellulose and lignin in a food is required. Studies of the crude-fiber procedure[35] have shown that a variable proportion of the cellulose is solubilized by the acid and alkali treatments and that some lignin is removed during the preliminary solvent extraction. The crude-fiber fraction, moreover, always contains nitrogenous material. In addition, there are technical problems of considerable complexity associated with the need for a rigid adherence to the provisions of the analytical protocol for the method, especially the control of the periods under reflux and the filtration stages.

The introduction of the normal-acid-fiber (NAF) procedure[94] resolved some of the technical problems but still gave a fraction that contained appreciable amounts of nitrogenous material.

An important advance in these methods was the observation of van Soest[82] that the nitrogenous contamination of the NAF fraction was considerably reduced if a detergent was added to the extracting acid. Using this observation, he was able to develop two procedures; the first was based on extraction of the food with acid-detergent to give an acid-detergent-fiber (ADF) fraction that was an accurate measure of the cellulose and lignin in a forage; the second, using a neutral-detergent extraction to give neutral-detergent fiber (NDF), provided a measure of the total cell-wall material in a forage.

The crude-fiber procedure is still widely used, and for this reason alone it is described below. The ADF procedure is rapidly becoming established as the method of choice for measuring fiber in animal feeds and in time must surely replace the crude-fiber procedure entirely.

A. Crude Fiber

The crude-fiber procedure is based on those described in the *Official Methods of Analysis* of the AOAC.[3] The principle of the method is that an air-dry sample of the food is extracted to remove lipid and that this dry, fat-free sample is then extracted successively with boiling acid and alkali. The residue is filtered off and washed thoroughly. After drying and weighing, the residue is ashed and residual inorganic matter measured (Figure 1).

Figure 1. Summary of stages in crude-fiber method.

The procedure is highly empirical, and strict adherence to the prescribed protocol is essential if reproducible results are to be obtained. For a detailed account of the specifications of the apparatus used in the official AOAC procedures, reference should be made to the appropriate section in the current *Official Methods of Analysis.*

1. Reagents

1. Sulfuric acid, 0.255 N, 1.25 g H_2SO_4/100 ml.
2. Sodium hydroxide, 0.313 N, 1.25 g NaOH/100 ml. This solution should be free or virtually so from carbonate. The strengths of these two reagents are critical and must be checked by titration. The alkali must be stored in such a way as to prevent absorption of carbon dioxide.
3. Asbestos. This is used as a filtering aid, and it must be ignited and extracted with acid and alkali before use. A blank determination on the prepared asbestos must give less than 1 mg of crude fiber.
4. Antifoaming agent.

2. Apparatus

Digestion Apparatus. A digestion apparatus is required with a condenser to fit a 600-ml beaker and a hot plate capable of bringing 200 ml water at 25°C to a rolling boil in 15 ± 2 min.

Filtration. A filtration device with a 200-mesh screen held in a special filter holder is also required.

Reagent Preheater. A preheater is necessary for preheating and maintaining the acid and alkali at the boiling point of water.

3. Procedure

A finely ground air-dry sample (2 g) from which the fat has been extracted with diethyl ether or petroleum spirit is used. If the fat content of the sample is less than 1%, the fat extraction can be omitted.

The fat-free sample is transferred to a 600-ml beaker, and about 1 g prepared asbestos and 200 ml boiling acid are added, together with a little antifoaming agent if necessary. The beaker is then heated on the digestion apparatus, with periodic rotation to keep the solids from adhering to the sides of the beaker.

The mixture is boiled for exactly 30 min and then filtered; in one operation and without breaking the suction, 50–75 ml boiling water are added to the filter. The beaker, filter mat, and residue are washed with three 50-ml portions of water and the residue sucked dry. The filter mat and residue are returned to the beaker and 200 ml boiling alkali added. The mixture is again boiled exactly 30 min and filtered as before.

Without breaking the suction, the filter is washed with 25 ml boiling acid, then with three 50-ml portions of boiling water and 25 ml alcohol.

The filter and residue are dried for 2 hr at 130°C, cooled in a desiccator and weighed. It is then ignited at 600 ± 15°C, cooled in a desiccator and reweighed.

Crude fiber ≡ loss on ignition − loss of asbestos blank

B. Normal-Acid Fiber

The normal-acid-fiber (NAF) procedure was proposed by Walker and Hepburn[94] as an alternative to the crude-fiber method and was developed following a critique of the crude-fiber method by Hallsworth.[36] The method involves the digestion of a solvent-extracted sample with normal sulfuric acid and the subsequent ignition of the residue; the loss in weight is taken as the normal-acid fiber in the sample. The philosophy behind this method was that the alkali stage in the crude-fiber method was a major cause of variability, probably because it was the most difficult to standardize and control. It was therefore hoped that elimination of this stage would result in an improved analytical procedure.[16,17] Since the method has not been as widely adopted as the crude-fiber method, and therefore neither the number of variants nor the precise protocol have been described at length, the method described is based on the original description (Figure 2).

1. Reagents

1. H_2SO_4, 1 N. The strength of the acid does not appear to be critical.
2. Teepol.

2. Procedure

The material to be analyzed is oven dried at 100°C and ground to pass a 0.8–1-mm sieve. A 1-g sample is then extracted for 8 hr with ethanol/benzene (1:2 v/v) in a Soxhlet apparatus. The solvents are driven off the residue, which is then transferred to a 500-ml conical or Kjeldahl flask; 200 ml of hot N H_2SO_4 is added followed by 0.2 ml Teepol, and the mixture is boiled gently under reflux for 1 hr with shaking at intervals.

The mixture is filtered through a 50-ml sintered glass crucible (porosity 1) and the residue washed three times with 50-ml portions of hot water. The first two portions are decanted from the flask, and the final wash is used to transfer the residue to the filter. The residue and filter are then washed with alcohol followed by diethyl ether, and the crucible is dried at 100°C overnight, cooled in a

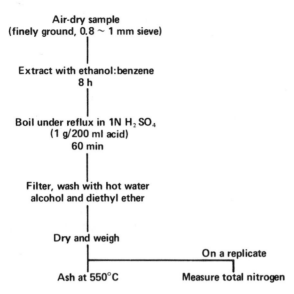

Figure 2. Summary of stages in normal-acid-fiber method.

desiccator and weighed. The crucible is ignited overnight at 550°C, cooled, and reweighed.

$$\text{Normal-acid fiber} \equiv \text{loss on ignition}$$

3. Correction for Nitrogen Content[93]

Another replicate is taken through the procedure and total nitrogen measured on the residue after acid treatment.

$$\text{Corrected normal-acid fiber} \equiv \text{loss on ignition} - (\text{N in residue} \times 6.25)$$

4. Comments on the Method[67]

Some variations in the conditions used are permissible because they appear to have little or no effect on the final result obtained. The time of hydrolysis is not critical within about 10 min, and the results are not affected by the size of the sample nor the rate at which the flask is heated. The degree of grinding of the sample is critical, and elevated values are obtained if the solvent stage is omitted. Extraction with ethanol/benzene gave lower results than with petroleum spirit and diethyl ether. The addition of a wetting agent such as Teepol greatly improved the reproducibility of the method.

C. Acid-Detergent Fiber

The use of detergents in fiber analysis was studied by van Soest,[82] and in what has become a classical paper, he examined the use of anionic, cationic, and nonionic detergents in modifications of a NAF type of method. In these studies the main aim was to produce an acid fiber with a low nitrogen content. On the basis of these studies, van Soest was able to describe a procedure for measuring a fraction that was closely correlated with the nutritive value of a forage.[83] The fraction measured by this procedure was also shown to be composed mainly of cellulose and lignin.

The method involves heating the sample with normal sulfuric acid containing cetyltrimethylammonium bromide (CTAB), the residue being filtered off and washed in the usual way (Figure 3).

Goering and van Soest give a detailed account of the procedure and the apparatus suggested for use in the method,[32] and the following description is based on their publication.

1. Reagents

Acid detergent, 20 g CTAB, dissolved in 1 liter N H_2SO_4. It is necessary to check the normality of the acid before adding the CTAB.

2. Procedure

The sample (1 g) should be air-dry and ground to pass a 1-mm sieve. If a wet sample is to be analyzed, the sample taken should contain around 1 g dry matter.

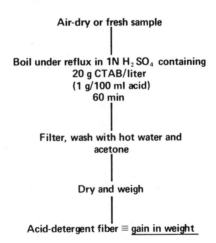

Figure 3. Summary of stages in acid-detergent fiber method.

The sample is weighed into a 600-ml beaker equipped with a suitable reflux device. 100 ml of cold (i.e., room temperature) acid-detergent reagent is added together with 2 ml decahydronaphthalene (reagent grade). The mixture is heated to boiling in 5–10 min, the heating being reduced as boiling commences to avoid excessive foaming. Boiling is continued at a slow even rate under reflux for 60 min.

The mixture is filtered through a tared 50-ml sintered glass crucible (coarse porosity) using gentle suction. The beaker and the residue on the filter are washed twice with hot water (90–100°C), rinsing the crucible in the process.

The filter is then washed with acetone until no further color is extracted, breaking up the residue on the filter to ensure complete extraction with the acetone.

The filter is sucked dry and the crucible heated at 100°C for 8 hr or overnight, cooled, and weighed.

$$\text{Acid-detergent fiber} \equiv \text{gain in weight}$$

The fraction is mainly composed of the cellulose and lignin in the sample and can be used for the subsequent measurement of these constituents. It also contains some silica.

D. Neutral-Detergent Fiber (Cell Wall)

This method was also developed by van Soest[82] and is thought to give an accurate measure of the cell-wall constituents in a vegetable foodstuff. It appears to divide the dry matter of feeds very nearly into those constituents which are nutritionally available by the normal digestive processes and those which depend on microbial fermentation. The method depends on the extraction of the food with a hot neutral solution of a detergent, sodium lauryl sulfate (Figure 4). The description given here is based on the very full account by Goering and van Soest.[32]

1. Reagents

Neutral-Detergent Solution. Dissolve 18.61 g ethylenediaminetetraacetic acid dihydrate (reagent grade) and 6.81 g sodium borate decahydrate (reagent grade) in some distilled water, heating the mixture to dissolve the reagents. Add this solution to a solution of 30 g sodium lauryl sulfate (USP) and 10 ml 2-ethoxy ethanol.

Dissolve 4.56 g disodium hydrogen phosphate (anhydrous reagent grade) in some distilled water with heating and add this to the solution of the other ingredients.

Adjust the volume to 1 liter with distilled water. The pH of the reagent should be between 6.9 and 7.1.

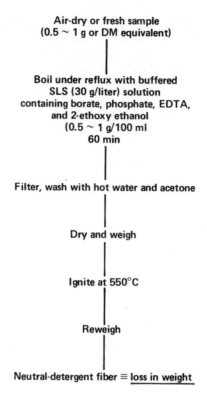

Figure 4. Summary of stages in neutral-detergent fiber method.

2. Procedure

The sample is prepared in the same way as in the acid-detergent method and weighed into a 600-ml beaker equipped for refluxing. 100 ml neutral-detergent solution is added followed by 2 ml decahydronaphthalene and 0.5 g sodium sulfite. The mixture is heated to boiling in 5–10 min and the heat adjusted to give a gentle even reflux. Heating is continued for 60 min from the time that boiling started.

The mixture is filtered through a tared 50-ml sintered glass crucible. The contents of the beaker are mixed by swirling, and the crucible is filled before suction is applied; only gentle suction is used at first. Finally, the contents of the beaker are rinsed into the crucible with the minimum of hot water (90–100°C). The suction is broken and the residue on the filter broken up; the crucible is filled with hot water and suction reapplied. This washing procedure is repeated. Finally, the residue is washed twice with acetone in the same way and sucked dry. The crucible is dried at 100°C for 8 hr or overnight and weighed.

Neutral-detergent fiber ≡ gain in weight

This residue invariably contains some inorganic material, which can be measured by igniting the residue for 3 hr at 500–550°C.

E. A Review of Fiber Methods

The primary aim of the methods described in this section is to provide estimates of the indigestible cell-wall material in a forage in connection with the nutrition of the ruminant. This factor must be borne in mind when considering the place of these methods in the study of dietary fiber in human nutrition.

The limitations of the crude-fiber method have been reviewed on many occasions; following is a summary of the deficiencies of the method.

First, it is highly empirical, and for this reason repeatability within the same laboratory and agreement between different analysts and different laboratories is only satisfactory if the defined procedure is followed precisely.

Second, it measures a fraction that has a variable composition from food to food. This is probably due to the solubilization of cellulose and lignin during the extractions. The crude-fiber method must therefore be regarded as an unsatisfactory method both technically and on the basis that it does not measure a chemically defined fraction in a food.

The normal-acid method (NAF) and the latest derivative of this method, the acid-detergent method (ADF), were developed to meet these major objections.

As measures of the indigestible material in a forage for the ruminant, the values obtained for using the NAF procedure are only marginally better than crude-fiber values. The main advantages of this method are its relative simplicity and the fact that in principle it should be easier to obtain reproducible and repeatable values using it. In practice the variability of the values obtained is not substantially better than with the crude-fiber method if the wetting agent is omitted. The NAF fraction contains substantial amounts of nitrogen (between 2 and 5% N × 6.25), and the deduction of N × 6.25 as a correction is not wholly satisfactory.

Viewed in the context of the fiber methods in general, the NAF procedure must be regarded as a stage in the evolution of the detergent-fiber procedures of van Soest.

One of these procedures, the ADF method, gives values that correlate well with the indigestible matter for the ruminant.[84] The method yields a fraction with a low nitrogen content (0.2–0.8%) and also gives an accurate measure of the cellulose and lignin in a forage.

From the technical viewpoint, the ADF method has been developed to a stage where it can be carried out routinely in a laboratory with a reasonable through-put of samples.

The method has a number of limitations when applied to human foodstuffs. The first concerns the size of sample; in a typical mixed diet the cellulose and lignin are of the order of 1–2% of the dry matter, and a sample of 1 g will therefore give a final residue of only 10–20 mg. Precision is therefore difficult to achieve when working with the method at these levels. Increasing the size of the sample leads to problems with foaming and the final filtration. The second is related and concerns the high levels of lipid in a mixed diet; as these tend to accentuate foaming and filtration problems, it is better, in general, to use a lipid-free, extracted sample when measuring ADF in human foods.

The third objection is, however, more fundamental. The plant cell wall contains a whole range of polysaccharides that are not hydrolyzed by human digestive secretions. Dietary fiber includes all these polysaccharides, whereas ADF measures only a fraction, a significant one but, nevertheless, only a portion of the total dietary fiber.

Neutral-detergent fiber (NDF) would therefore seem to be the best procedure yet developed in the "fiber school" for the measurement of total dietary fiber. In principle, it provides a measure of all the cell-wall material in a food. There are, however, a number of technical difficulties associated with the application of the unmodified method to human foodstuffs.

As for the ADF, the concentration of NDF in most foodstuffs is quite low, and this means that with a sample of 0.5–1 g, precision is difficult to achieve.

The proportion of starch and lipid, moreover, is much higher in most human foods than in the forages for which the method was developed. These constituents are responsible for technical difficulties at two points in the method: foaming during extraction and difficulties with filtration.

In the case of foods that are rich in starch, not only is the filtration stage difficult, but in many cases some starch remains insoluble in the hot detergent and is measured with the residual NDF.[80]

These difficulties can be largely resolved by suitable modifications of the procedure; these involve the insertion of a lipid-extraction stage and the removal of starch by enzymatic hydrolysis.[33] A fundamental objection, however, still remains in that the method will underestimate total dietary fiber because the water-soluble polysaccharides are lost during the extraction, these losses being quite substantial with some materials.

III Measurement of Constituents of the Plant Cell Wall

A. Analytical Procedures Based on Fractionation of the Plant Cell Wall

In general, the methods described in this section have been generated by two fields of interest. The first of these lies among workers who are concerned with the chemistry of the plant cell wall and the second among those who were primarily interested in the nutrition of the ruminant and were disenchanted with the information obtained by the standard analytical methods.

A number of analytical protocols have been described (for example, see refs. 4,65,72), and most workers in this field have found that each type of plant tissue requires some modification of the analytical procedures. Siegel[73] feels that a unified scheme, which is applicable to all types of tissue, may represent an unattainable goal.

An outline of the fractionation scheme for the isolation of the components of the plant cell wall was described in Chapter 2, and the quantitation of the various stages provides the basis of the analytical procedures developed in this field of study.

As an analytical procedure, the fractionation scheme can be considered to be divided into six main stages (Figure 5). In this section the principles that have been used in these stages will be described, as it is not possible in the space available to describe the many variants in any detail.

1. Preparation of the Sample

Many studies of the composition of the plant cell wall have been made on woody tissues, which, as a rule, have a relatively low water content and for which drying is often unnecessary. Other plant tissues, however, required dehydration; oven drying at an elevated temperature was used in many early studies, but it is now generally felt that the heating produces artifacts formed by condensation between proteins and other constituents.[86] Freeze-drying can be seen as a suitably mild alternative, and acetone-drying also provides a suitable procedure for the preparation of the sample.

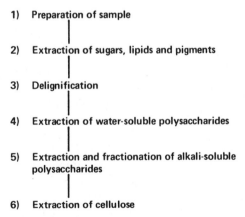

Figure 5. Stages in the fractionation of the plant cell wall.

Some degree of subdivision of the sample is desirable, and the production of an homogeneous sample of the tissue is essential if relatively small subsamples are used later. The intrinsic inhomogeneity of many plant tissues provides a challenge to the analyst at this stage. Ball-milling will reduce most samples to a very homogeneous powder but should be used with caution as intense local heating can occur, and prolonged ball-milling of cellulose preparations is known to cause depolymerization.

2. Extraction

An extraction stage which removes pigments, lipids, and soluble sugars is a usual feature of the methods under review.

Technically it provides a sample of plant tissue that is easier to manipulate, and the removal of pigments and a range of extractives prevents their interfering during the later stages of the fractionation. Care is necessary to ensure that some protein–polysaccharide complexes are not extracted; in addition, prolonged extraction with some lipid solvents will slowly solubilize lignin or lignin-like material.[62]

No general rules can be expounded that apply to all materials, and the intending analyst must verify that polysaccharides are not extracted during the extraction used.

Hot aqueous alcohols are frequently used in the extraction; ethanol, methanol, and isopropanol have been used in concentrations around 750–900 ml/liter. At lower concentrations some polysaccharides are partially soluble.[27] These hot aqueous alcohols dissolve waxes, some lipids, and most of the chorophyll pigments. A typical procedure is that of Gaillard.[27] In this procedure, the dried plant tissue is extracted with 90% v/v ethanol by boiling under reflux for 1 hr and repeating this treatment after removing the first extract by filtration.

3. Delignification

In woody tissues the polysaccharides in the cell wall cannot be extracted completely unless the lignin is first removed. This also appears to be true for grasses where the level of lignification can be quite high.[95] In the case of herbaceous plants (which includes most plant foods in the human diet), it is not certain whether this stage is essential.[98]

The classical procedure for delignification involves treatment of the tissue with chlorine or sulfite. The course of the reactions involved is not completely established, but the chlorine treatment has a substituting and oxidative effect on the aromatic material leading to depolymerization.

Some functional groups are introduced into the cell-wall polysaccharides,[12] and a fraction of the noncellulosic polysaccharides dissolves during the treatment.[11] In view of the potential changes associated with delignification, it should only be used if complete extraction is not possible in the presence of the lignin in the actual samples being examined. Many modifications have been described, but that of Whistler et al.[95] is fairly typical and has been used to prepare material for structural studies on corncob xylans. The delignified cell-wall material is frequently called holocellulose.

Preparation of Holocellulose from Corncobs.   A sample (30 g) of ground material is suspended in 680 ml water in a round-bottomed flask, heated to 74°C and maintained at that temperature ± 1°C. Acetic acid (2.5 ml) and sodium chlorite (7.5 g) are then added with stirring. At the end of 15 min, these additions are repeated and again at 15-min intervals until four additions have been made. During this process the flask is flushed with carbon dioxide to prevent the accumulation of the explosive oxides of chlorine. The reaction must, of course, be carried out in a fume hood.

After another hour, during which the pH rises from 4.2 to around 4.7, the mixture is cooled and the holocellulose filtered off rapidly and washed thoroughly with water and acetone.

4. Extraction of Water-Soluble Polysaccharides (Pectic Substances)

This is usually carried out on an undelignified sample. The pectic substances in many tissues are extractable with hot water alone. However, the pectic material in the wall is frequently deposited as salts (of calcium particularly),[20] and the addition of ammonium oxalate or EDTA to the extracting medium results in more complete extraction.

The other water-soluble noncellulosic polysaccharides also dissolve at this stage, and it is usual either to reextract these substances from the crude pectic fraction or to precipitate the pectic fraction as the calcium salt.

Gaillard[27] describes a typical procedure.

Procedure.   The residue insoluble in alcohol is extracted with hot (near boiling) 0.5% w/v aqueous ammonium oxalate solution. After cooling, the extract is slightly acidified and the pectic substances precipitated by the addition of four volumes of ethanol.

5. Extraction of the Alkali-Soluble "Hemicellulose" Fraction

The extraction can be carried out with a range of strengths of alkali, and the temperature of extraction has been used to provide an additional fractionating condition.[57] It is essential to carry out the extraction in the absence of oxygen if very extensive changes in the polysaccharides are to be avoided, and it is usual to perform the extraction under nitrogen.

The treatment of the alkali extracts still owes much to the classical studies of O'Dwyer.[63] The first stage of the treatment is neutralization with acetic acid and the recovery of the hemicellulose-A fraction. The hemicellulose-B fraction precipitates on the addition of ethanol to the neutralized extract.[8,9] The procedure described incorporates Gaillard's[28-30] fractionation of the hemicelluloses by the formation of iodine complexes, which separates linear and branched polysaccharides.

Procedure. After delignification and extraction with ammonium oxalate, the residue is extracted first with 5% w/v KOH and then with 24% w/v KOH. The extractions are carried out by shaking the mixture in a closed vessel which has been flushed with nitrogen.

Extraction is usually carried out over 24 hr, after which the residue is filtered off on a sintered glass crucible. Three extractions are usually necessary.

The combined filtrates are made slightly acid by the addition of 50% v/v acetic acid, and the precipitated hemicellulose-A fraction is collected by filtration or centrifugation. Four volumes of ethanol are added and the precipitated B fraction is collected.

The precipitated fractions are further fractionated by dissolving them in 20 ml of aqueous calcium chloride (S.G. 1.3) and adding 2 ml of iodine solution (3 g I_2 and 4 g KI in 100 ml water). The linear hemicelluloses form an insoluble complex with iodine.

The precipitates of hemicelluloses from many human foods are very fine and gelatinous and do not filter easily; centrifugation is usually the easiest method of collecting them after precipitation.

6. Extraction of Cellulose

The residue insoluble in strong alkali is often designated cellulose, and in the residue from many tissues the $\beta(1\text{-}4)$ glucan component is the major component. Some xylose is almost invariably detectable in hydrolysates of α cellulose fractions, and frequently traces of other sugars are present. Whether these represent contaminants or are a measure of the inhomogeneity of the cellulose fibrils in the particular tissues is uncertain. It is, however, reasonable to suppose that these contaminants are derived from polysaccharides that are intimately associated with the cellulose fibrils, and from the point of view of cell-wall structure they form part of the fibrillar structure. Some residual lignin and inert material derived from cuticular matter may be present along with some inorganic matter.

The cellulose in the residue may be dissolved out with 72% w/w H_2SO_4; other cellulose solvents such as cuprammonium sulfate do not appear to have been very widely used in cell-wall studies.[19]

B. Measurement of Individual Components

The fractionation procedures described in the previous section yield the components of the cell wall in the form of precipitates (usually from aqueous alcohol) or residues. In many studies a gravimetric procedure has been used to quantitate the fractionation; however, this approach is not entirely satisfactory for two reasons. The first is purely technical and concerns the isolation of the precipitates in a form suitable for accurate gravimetric work. The precipitates are usually gelatinous and often extremely difficult to filter. A second and more important factor is that the precipitates are frequently not pure and contain protein and other contaminants. This is especially noticeable in the fractions isolated by alkali extraction, where protein and in some cases polyphenolic substances are precipitated along with the polysaccharides.

An alternative approach that frequently accompanies the gravimetric method is the measurement of the component sugars. Provided that some estimate of the hydrolytic losses of the component sugars is made, summation of the sugars present in the fractions can be a better measure of the total polysaccharide in a fraction. This summation also provides more information about the nature of the polysaccharide in the fraction.

All plant tissues will give mixtures of polysaccharides, and a range of appropriate fractionation techniques must be used if defined polysaccharides are to be measured.[8,78] A description of these is, however, outside the scope of this chapter, and reference should be made to the original literature.

This section is concerned with the measurement of the isolated polysaccharide fractions. If a delignification procedure is used, the fractionation provides no measure of the lignin present, necessitating alternative procedures, which are also described.

1. Measurement of Component Monosaccharides

Hydrolysis. Complete hydrolysis of many noncellulosic polysaccharides can be achieved with dilute mineral acid; nevertheless, hydrolytic losses of the component monosaccharides can occur in the presence of oxygen or other constituents, notably protein. The more labile furanose sugars seem to be more susceptible to degradation. The glycosidic bonds involving uronic acids are particularly resistant to hydrolysis, and most uronic acid-containing polysaccharides yield aldobiuronic acids on hydrolysis. If the hydrolytic conditions are sufficiently severe to hydrolyze these bonds, then extensive destruction of both uronic acids and other components occurs. An alternative approach is therefore necessary to measure the uronic acid components.[7,23]

No one set of hydrolytic conditions has yet been proposed which will guarantee the complete hydrolysis of all noncellulosic polysaccharides with the minimum of degradation. In most cases the analyst has to establish the optimum conditions for the material in hand.

Hydrolysis with 1 N H_2SO_4 for 1–2 hr at 100°C either under reflux or in a sealed tube is a relatively reliable procedure and makes a convenient starting point in the assessment of hydrolytic conditions.[37] The sulfuric acid is conveniently removed by treatment of the hydrolysate with barium carbonate.

Hydrolysis with trifluoroacetic acid has been widely used by Albersheim and his coworkers (2) and has the advantage that it is volatile and can be conveniently removed prior to preparation of derivatives for gas-liquid chromatography.

Cellulose is not hydrolyzed to any appreciable extent by dilute mineral acid but can be conveniently hydrolyzed after it has been brought into solution in 72% w/w H_2SO_4. It is usual to allow the initial stage to take place at a low temperature (0–4°C), followed by dilution to give an acid strength of 2–3 N and then boiling under reflux.[71,81]

2. Analysis of Sugar Mixtures Produced by Hydrolysis

The hydrolysates of the noncellulosic polysaccharides and, to a lesser extent, those of the cellulosic fraction contain mixtures of monosaccharides. Until chromatographic procedures were available, the analysis of these mixtures was extremely difficult and in many cases virtually impossible. Now a range of sensitive and reliable procedures is available.

Paper Chromatography. A large number of chromatographic systems for the separation of mixtures and their quantitative analysis have been used.[43,69,96] Hough and Jones[39] have provided a reasonably concise account, and most textbooks on chromatographic analysis contain extensive accounts.[46]

Gas–Liquid Chromatography.[5,68,79] The main difficulty with the separation of sugars has been the preparation of suitable volatile derivatives. Separation of the alditol acetates appears to be a most versatile and reliable procedure and has been widely used in studies of the chemistry of the plant cell wall by Albersheim and his colleagues.[2]

In their procedure 20 mg of cell-wall material is hydrolyzed by heating in a sealed tube with 2 ml of 2 N trifluoroacetic acid for 1 hr at 121°C. The acid is then removed by evaporation at 50°C in a stream of air. The sugar mixtures are reduced to the alditols with sodium borohydride in ammonia (10 mg in 0.5 ml N ammonia). The excess of borate is decomposed with acetic acid and the borate in the mixture removed by evaporation with methanol. The alditols are then acetylated with acetic anhydride and this solution injected directly onto the column. The procedure gives a good separation of all the sugars encountered in the cell-wall polysaccharides. An internal standard, *myo*-inositol, added at the hydrolysis stage, permits quantitation of the peaks in terms of the original starting material.

Ion-Exchange Chromatography. Of a number of ion-exchange systems for the analysis of sugar mixtures available,[23,44,45,90,91,92] the procedure described by Floridi[21] using a Technicon system is a useful and reliable one.

Chemical Methods. Several colorimetric procedures are available for measuring sugars in mixtures.[14] These do not have the specificity of the chromatographic methods, and care is necessary in applying the corrections for interference. Southgate[75] used the anthrone method for hexoses,[18,70] a ferric chloride/orcinol procedure for pentoses,[1,52] and a carbazole method[6] for uronic acids.

3. Specific Methods for Uronic-Acid-Containing Polysaccharides

Complete acid hydrolysis of these polysaccharides is rarely possible without some decarboxylation or the formation of aldobiuronic acids. The carbazole reaction[6,15,31,34] is given by uronic acid residues in many intact polysaccharides, but the color yield can be influenced by the nature of neighboring residues. For accurate determination, calibration of the color yield against an alternative procedure is necessary.

Quantitative decarboxylation formed the basis of the classic procedure for the measurement of uronic acids, and semimicro procedures have been developed.[97] The carboxylation methods are relatively slow and technically exacting, and the availability of polygalacturonase preparations has provided more specific procedures for glycuronans of the pectic type. Dekker and Richards[13] have described a procedure where 5 mg of the polysaccharide dissolved in 10 ml 0.1 M acetate buffer, pH 4.4–4.5, was incubated with 1 mg of a polygalacturonase preparation (Sigma); the uronic acids released were measured with the carbazole method.

4. Lignin

All procedures for lignin are semiempirical because of the ill-defined character of the polymer itself. In many cases the so-called lignin fraction reported in the literature is contaminated with nitrogenous and other materials, and the reported values are almost without exception too high.[10,58,99] Furthermore, any heat treatment of a food will almost invariably increase the apparent yield of lignin because of the formation of ill-defined condensation products.[86]

The classic procedure for the measurement of lignin measures the residual matter insoluble in 72% w/w H_2SO_4 (640 g H_2SO_4/liter) or Klason lignin. The method described is that of Goering and van Soest[32] from acid-detergent fiber.

Procedure for Acid-Detergent Lignin. The ADF is filtered into a crucible as described previously (p. 80), and the filter mat is covered with cooled (15°C) 72% w/w H_2SO_4 and stirred, breaking up all the lumps with a glass rod. More acid is added to the crucible as the solution filters. After 3 hr, suction is applied, and the contents of the crucible are washed with hot water until the washings are free from acid. The crucible is dried at 100°C, cooled and weighed; it is then ignited at 550°C for 3 hr, cooled, and weighed.

$$\text{AD lignin} \equiv \text{loss in weight}$$

This lignin may be contaminated with cutin and nitrogenous material. In the more conventional methods for Klason lignin,[58] many workers have found that a preliminary peptic digestion reduces nitrogenous contamination, but ADF is usually so low in nitrogen that this is unnecessary with the procedure described.

Permanganate Lignin.   As an alternative to acid-detergent lignin, van Soest and Wine[88] developed a method in which the lignin was removed from the ADF by oxidation with permanganate. The values obtained by this method are less affected by heat-induced artifacts and are thought to be closer to the "true lignin" content. Complete technical details of the procedure are given by Goering and van Soest.[32] The principle of the method is as follows.

The ADF is prepared in the normal way and is then treated with acetic acid-buffered potassium permanganate in *tert*-butyl alcohol containing ferric iron and silver as catalysts. The manganese and iron oxides deposited on the residue are dissolved out with an alcoholic solution of oxalic and hydrochloric acids, leaving a residue of cellulose and silica. Lignin is taken as the loss in weight. Cellulose is measured in the residue as the loss in weight on ignition.

The control of the permanganate oxidation stage is critical, and samples with a high lignin content require more reagent; conversely, samples with a low lignin content are rapidly oxidized, and losses of cellulose can occur if the rate of flow of permanganate through the sample is too high.

Comments on Lignin Procedures.   In general, the methods for the measurement of lignin are the least satisfactory of the methods available for any one component of the plant cell wall. This is primarily due to the nature of lignin itself which, on the basis of present knowledge, is a group of polymeric substances. Furthermore, the composition of the polymers is not fixed and probably varies during the course of lignification in one plant tissue and from species to species. However, sufficiently detailed studies of the structure of the lignin polymers which could enable one to assess the possible analytical significance of these changes are not available. It is therefore difficult to devise a method that is specific for lignin on the basis of, for example, a functional group or a fragment of the molecule.

Morrison[59-61] has devised a procedure of this kind based on the absorption in the ultraviolet of the products formed by the reaction with acetyl bromide; it was necessary, however, to calibrate the method by reference to a Klason lignin preparation, and the yield of products absorbing in the UV depended on the source of the lignin. Since this means that different calibrations are required for lignin measurements on different materials, a method of this kind is difficult to apply to mixtures of plant foods.[51]

The main problems in the measurement of lignin may be summarized as follows. First, because of the ill-defined and variable composition, a specific method is not at present available. Second, the methods based on the progressive removal of other constituents and measurement of a residue are unsatisfactory in many ways because they depend on the balance between losses of lignin during the extractions and contamination with artifacts. Treatment of foods with strong acid usually results in the formation of humin, and this material is often measured in the lignin residue.

Probably the best available method is the permanganate oxidation method.[32] However, the control of the oxidation is critical, and losses of cellulose and consequential overestimation of lignin can occur. For the most accurate use of this method with foods having a low lignin content (that is, most plant foods eaten by man), a careful investigation of the oxidation conditions is necessary and should be undertaken with the same attention to detail as that used by van Soest and Wine[87] for forages.

C. The Place of Methods for the Cell-Wall Constituents in Studies of Dietary Fiber

The methods described in this section have, in the main, been developed by workers studying the plant cell wall. The approach of these workers has been different from those concerned with ''fiber'' methods as such, in that most of this work has been conducted as part of a research program on the plant cell wall and not primarily because of any nutritional interests. Among the exceptions to this generalization are the studies of Gaillard[27,29] and Waite and Gorrod,[89] which were primarily concerned with gaining a greater insight into the nutritional value of the cell walls of forage plants. Nevertheless, the procedures are research-oriented, time-consuming, and not readily applicable as such to the routine analysis of human foodstuffs. They have been described at some length because they provide the most detailed methods available for the measurement of the major components of dietary fiber and thus must form the basis of any comprehensive method for dietary fiber.

The direct application of these methods to human diets raises many technical problems. Many of these stem from the relatively low concentrations of dietary fiber in human foods and the high concentrations of sugars, starches, and lipids compared with the plant materials for which the methods were devised. In practice, this means that the protein contamination of many fractions is often high and the direct application of an alkaline fractionation is difficult. Delignification prior to extraction of the noncellulosic polysaccharides is, however, probably unnecessary.[98] And in any case great care is necessary when applying the chlorite delignification developed for plant tissues to the complex mixture in human foods.

The measurement of lignin is especially difficult because of the low concentrations and its uneven distribution in the diet. The human diet also includes many cooked items, which will contain lignin-like artifacts produced by heat. The carbohydrates in many baked foods have undergone some degree of caramelization; in the crust of bread, for example, the dry heating may degrade some starch virtually to the carbon stage. The behavior of these carbohydrate degradation products is ill-defined, but experience suggests that they tend to be measured as lignin.

A final technical consideration that it is often difficult to discount is the state of division of much plant cell-wall material in a human diet. This is frequently very fine indeed, as a considerable amount is derived from thin-walled material from cereal endosperm cells and the parenchymatous cells of fruits and vegetables. These materials are extremely difficult to filter properly, as losses of cellulosic material and lignin can occur under conditions where most of the corresponding material in a typical animal foodstuff would be retained on the filter.

IV. Unified Procedures for the Measurement of Unavailable Carbohydrates, Indigestible Residue, and Dietary Fiber

The procedures in this section were developed by workers in the field of human nutrition. As was the case with the most recent methods described in Section II, the initial impetus in their development came from dissatisfaction with the crude-fiber method.

For the nutritional work concerned with man, the crude-fiber method had a double disadvantage. First, there were the empirical nature of the method and the generally unsatisfactory analytical aspects. Second, and more important, was the fact that the estimates of cellulose and lignin represented only a portion of the indigestible polysaccharides in the human diet.[76]

A. Total Unavailable Carbohydrate

The study of carbohydrates in foods by McCance and Lawrence[49] provides a convenient point from which to trace the development of methods designed to measure the indigestible plant polysaccharides and lignin in human foods. The primary interest of these two workers was the available carbohydrates,[26,64,74,100] but in providing the impetus for the development of procedures for measuring this fraction directly, they also provided a means for measuring the unavailable carbohydrates. Their use of this term for the sum of the polysaccharides and lignin not digested by the endogenous secretions of man is equivalent to the present use of the term dietary fiber. In some ways the older terminology is preferable because many of the polysaccharides making up dietary fiber are in no way fibrous.

McCance et al.[50] measured total unavailable carbohydrates in a range of fruits, nuts, and vegetables by determining the residue insoluble in 80% v/v

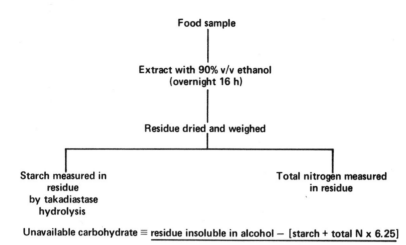

Figure 6. Stages in the measurement of total unavailable carbohydrate.

ethanol and then measuring the starch in the residue after enzymatic hydrolysis with a takadiastase preparation. They also made a correction for the protein in the residue (Figure 6).

B. Indigestible Residue

Williams et al.[101] adopted a a different approach. They had found that the composition of the crude fiber varied from food to food, and the amount of crude fiber was not a measure of much of the indigestible carbohydrate in human foods as judged by fecal analysis. The principle of their procedure, summarized in Figure 7, was the removal of protein by a prolonged incubation with pancreatin, followed by filtration of the residue and the subsequent removal of lipid by solvent extraction. The residue was then treated with 21. 4 N H_2SO_4 (=72% w/w) and left at $0°C$ for 24 hr. The mixture was rapidly diluted and boiled under reflux for 3 hr. After neutralizing the hydrolysate, reducing sugars were measured before and after removal of fermentable sugars with yeast.

The residue insoluble in acid was taken as lignin; the nonfermentable sugars, as pentose, were taken as a measure of the water-insoluble hemicelluloses (pentosans), and the fermentable sugars, as glucose, provided a measure of the cellulose. The limitations of this procedure can easily be seen, and in light of what is known about the structural polysaccharides, could only give an approximate idea of the composition of the indigestible material. Although the method was originally described as a method to determine the indigestible residue in feces, it was also used for foodstuffs. Macy[54] gives a detailed account of the application

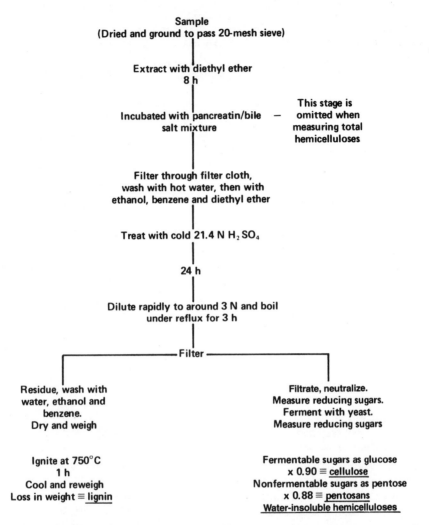

Figure 7. Stages in the measurement of indigestible residue.

of the method in her study of children.[55] In this work total hemicelluloses were measured by omitting the pancreatin digestion, during which some hemicelluloses dissolved, and measuring these after total acid hydrolysis and removal of fermentable sugars.

C. Cellulosic Fraction of Cereals

Further interest in the measurement of indigestible carbohydrates in human foods was stimulated by a report of the FAO[22] which stressed the urgent need for a better knowledge of the physiological value of the carbohydrate fraction of foods and, in particular, for the development of direct methods for the analysis for food carbohydrates.

Fraser et al.[25] took up this topic with special reference to cereals and, using a combination of existing procedures, proposed and tested a scheme for the analysis of the carbohydrates in wheat flours. Two parts of the procedures, which covered all the carbohydrates present in flours, are relevant to the present discussion.

The first of these measured a cellulosic fraction (Figure 8) and was based on enzymatic treatments of the sample, first with a takadiastase preparation and,

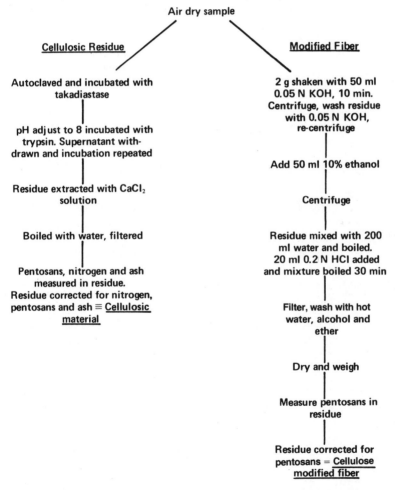

Figure 8. Summary of stages in the measurement of cellulosic material and modified fiber.

after adjustment of the pH, with trypsin. The tryptic digestion was repeated and the residue washed with calcium chloride to remove residual starch. Finally, the residue was washed with hot water, alcohol, and ether, then dried and weighed. The residue contained protein, some pentosan and ash; the value for cellulosic material was derived by correcting for the amounts of these constituents present.

The second procedure was a "modified fiber" method. The sample was treated with dilute alkali (0.05 N KOH) by shaking at room temperature and the alkaline extract discarded after centrifugation. The residue was washed with 10% v/v ethanol and suspended in water (200 ml) and boiled. Acid (20 ml 0.2 N HCl) was added to the boiling solution and the mixture boiled for 30 min. At the end of this time the residue was filtered off, washed with hot water, alcohol, and ether, and dried and weighed. The residue contained some pentosan but virtually no ash or protein; the value for cellulose was obtained after correction for the amount of pentosan present.

These two procedures gave identical results for the cellulosic fraction in wheat flours. Since the pentosans in wheat account for the major part of the noncellulosic polysaccharides, these procedures could be said to give a reasonable measure of total dietary fiber in wheat flours. Some of the values obtained with four types of flour are shown in Table I, with the values for crude fiber shown for comparison.[24] In all cases the crude fiber values are significantly lower than the cellulosic fraction, and the pentosan fraction is, of course, not measured at all in the crude-fiber method.

The scheme of analysis was developed specifically for wheat flours where the major noncellulosic polysaccharides are mainly arabinoxylans. In other foods this is not usually the case, and the measurement of pentosans is not wholly appropriate. There are, in addition, technical problems involved in the manipulation of the sample when foods richer in lipid and protein are being analyzed. The distillation method used for pentosans is rather time-consuming and is interfered with by other carbohydrates.

TABLE I. Pentosans, Cellulosic Material, and Crude Fiber
in Wheat Flours of Different Extraction Rate (g/100 g)

	Pentosan	Cellulosic	Crude fiber
Low extraction, 70%	2.0	0.4	0.12
Brown flour			
<93%	4.9	2.8	1.44
>93%	5.2	3.0	1.60
Wholemeal, 100%	5.8	4.4	1.99

D. Individual Components of Unavailable Carbohydrate

The same FAO report stimulated another type of study by Southgate and Durnin,[77] who set out to reinvestigate the use of energy-conversion factors in the calculation of the energy value of mixed diets. This study required the measurement of the unavailable carbohydrates in the food and feces of subjects receiving mixed diets as part of the study. The methods developed during this work[74,75] were designed to measure the various components of the unavailable carbohydrate fraction and thus form a basis for the measurement of dietary fiber. The principle adopted at the start of the work was that, if possible, specific methods for carbohydrates would form the basis for the measurements and that, as far as possible, the results obtained should be comparable with the results from detailed studies of the plant cell wall. The procedure, which has been modified since it was first described, is here described in some detail in its latest form.

1. Procedure

An outline of the procedure is given schematically in Figure 9. The sample (which can be fresh or dried and if the latter, preferably freeze-dried) is extracted with boiling 85% v/v methanol. The extraction with hot methanol is repeated three more times allowing the filtrate to drain from the residue between each extraction. The residue is then extracted on the filter with three portions of warm diethyl ether and allowed to dry in air at room temperature. The residual solvent can be driven off in an oven if necessary. The residue is weighed and then finely ground (ball-milling provides a very homogeneous material, but the technique should be used with caution as excessive milling can degrade cellulose).

Portions of this sugar-and fat-free residue are carried through the fractionation. Starch is first removed by enzymatic hydrolysis. In the original description of the method, takadiastase (Parke Davis, for analysis, on talc) was used, but this is no longer commercially available and an amyloglucosidase (Sigma or Boehringer) is now used.

After an overnight incubation at 37°C, 4 vol ethanol are added to precipitate any polysaccharides soluble in the buffer.

The residue is extracted with hot water, the extract is cooled, and the extracted polysaccharides are precipitated by the addition of 4 vol ethanol. The precipitate is then hydrolyzed with 1 N H_2SO_4 for 2½ hr at 100°C in a boiling water bath, and the hydrolysate analyzed as described below.

The residue, after water extraction, is similarly hydrolyzed with 1 N H_2SO_4. The hydrolysate is diluted with an equal volume of ethanol and filtered or centrifuged and the supernatant removed. The residue is washed with 50% v/v ethanol and the supernatants and washings are combined for analysis.

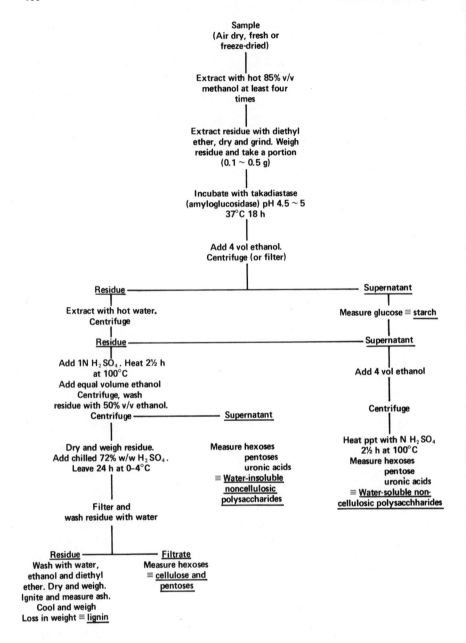

Figure 9. Schematic outline of the method for measuring the components of the unavailable carbohydrates or dietary fiber.

The two hydrolysates (of the water-soluble and water-insoluble polysaccharides) are analyzed for hexoses,[70] pentoses,[1] and uronic acid.[6]

The residue after dilute acid hydrolysis is dried by treatment with ethanol and diethyl ether and treated with chilled 72% w/w H_2SO_4 and left at $0°-4°C$ for at least 24 hr. During the first few hours of this treatment, the residue is stirred and any lumps broken up. At the end of 24 hr the mixture is filtered through a sintered glass crucible (porosity 1) and the crucible and residue washed with water. The acid filtrate and washings are combined and analyzed as above. Finally, the crucible and residue are washed with ethanol and diethyl ether, dried in an air oven at $90°-100°C$ for 10 min, cooled and weighed.

2. Interpretation of Results

This procedure gives values for the unavailable carbohydrate in terms of the sugar components. Quantitative paper chromatography, ion-exchange chromatography, or gas–liquid chromatography can alternatively be used to measure the component sugars in the hydrolysates.

The sugars found in the 72% w/w H_2SO_4 hydrolysates almost invariably include some pentoses, usually xylose,[66] which are probably derived from xylans intimately associated with the lignocellulose complex.[51]

The residue insoluble in 72% w/w H_2SO_4 was taken as lignin in the original method. This value may be too high for the reason discussed earlier (p. 92). The preliminary treatment of the sample should remove much of the cuticular matter. In practice the enzymes used seem to possess some proteolytic activity, and protein contamination is quite low, possibly because the initial treatment of the samples does not involve excessive heating.

In the application of this method, technical problems do arise in connection with the filtration of the finer cellulose and lignin particles; in these cases centrifugation is virtually obligatory.

The earlier criticisms regarding the use of a gravimetric procedure for lignin still apply. In many cases the concentration of lignin is so low that it is necessary to carry larger samples than are necessary for the other stages through the method in order to recover sufficient lignin to weigh with any degree of accuracy.

E. Review of Unified Procedures

In general, procedures of this type provide a better approach to the measurement of dietary fiber. For practical and theoretical reasons, any analytical scheme for the complex mixture making up dietary fiber must be a compromise between what is practicable in terms of time and equipment and the ideal procedure in which all the various classes of polysaccharide are separated and analyzed by specific procedures.

The analytical schemes described in this section represent an evolutionary development and should not be considered as alternatives. The procedure of McCance et al.[50] is possibly the simplest and could still be used to estimate total dietary fiber, although it would tell one nothing about its compositions; a similar estimate is obtained early in the Southgate method.

Williams and Olmsted's method provided an advance in that some division into cellulose, lignin, and hemicelluloses was obtained. The techniques of sugar analysis available to them limited the quantitative interpretation of their scheme so that reliance on a pentose method for measuring total hemicellulose probably resulted in an underestimation of this fraction.

The final procedure described in this section gives values for the noncellulosic polysaccharides in two fractions: one, those that are water-soluble and hydrolyzable with dilute acid, and the other, the cellulose and any associated noncellulosic pentosans and lignin. The polysaccharides are measured as their component sugars, and these values give some indication of the types of polysaccharide present in the foods; these values can be extended if a detailed analysis of the hydrolysates is performed.

Comparison of the values obtained for cellulose and lignin by the procedures of van Soest and Wine[87] and by the Southgate procedure gave reasonable agreement for a limited range of foods and mixed diets.[33,51] The results obtained also agree quite well with the values obtained in detailed studies of the plant cell wall. Unfortunately the cell walls of many food plants have not been studied in any great detail, and until these comparisons are available, the significance of the approximations in the procedure cannot be assessed.

To a greater or lesser degree, all the unified procedures make an assumption about the division between available carbohydrates, that is, those which are digested by the endogenous secretions of the human digestive tract and absorbed, and the unavailable carbohydrates. In most cases the division is effected by an enzymatic procedure[53,74] and, in the latest modification of the Southgate method, by the use of an enzyme specific for $\alpha 1 \rightarrow 4$ and $1 \rightarrow 6$ glucosidic bonds. This enzyme converts starch quantitatively to glucose and appears to have little or no activity towards pectic substances, cellulose, and some arabinoxylans, but does seem to have slight activity towards the galactomannans in guar gum.[41] It would be difficult, on the present evidence regarding the structure of starch, to dispute that the amyloglucosidase is specific for starch, but there is evidence that some chemically modified starches[40] are not hydrolyzed by this enzyme. However, whether this reflects unavailability *in vivo* remains to be seen, although observations with rats seems to suggest that some modified starches are unavailable.[47,48]

Undoubtedly the ideal procedure for measuring the polysaccharides in foods would involve a detailed fractionation prior to hydrolysis. Such an ideal would be practicable only in connection with a small project. The procedures available at present do provide a reasonably practicable method for obtaining values that both characterize the types of polysaccharide making up dietary fiber and give quantitative estimates of the amounts present. The Southgate method is applicable to both foods and feces and can be used in studies of the availability of dietary fiber in man.[77]

V. Conclusions: Choice of Analytical Procedures in Studies of Dietary Fiber

Although it is clearly desirable that more methodological studies be made of the analysis of dietary fiber, it is possible to give some guidance as to the choice of analytical methods in studies of dietary fiber.

The method for crude fiber has little or no place in studies with man; indeed, its continued use can only lead to confusion.[42]

In studies where only cellulose and lignin values are of interest, the measurement of acid-detergent fiber represents the method of choice, as it provides a small saving in time over other methods. However, there are manipulative problems, and the low concentrations of these two constituents in mixed diets can make accurate measurements difficult.

The neutral-detergent fiber procedure, in our experience, cannot be used in an unmodified form on human foods that contain starch or significant amounts of lipid. A preliminary treatment to remove lipid and an enzymatic hydrolysis of starch are essential if reliable values are to be obtained. The method does not measure any water-soluble polysaccharides and does not provide any evidence regarding the types of noncellulosic polysaccharides present.

A very simple procedure of this type used by McCance et al.[50] or the first stages of the Southgate method[75] provide a more convenient procedure for total dietary fiber.

In conclusion, it must be recognized that the physiological significance of the various components of dietary fiber are, as yet, poorly understood. At the present stage of the investigation of these physiological properties, it is important to have as much information as one can about the composition of the dietary fiber being studied.

References

1. ALBAUM, H. G., and W. W. UMBREIT. Differentiation between ribose-3-phosphate and ribose-5 phosphate by means of the orcinol-pentose reaction. *J. Biol. Chem.* **167:**369, 1947.
2. ALBERSHEIM, P., D. J. NEVINS, P. D. ENGLISH and A. KARR. A method for the analysis of sugars in plant cell-wall polysaccharides by gas liquid chromatography. *Carbohydr. Res.* **5:**340, 1967.

3. AOAC. *Official Methods of Analysis of the Association of Official Analytical Chemists*, 11th ed., W. Horwitz (ed.). Washington, D. C., AOAC, 1970, p. 129.
4. BATH, I. H. Analysis of the structural carbohydrates of grasses. *J. Sci. Food Agric.* **11**:560, 1960.
5. BISHOP, C. T. Gas-liquid chromatography of carbohydrate derivatives. *Adv. Carbohydr. Chem.* **19**:95, 1964.
6. BITTER, T., and H. M. MUIR. A modified uronic acid carbazole reaction. *Anal. Biochem.* **4**:330, 1962.
7. BLAKE, J. D., and G. N. RICHARDS. Problems of lactonisation in the analysis of uronic acids. *Carbohydr. Res.* **8**:275, 1968.
8. BLAKE, J. D., and G. N. RICHARDS. An examination of some methods for fractionation of plant hemicelluloses. *Carbohydr. Res.* **17**:253, 1971.
9. BLAKE, J. D., P. T. MURPHY and G. N. RICHARDS. Isolation and A/B classification of hemicelluloses. *Carbohydr. Res.* **16**:49, 1971.
10. BONDI, A., and H. MEYER. Lignins in young plants. *Biochem. J.* **43**:248, 1948.
11. BUCHALA, A. J., C. G. FRASER and K. C. B. WILKIE. Extraction of hemicellulose from oat tissue during the process of delignification. *Phytochemistry* **10**:1249, 1972.
12. COLE, E. W. Note on the reaction of gaseous chlorine with a wheat flour hemicellulose. *Cereal Chem.* **47**:696, 1970.
13. DEKKER, R. F. H., and G. N. RICHARDS. Determination of pectic substances in plant materials. *J. Sci. Food Agric.* **23**:475, 1972.
14. DISCHE, Z. New color reactions for determination of sugars in polysaccharides. *In: Methods of Biochemical Analysis*, D. Glick (ed.). New York, Interscience Publishers, 1955, Vol. II, p. 313.
15. DISCHE, Z., and C. ROTHSCHILD. Two modifications of the carbazole reaction of hexuronic acids for the differentiation of polyuronides. *Anal. Biochem.* **21**:125, 1967.
16. DOUGALL, H. W. Observations on 'crude fibre' estimated by acid digestion. *J. Sci. Food Agric.* **9**:1, 1958.
17. DOUGALL, H. W. The 'conventional' crude fibre content of roughages in relation to their acid fibre content. *J. Sci. Food Agric.* **9**:342, 1958.
18. DREYWOOD, R. Qualitative test for carbohydrate material. *Indr. Eng. Chem., Anal. Ed.* **18**:499, 1946.
19. EDWARDS, C. S. Determination of lignin and cellulose in forages by extraction with triethylene glycol. *J. Sci. Food Agric.* **24**:381, 1973.
20. EPSTEIN, E. Mineral Metabolism. *In: Plant Biochemistry*, J. Bonner and J. E. Varner (eds.). New York, Academic Press, 1965, p. 460.
21. FLORIDI, A. An improved method for the automated analysis of sugars by ion-exchange chromatography. *J. Chromatogr.* **59**:61, 1971.
22. Food and Agricultural Organization of the United Natiions. *Energy-Yielding Components of Food and Computation of Calorie Values*, Washington, 1947.
23. FRANSSON, L., L. RODEN and M. SPACH. Automated ion-exchange chromatography of uronic acids and uronic acid containing oligosaccharides. *Anal. Biochem.* **21**:317, 1968.
24. FRASER, J. R., and D. C. HOLMES. Proximate analysis of wheat flour. 2. The analysis of the carbohydrate fractions of different flour types. *J. Sci. Food Agric.* **7**:589, 1956.
25. FRASER, J. R., M. BRENDON-BRAVO and D. C. HOLMES. The proximate analysis of wheat flour carbohydrates. I. Methods and scheme of analysis. *J. Sci. Food Agric.* **7**:577, 1956.
26. FRIEDEMANN, T. E., N. F. WITT, B. W. NEIGHBORS and C. W. WEBER. Determination of available carbohydrates in plant and animal foods. *J. Nutr.* **91**(2):10, 1967.
27. GAILLARD, B. D. E. A detailed summative analysis of the crude fibre and nitrogen-free extractives fraction of roughages. I. Proposed scheme of analysis. *J. Sci. Food Agric.* **9**:170, 1958.
28. GAILLARD, B. D. E. Separation of linear from branched polysaccharides by precipitation as iodine complex. *Nature* **191**:1295, 1961.

29. GAILLARD, B. D. E. Comparison of the hemicelluloses from plants belonging to two different plant families. *Phytochemistry* **4:**631, 1965.

30. GAILLARD, B. D. E., and N. S. THOMPSON. Interaction of polysaccharide with iodine. Part II. The behaviour of xylans in different salt solutions. *Carbohydr. Res.* **18:**137, 1971.

31. GALAMBOS, J. T. The reaction of carbazole with carbohydrates. I. Effect of borate and sulfamate on the carbazole color of sugars. *Anal. Biochem.* **19:**119, 1967. II. Effect of borate and sulfamate on the ultraviolet absorption of sugars. *Anal. Biochem.* **19:**133, 1967.

32. GOERING, H. K., and P. J. VAN SOEST. *Forage Fiber Analyses* (Apparatus, Reagents, Procedures and Some Applications). U. S. Dept. Agric. Agricultural Handbook No. 379, U. S. Government Printing Office, Washington, D. C., 1970.

33. GREENBERG, C. J. Personal communication.

34. GREGORY, J. D. The effect of borate on carbazole reaction. *Arch. Biochem. Biophys.* **89:**157, 1960.

35. HALLAB, A. H., and E. A. EPPS. Variables affecting the determination of crude fiber. *J. Assoc. Off. Agric. Chem.* **46:**1006, 1963.

36. HALLSWORTH, E. G. The crude fibre determination and its alternatives. *Agric. Progr.* **25:**39, 1950.

37. HARWOOD, V. D. Analytical studies on the carbohydrates of grasses and clovers. V. Development of a method for estimation of cell wall polysaccharides. *J. Sci. Food Agric.* **5:**270, 1954.

38. HENNEBERG, W., and F. STOHMANN. Beitr. zur Begrundung einer rationellen Futterung der Wiederkauer I & II. Braunsweig, 1860–1864.

39. HOUGH, L., and JONES. Chromatography on paper. *In: Methods in Carbohydrate Chemistry,* R. L. Whistler and M. L. Wolfrom (eds.). New York, Academic Press, 1962, Vol. I., p. 21.

40. HOWLING, D. Modified starches for the food industry. *Food Technol. Aust.* **26:**464, 1974.

41. HUDSON, G. J. Personal communication.

42. JAMES, W. P. T., and D. A. T. SOUTHGATE. Bran and blood lipids. *Lancet* **1:**800, 1975.

43. JERMYN, M. A., and F. A. ISHERWOOD. Improved separation of sugars on the paper partition chromatogram. *Biochem. J.* **44:**405, 1949.

44. JONSSON, P., and O. SAMUELSON. Automated chromatography of sugars on cation exchange resins. *Anal. Chem.* **39:**1156, 1967.

45. KIESLER, R. B. Rapid quantitative anion-exchange chromatography of carbohydrates. *Anal. Chem.* **39:**1416, 1967.

46. LEDERER, E., and M. LEDERER. *Chromatography. A Review of Principles and Applications,* 2d ed. New York, Elsevier, 1957. p. 37.

47. LEEGWATER, D. C., and J. W. MARSMAN. Identification of a (6-0-[hydroxypropyl])D glycosyl D glucose as a faecal metabolite of 0-(hydroxypropyl) starch in the rat. *Carbohydr. Res.* **29:**271, 1973.

48. LEEGWATER, D. C., M. C. TEN NOEVER DER BRAUW, A. MACKOR and J. W. MARSMAN. Isolation and identification of an O(2 hydroxy propyl) maltose from faeces of rats fed with 0-(hydroxypropyl starch). *Carbohydr. Res.* **25:**411, 1972.

49. McCANCE, R. A., and R. D. LAWRENCE. The carbohydrate content of foods. *Med. Res. Counc. Spec. Rep. Ser.* **135** (HMSO), 1929.

50. McCANCE, R. A., E. M. WIDDOWSON and L. R. B. SHACKLETON. The nutritive value of fruits, vegetables and nuts. *Med. Res. Counc. Spec. Rep. Ser.* **213** (HMSO):32, 1936.

51. McCONNELL, A. A., and M. A. EASTWOOD. A comparison of methods of measuring 'fibre' in vegetable material. *J. Sci. Food Agric.* **25:**1451, 1974.

52. McKAY, E. Pentose estimation by the orcinol method, with particular reference to plasma pentose. *Clin. Chim. Acta* **10:**320, 1964.

53. MacRAE, J. C., and D. G. ARMSTRONG. Enzyme method for the determination of α linked glucose polymers in biological materials. *J. Sci. Food Agric.* **19:**578, 1968.

54. MACY, I. G. *Nutrition and Chemical Growth in Childhood.* Springfield, Illinois; C. C. Thomas, 1942, p. 251.

55. MACY, I. G., F. C. HUMMEL and M. L. SHEPHERD. Value of complex carbohydrates in diets of normal children. *Am. J. Dis. Child.* **65:**195, 1943.
56. MANGOLD, D. E. The digestion and utilisation of crude fibre. *Nutr. Abstr. Rev.* **3:**647, 1934.
57. MONRO, J. A., R. W. BAILEY and D. PENNY. Differential alkali-extraction of hemicellulose and hydroxyproline from nondelignified cell walls of lupin hypocotyls. *Carbohydr. Res.* **41:**153, 1975.
58. MOON, F. E., and A. K. ABOU-RAYA. The lignin fraction of animal feedingstuffs. I. Preliminary examination of lignin determination procedures. II. Investigation of lignin determination procedures and development of a method for acid-insoluble lignin. *J. Sci. Food Agric.* **3:**399, 407, 1952.
59. MORRISON, I. M. A semi-micro method for the determination of lignin and its use in predicting the digestibility of forage crops. *J. Sci. Food Agric.* **23:**455, 1972.
60. MORRISON, I. M. Improvements in the acetyl bromide technique to determine lignin and digestibility and its application to legumes. *J. Sci. Food Agric.* **23:**1463, 1972.
61. MORRISON, I. M. Some investigations on the lignin-carbohydrate complex of *Lolium perenne*. *Biochem. J.* **139:**197, 1974.
62. MURPHY, R. P. A method for the extraction of plant samples and the determination of total soluble carbohydrates. *J. Sci. Food Agric.* **9:**714, 1958.
63. O'DWYER, M. H. The hemicelluloses of beech wood. *Biochem. J.* **20:**656, 1926.
64. OLMSTED, W. H. Availability of carbohydrate in certain vegetables. *J. Biol. Chem.* **41:**45, 1920.
65. PAECH, K., and M. V. TRACEY. Modern Methods of Plant Analysis. Berlin, Springer Verlag, 1955, Vol. III.
66. PRITCHARD, P. J., E. A. DRYBURGH and B. J. WILSON. Carbohydrates of spring and winter field beans *(Vicia faba L.) J. Sci. Food Agric.* **24:**663, 1973.
67. RAYMOND, W. F., E. C. JONES and C. E. HARRIS. Factors affecting the accuracy of normal-acid fibre determinations and its use as a faecal index. *Agric. Progr.* **23:**120, 1955.
68. REID, P. E., B. DONALDSON, D. W. SECRET and B. BRADFORD. A simple, rapid, isothermal gas chromatographic procedure for the analysis of monosaccharide mixtures. *J. Chromatogr.* **47:**199, 1970.
69. ROBYT, J. F. Paper chromatographic solvent for the separation of sugars and alditols. *Carbohydr. Res.* **46:**373, 1975.
70. ROE, J. H. The determination of sugar in blood and spinal fluid with anthrone reagent. *J. Biol. Chem.* **212:**335, 1955.
71. SAXENA, K. K., R. MUKHERJEE and V. MAHADEVAN. A new micromethod for the estimation of cellulose in biological materials using anthrone reaction. *J. Sci. Ind. Res.* **20C:**186, 1961.
72. SELVENDRAN, R. R. Analysis of cell wall material from plant tissue; extraction and purification. *Phytochemistry* **14:**1011, 1975.
73. SIEGEL, S. M. The biochemistry of the plant cell wall. *In: Comprehensive Biochemistry,* M. Florkin and E. H. Stotz (eds.). Amsterdam, Elsevier, 1968, Vol. 26A, p. 1.
74. SOUTHGATE, D. A. T. Determination of carbohydrates in foods. I. Available Carbohydrates. *J. Sci. Food Agric.* **20:**326, 1969.
75. SOUTHGATE, D. A. T. Determination of carbohydrates in foods. II. Unavailable carbohydrates. *J. Sci. Food Agric.* **20:**331, 1969.
76. SOUTHGATE, D. A. T. Problems in the analysis of polysaccharides in foodstuffs. *J. Assoc. Pub. Anal.* **12:**114, 1974.
77. SOUTHGATE, D. A. T., and J. V. G. A. DURNIN. Calorie conversion factors. An experimental reassessment of the factors used in the calculation of the energy value of human diets. *Br. J. Nutr.* **24:**517, 1970.
78. STEPHANENKO, B. N., and L. B. UZDENIKOVA. Precipitation of neutral polysaccharides and separation of their mixtures by use of various quaternary salts. *Carbohydr. Res.* **25:**526, 1972.

79. Sweet, D. P., P. Albersheim and R. H. Shapiro. Partially ethylated alditol acetates as derivatives for elucidation of the glycosyl linkage-composition of polysaccharides. *Carbohydr. Res.* **40:**199, 1975.

80. Terry, R. A., and G. E. Outen. The determination of cell-wall constituents in barley and maize. *Chem. Ind.* **23:**1116, 1973.

81. Updegraff, D. M. Semimicro determination of cellulose in biological materials. *Anal. Biochem.* **32:**420, 1969.

82. van Soest, P. J. Use of detergents in the analysis of fibrous feeds. I. Preparation of fiber residues of low nitrogen content. *J. Assoc. Off. Agric. Chem.* **46:**825, 1963.

83. van Soest, P. J. Use of detergents in the analysis of fibrous feeds. 2. A rapid method for the determination of fiber and lignin. *J. Assoc. Off. Agric. Chem.* **46:**829, 1963.

84. van Soest, P. J. The uniformity and nutritive availability of cellulose. *Fed. Proc.* **32:**1804, 1973.

85. van Soest, P. J. Personal communication.

86. van Soest, P. J., and R. W. McQueen. The chemistry and estimation of fibre. *Proc. Nutr. Soc.* **32:**123, 1973.

87. van Soest, P. J., and R. H. Wine. Method for determination of lignin, cellulose and silica. *J. Anim. Sci.* **26:**940, 1967.

88. van Soest, P. J., and R. H. Wine. Determination of lignin and cellulose in acid-detergent fiber with permanganate. *J. Assoc. Off. Agric. Chem.* **51:**780, 1968.

89. Waite, R. and A. R. N. Gorrod. The comprehensive analysis of grasses. *J. Sci. Food Agric.* **10:**317, 1959.

90. Walborg, E. F., Jr., L. Christensson and S. Gardell. An ion-exchange-column chromatographic method for the separation and quantitative analysis of neutral monosaccharides. *Anal. Biochem.* **13:**177, 1965.

91. Walborg, E. F., Jr., and R. S. Lantz. Separation and quantitation of saccharides by ion-exchange chromatography utilizing boric acid/glycerol buffers. *Anal. Biochem.* **22:**123, 1968.

92. Walborg, E. F., Jr., D. B. Ray and L. E. Ohrebearg. Ion-exchange chromatography of saccharides. An improved system utilising boric acid/2:3 butanediol buffers. *Anal. Biochem.* **29:**433, 1969.

93. Walker, D. M. A note on the composition of normal acid fibre. *J. Sci. Food Agric.* **10:**415, 1959.

94. Walker, D. M., and W. R. Hepburn. Normal-acid fibre: A proposed analysis for the evaluation of roughages. I. The analysis of roughages by the normal-acid fibre method and its use for predicting the digestibility of roughages by sheep. *Agric. Progr.* **30:**118, 1955.

95. Whistler, R. L., J. Bachrach and D. R. Bowman. Preparation and properties of corncob holocellulose. *Arch. Biochem.* **18:**25, 1948.

96. Whistler, R. L., and J. N. Bemiller. Quantitative paper chromatography. *In: Methods in Carbohydrate Chemistry*, R. L. Whistler and M. L. Wolfrom (eds.). New York, Academic Press, 1962, Vol. I, p. 395.

97. Whistler, R. L., and M. S. Feather. Carboxyl determination. *In: Methods in Carbohydrate Chemistry*, R. L. Whistler and M. L. Wolfrom (eds.). New York and London, Academic Press, 1962, Vol. 1, p. 464.

98. Whitehead, D. L., and G. V. Quicke. A comparison of the six methods of estimating lignin in grass hay. *J. Sci. Food Agric.* **15:**417, 1964.

99. Widdowson, E. M., and R. A. McCance. The available carbohydrate of fruits. Determination of glucose, fructose, sucrose and starch. *Biochem. J.* **29:**151, 1935.

100. Williams, R. D., and W. H. Olmsted. A biochemical method for determining indigestible residue (crude fiber) in faeces, lignin cellulose and nonwater-soluble hemicelluloses. *J. Biol. Chem.* **108:**653, 1935.

101. Williams, R. D., L. Wicks, H. R. Bierman and W. H. Olmsted. Carbohydrate values of fruits and vegetables. *J. Nutr.* **19:**593, 1940.

4

Physical Properties of Fiber: A Biological Evaluation

Martin A. Eastwood and W. D. Mitchell

I. Introduction

It is important in nutrition to distinguish between chemical and biological evaluation of a dietary component. Such a distinction is well demonstrated for protein, in which the essential amino acids affect the biological quality;[1] vitamin D is another example.[30]

This chapter attempts to evaluate dietary fiber biologically. Fiber is a very imprecise term and is generally used to describe the mixture of cell-wall polysaccharides and lignins.[7] It is hoped to describe characteristics with physiological relevance that would be capable of being tested in nutritional studies.[11] Chemical analysis of the polysaccharides and lignins that constitute the heterogeneous complex of vegetable dietary fiber would then have more interpretive value.

In the plant, polysaccharides and lignin molecules intermesh for anatomical and physiological function. The anatomy is destroyed by cooking, mastication, and digestion, although the original functional characteristics are retained, modified by the particular environment in the gastrointestinal tract. The physiological effects of these characteristics are quite different from those observed in *in vitro* studies. Such factors as osmolality, the presence of other macromolecules, the possibility of adsorption, and the presence of water and bacteria will markedly influence these characteristics.[28]

MARTIN A. EASTWOOD • Consultant Physician, Wolfson Gastrointestinal Laboratories, Department of Medicine, Western Hospital, Edinburgh, Scotland. W. D. MITCHELL • Wolfson Gastrointestinal Laboratories, Department of Medicine, Western General Hospital, Edinburgh, Scotland.

II. Physical Properties of Living Plant Fiber

A. Polysaccharides and Lignins

The primary cell wall of a plant is essentially a pectin-layer cellulose. As secondary thickening progresses, there is a formation of fibers which run through the cell matrix. Cellulose, either wholly or partly in a crystalline form, is distributed at various axes through the structure, and the interspaces filled with lignin, hemicellulose, and pectic material are held in a network.[41]

Cellulose is essentially a straight-chain polymer. The cellulose molecules in the plant form lengthwise chains, which determine the shape and strength of the structure.[33] Their distribution is stabilized by hydrogen bonds between hydroxyl groups, which form strong intermolecular bonds.

Pectin is found in the primary cell walls and intracellular layers. It is a uronic acid polymer, which is maximally branched in growing tissues and becomes less branched as it becomes a supporting structure. These pectic substances are hydrophilic and adhesive, and the pectin changes from an insoluble material in the unripe fruit to a much more water-soluble substance in the ripe fruit.[54] This highlights a variation in physical characteristics which is not immediately apparent from the general term pectin.

Hemicelluloses are a mixture of linear and highly branched polysaccharides which contain various sugar residues: xylose, arabinose, glucose, and mannose. Hemicelluloses act as plasticizers and intertwine with lignin between the cellulose fibers.[2] Mucilages are polymers that contain mainly uronic acid and, having a high capacity for holding water, form potential reservoirs for water retention in the plant.

Lignin is laid down with cellulose and hemicelluloses to produce structural material. It reinforces fiber and provides a strong support structure, forming approximately 12% of the structural material of annual plants. It consists of phenylpropane polymers and is therefore markedly different from the polysaccharide material with which it is so closely associated. There is continuous polymerization of lignin which is closely related to the age of the plant.[43] It is believed that there is association between the ferulic and coumaric acid components of the lignin and the hemicelluloses.[22]

B. Primary Variations in Vegetable Fiber

Perhaps the main reason for the difficulty in defining dietary fiber is that it is part of a living organism and therefore undergoing continuous change. The nutritional, physical, and chemical properties of its constituents change not only with the age of the plant, but also with the anatomical source of the fiber; e.g., cauliflower is a flower, celery is a stalk, and turnip a root. Grasses differ chemically throughout the season in that the amount of cellulose and hemicellulose will change during the growing period. The fully developed stem is less easily digested by ruminants than is the leaf. Spring grass contains more soluble carbohydrate and a higher ratio of sugar to protein than grass eaten later in the year. The fructosan and total soluble-carbohydrate content of spring grass is related to the low growth rate and ground temperature. It is interesting that if a comparison is made between grass in Scotland and that in the warmer parts of the United States, such differences are even more accentuated.[47] Similar variations may be assumed to occur in fruits and vegetables. It is possible that the development of varieties of fruits and vegetables for speed of growing, succulence, and suitability for canning and freezing may coincide with a reduction in fiber content.

To an extent, the chemical descriptions of plant fiber polysaccharides are artificial. These descriptions are based on a fractionation procedure that is dependent on the solubility of the constituents in different solutions, whereas within the plant these polysaccharides and lignins are part of an intertwined heterogeneous complex, the properties of which will be peculiar to that complex. In our studies we have regarded plant fiber as a composite entity, which will have physicochemical characteristics dependent on its constituents. Thus we have defined plant dietary fiber according to its physical characteristics rather than its chemical constituents.

III. Preparation of Vegetable Fiber for Physical Characteristics Studies

The nature of the fiber used in these studies is variable. Plant material should be prepared in a manner that allows biological comparisons. To achieve cellular disruption, the plant material may be macerated in a blender in the presence of water and the homogenate frozen at $-20°C$ at least overnight before thawing. The material must then be washed repeatedly with water to remove water-soluble materials including protein. "Never-dried fiber" is prepared in this way. The fiber may be dried either by freeze-drying or by washing with acetone. The resultant powder may be left in such a prepared form or ground in a mill. These preparations differ from each other in physical and, presumably, biological activity. There is a strong suspicion that polysaccharides and possibly lignin material may be washed from the general mass of fiber at each stage of preparation.[35,36] It is possible that a similar leaching occurs along the gastrointestinal tract. Here leached materials might move along the tract at various rates and could even be metabolized by bacteria at differing rates and to differing extents. The definition of fiber that is based on methods used in the nineteenth century for legislation against the adulteration of cattle feed[46] does not highlight such complexities.

IV. Water Adsorption

A. Fiber and Water

Polysaccharides swell in water to form gels with a high water content, depending on their chemistry, species, and anatomical sources. Pectic substances and hemicelluloses also form gels, but celluloses are more insoluble. Plant cell-wall materials swell in water to a coherent structure. When a material is exposed to water, there is initial adsorption of surface water onto the fiber, while the interstitial spaces of the fibrous material are filled with water later.

Fiber has a finite capacity to hold water, known as the fiber saturation capacity,[44] which has been reached when the surface and the interstices of the fibrous material are saturated with water. Any water subsequently added will be free water superfluous to the plant matrix (Figure 1). This capacity is determined not only by the chemistry and shape of the macromolecules but also by the electrolyte concentration and pH of the surrounding liquid. Lignins are much less hygroscopic than the polysaccharides and are relatively insoluble in water. The hydrophilic nature of lignin is determined by the hydroxyl, carbonyl, and carboxyl moieties.[32]

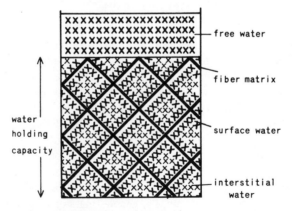

Figure 1. A representation of water phases in relation to the fiber matrix, (a) held water (surface and interstitial) which is measured by the water-holding capacity and (b) free water excess to (a).

B. Methods of Water Adsorption

Water-holding capacity means the extent to which a fibrous material is capable of holding water within its interstices in the presence of excess water. Water outside the interstices or matrix is expressable and is free water. Water-holding capacity is defined as the amount of water that can be taken up by unit weight of dry fiber and is the point at which there is no free water. There are several methods of measurement. It is essential to state exactly how the fiber was prepared and under what conditions the water-holding capacity is measured, since various factors can influence the water-holding capacity. For example, the water-holding capacity of materials that have been dried differs from that of materials that have not. This is particularly the case in the comparison of materials that have been washed with hydrochloric acid to convert all the ionic groups to the H^+ form. Particle size may also materially affect the water-holding capacity (Table I). The presence of electrolytes and also the pH of the solution can affect water-holding capacity.

Methods of determining the water-holding capacity require the separation of the free or unadsorbed water phase from the entrapped or interstitial phase of water. One method is centrifugation, which produces two defined phases: a supernatant and a pellet. There are, however, limitations in that too severe centrifugation can result in the collapse of the matrix produced by the fibrous material. Furthermore, the matrices from different plant sources vary in their physical and chemical characteristics so that it may well be wrong to use constant centrifugation conditions for a range of different vegetable sources. An alternative method of studying water-holding capacity is to equilibrate the fiber with water and to explore the gel-filtration characteristics of the swollen material. Polyethylene glycol of varying molecular weight[45] or dyes such as dextran blue

TABLE I. Effect of Preparation on Fiber Water-Holding Capacity
(g water/g powder)

Source	Natural[a]	H^{+b}
Celery		
Celery macerated, frozen		
at $-20°C$, ground 1 mm	42	76
Water-washed preparation A	52	82
Acetone-dried preparation B	35	30
Freeze-dried preparation B	41	28
Wheat bran		
Unground	5.0	13.0
Ground 1 mm	3.7	7.1

[a]Natural — washed only in water.
[b]H^+ — acidic groups converted to H^+ form by exposure to 2 N HCl overnight; washed to neutrality.

(Pharmacia, Uppsala) may be used. In our experience, however, there have been problems of adsorption of the material to the fiber that have precluded our further use of this method.[36] It might be possible, by using different salt concentrations, to avoid this adsorption, particularly of dextran blue. This, however, causes further problems, as the water-holding capacity is influenced by the concentration of electrolytes in the solution.

As our interest in fiber is ultimately a biological one, it has seemed appropriate to use 14,000 g centrifugation.[36] This speed has proved useful in separating the normal from a diarrheal stool.[16] A normal stool does not separate to a supernatant and pellet under these circumstances, whereas a diarrheal stool will give a water phase and a pellet. The water phase represents free water and the water associated with the pellet is the interstitial water of the stool.

C. Water-Holding Capacity

The main source of variation in water-holding capacity measured under constant analytical conditions is the variation in the type of plant material. There is an almost 10-fold range (Table II). The fiber content of different vegetables is

TABLE II. Vegetable Fiber and Water-Holding Capacity

	Fiber (% raw material)	Water-holding capacity (g water/g fiber)	Capacity of fiber in 100 g of raw vegetable to adsorb water (g water)
Turnip	4.0	9.0	37
Potato	19.5	2.0	41
Rhubarb	4.2	14.4	60
Banana	22.7	2.9	68
Cauliflower	11.6	5.9	68
Tomato	6.6	10.8	71
Broad bean	18.8	4.1	77
Cucumber	3.7	20.9	77
Celery	6.0	16.2	97
Pea	21.6	4.6	99
Lettuce	4.2	23.7	99
Green bean	12.4	8.1	100
Pear	15.3	7.4	113
Orange	9.9	12.4	122
Maize	86.1	1.5	129
Aubergine	7.5	17.3	129
Apple	14.6	12.1	177
Carrot	8.9	23.4	208
Mango	15.3	20.4	312
Bran	89.3	3.0	447

considerably varied. The fiber content of cucumber is 4%, whereas the fiber content of bran is 90%. In this context the term fiber is applied to the freeze-dried dry weight of the vegetable obtained in the way described in Section III.

The water-holding capacity of a plant source is a function both of the fiber content and of the water-holding capacity of the particular fiber. Table II shows a comparison of the fiber content of a variety of fruits and vegetables and the water-holding capacity of the fiber that has been obtained from these fruits and vegetables. If the original fiber content of the vegetable source and the water-holding capacity of that fiber source are taken into account, it is possible to estimate the water-holding capacity of the fiber present in the fresh vegetables. Table II shows the capacity of the fiber in 100 g of raw fruits and vegetables to adsorb water. It will be seen that bran, while it has a modest water-holding capacity because of its high fiber content, is the most potent water-holding material. However, mango, carrot, apple, and Brussels sprouts each have a good water-holding capacity and for many people are more palatable than bran. Thus 50 g of bran, which is capable of holding 220 g of water, is functionally equivalent to 100 g of raw carrot, 150 g of apple, or 200 g of peeled orange. It would be necessary, however, to eat 600 g of potatoes to obtain the same overall water-holding capacity. Such a ranking as presented in Table II enables a theoretical rationale for a high-fiber-containing diet, either for treatment or for prophylaxis.[36] However, these data do not take into account the effect of intestinal digestive function or bacterial degradation of the fiber on water-holding capacity. Such changes may not parallel the order that has been described.

Water-holding capacity may well vary with age and variety of fruits and vegetables because of differences in the hydrophilic polysaccharide content.

V. Ion-Exchange Capacity

A. Cation Exchange

Polysaccharide material may, by virtue of acidic sugars, have a cation-exchange capacity. The cation-exchange capacity is constant for fiber from any particular plant source irrespective of the anatomical source. Species of plants have a characteristic cation-exchange capacity that is relatively independent of external nutritional factors in the soil.[29] It has been shown, for example, that calcium may be adsorbed by dietary fiber to a degree dependent on the presence of unsubstituted uronic acid groupings.[5] The physical properties of fiber may be affected by the presence of ions. It has also been shown that divalent metals have a considerable effect on the gel formation and precipitation properties of sodium alginate.[25]

The ion-exchange capacities of vegetable fiber may be measured by titration with sodium hydroxide after conversion of the fiber to the H^+ form by treatment

with excess hydrochloric acid.[26] The capacity measurements depend on a complete conversion to the H^+ form, and a 24-hr immersion in hydrochloric acid was
found to give consistent results.[36] In the natural form most of the carboxyls are in
the salt form. There are, however, problems associated with the use of hydrochloric acid, which leaches out polysaccharide material that we have yet to characterize. Less strong acids might, therefore, be used in preference. If the cation-
exchange capacity is to be measured, the fiber must first be washed to neutrality
and then reconstituted in water or sodium chloride solution. The measurement of
pK (the mean dissociation constant of the fixed ionogenic moieties of the
polysaccharide) requires titration to be made with the fiber in a sodium chloride
solution of sufficient strength for complete conversion of the nonionized to the
ionized form. In addition, in order to calculate pK, the water-holding capacity
must be known. This will vary with the salt concentration and the length of time
of immersion. The titration must be made slowly; a period of 4–8 hr is
recommended.[36]

Most of the vegetable fibers that we have studied act as monofunctional,
weak cation-exchange resins. Maize, oatmeal, bananas, cereal bran, and new
potatoes act as very weak polyfunctional exchangers. A polyfunctional ion-
exchange characteristic is the formation of steps at different pH levels in the
titration curve. This is due to different pH levels in the titration curve, which in
turn is due to different ionogenic groups and a spectrum of pK values. Figure 2
shows typical pH titration curves of monofunctional and polyfunctional cation
exchangers. The cation-exchange capacity of acetone-dried powder was shown

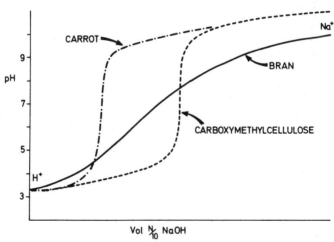

Figure 2. Titration of a strong monofunctional cation exchange resin (carrot;
sodium carboxymethylcellulose as a standard) and a weak polyfunctional cation
exchange resin.

to range from 0.6 mEq/g for pear through orange and lettuce, which were medium cation exchangers, to over 2.3 mEq/g for turnip, carrot, and spring cabbage. A strong exchange resin is one where there is a vertical change in the sigmoid titration curve; a medium exchanger is one where the range is greater than 45° but less than 90° (Figure 2). The effect of cooking on the cation-exchange capacity is variable in that winter cabbage appears to be increased in cation-exchange capacity, whereas spring cabbage is decreased.

There is a close relationship between acid-detergent-fiber (ADF) content[46] of the vegetable dietary fiber and the water-holding capacity, but there is no such relationship between the lignin content and the water-holding capacity. This suggests that it is the polysaccharide content of the plant that determines the ability to hold water. There is a linear relationship between the water-holding capacity and the cation-exchange capacity but not between the cation-exchange capacity and the ADF content of the vegetable dietary fiber or the lignin content.

B. Anion Exchange

There is a ready hydrolysis of polysaccharide in sodium hydroxide solution that creates difficulties in the conversion of the acidic groups to the salt. We have therefore used other methods but have been unable to detect any anion-exchange capacity in any of the fruit and vegetable sources studied.[36]

VI. Adsorption of Organic Materials to Vegetable Dietary Fiber

Vegetable dietary fiber can adsorb bile acids. Those bile acids conjugated with taurine found in the bile and upper duodenum are only weakly adsorbed to fiber, but unconjugated bile acids, e.g., those found in the colon, are strongly adsorbed. Such adsorption is enhanced by the chemical transformation of the bile acids by intestinal microorganisms. These bacterial changes in the physical state of the bile acids result in the formation of poorly soluble derivatives that are easily adsorbed to the residue.[4,10,31] The adsorption is pH-dependent, being greatest at an acidic pH, and is possibly influenced by methyoxylation of acidic hydroxyl moieties in the fiber. There is good evidence that it is the lignin component of the fiber that is important in the adsorption of bile acids to the fiber,[10] but there is little information about the adsorption of other colonic contents. Crude dietary homogenates and certain nondietary fibers can adsorb pteroyl monoglutamates, but not pteroyl polyglutamates.[34]

VII. Surface Properties

A. Surface Area, Pore Size

Fiber has an external surface into which there are intrusions of pores, the surface area of which contributes to the internal surface of the fiber. These pores have a volume and also an average radius. It is possible to measure the internal volume area and also the average radius in the dry material. None of the surface areas are high.[8] Table III shows such measurements in some of the fruits and vegetables which have been discussed in Table II. It will be seen that there is only a modest relationship between external surface area and water-holding capacity.

B. Filtration Properties

A gel-filtration system with a molecular-exclusion capacity may be provided by vegetable dietary fiber. This could affect the entrance of substrate, enzymes, and bacteria to the fiber interstices and hence the susceptibility of the

TABLE III. Vegetable Fiber Surface Area Data

	External surface area (M^2/g)	Pore area (M^2/g)	Pore volume (cm^3/g)	Pore radius ($\mathring{A}$)
Oatmeal (ground)	1.6	1.61	0.0026	500–1000, 200–300 20–30, 12–8
Banana (ground)	1.7	1.92	0.0034	100–150, 50–70, 20–30, <12
Broad bean	2.8	2.18	0.0025	50–60, 20–25, <12
Cauliflower,				
Uncooked	1.1	0.94	0.0014	Not available
Cooked	0.6	0.66	0.0008	Not available
Brussel sprouts	2.4	2.4	0.0033	500–1000, 200–300, 100–150, 20–25, <12
Apple				
Unground	1.7	2.05	0.0026	20–30, 12–18, <12
Ground	3.3	3.26	0.0038	500–1000, 100–150, 20–25, <12
Aubergine (ground)	1.7	1.59	0.0017	
Carrot (ground)	4.7	4.67	0.0061	100–150, 50–60, 20–25, <12
Rhubarb	1.6	1.59	0.0018	Not available

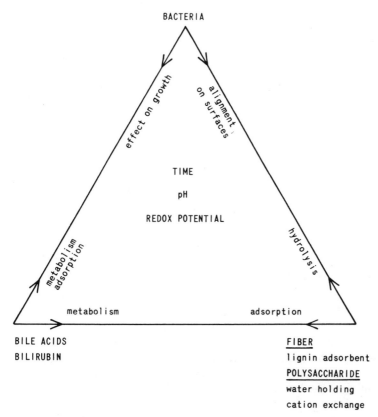

Figure 3.   Interrelationship among bacteria, fiber, and substrates in the colonic contents.

fiber to enzyme hydrolysis. This filtration capacity will dictate the permeability of molecules of differing water solubility and size into the fibrous mesh.

VIII. Relationship between Fiber and Bacteria

In the colon there are complex relationships between the fiber matrix, solute, and bacteria. There may be interaction between matrix and bacteria, bacteria and solute, or solute and bacteria, or all three may interact (Figure 3).

A. Effect of Bacteria on Fiber

In the intestine and particularly in the colon, vegetable dietary fiber will form a matrix which has a potential for biological sieving. Very little information is available about such sieving in the cecum. There is information about cellulose-degrading bacteria and their physical relationship to cellulose fiber during hydrolysis. Many bacteria can produce extracellular enzymes that can permeate the gel and hydrolyze the polysaccharide matrix. The rate of degradation has been shown to be dependent on the degree to which the fibrils of cellulose are opened and may be complicated by the coincidental presence of lignin, which can inhibit the passage of bacteria into the mesh.[3]

It has been shown that polysaccharides can be metabolized by colonic bacterial microflora; e.g., if pectic materials are taken by mouth, only 10% of the ingested pectin is recovered in the feces.[52] When cellulose fiber is degraded by bacteria, the bacteria penetrate into the lumen of the fiber and accumulate in large numbers. Thus the structure of the fiber greatly influences the rate at which different cellulolytic bacteria affect decomposition. This means that the fiber matrix may disintegrate by bacterial hydrolysis attack.[3]

B. Influence of Fiber on Bacteria

The hydrated fiber after passage to the cecum will have a cellular interior or mesh rather like a sponge, and the interior of the fiber will create different environments. Thus the substrate, enzyme, and bacteria inside and outside the fiber matrix may well behave differently as a result of passing down the intestine. The reactions between the substrate enzymes and bacteria will be complicated by the surface effects, which will themselves be complicated by some bacteria that can hydrolyze the fiber matrix, particularly the hemicelluloses. If bacteria in the cecum behave like bacteria in contact with polysaccharides in other situations, the bacteria may well align themselves on the surface of the fiber. However, little is known about the effects on bacterial physiology of their adsorption to fiber. Bacteria adsorbed to a Dowex 1 chloride column grow as well as those in free solution.[24] The optimum pH for oxidation metabolism by the adsorbed bacterial cells is $1-1.5$ units higher than that for free cells, and oxygen consumption of the adsorbed cells is low. Gram-negative $E.\ coli$ oxidize some organic materials more slowly when they are adsorbed to the resin than when they are in suspension.[23] In effect, the bacteria on the resin surface are growing and metabolizing in an environment different from that of the bacteria in a free solution. The surface characteristics of a polysaccharide cation exchanger such as fiber must be different from those of an anion-exchange resin. The possibilities exist for profound differences in the spectrum of metabolites produced from substrates present in the intestinal contents as a result of bacterial activity in the presence of fiber compared with the activity in free solution (Figure 4).

The interposition of bacteria into the fibrous matrix of the feces will be affected by cross-linkages in the fiber, and these will alter as a result of changes

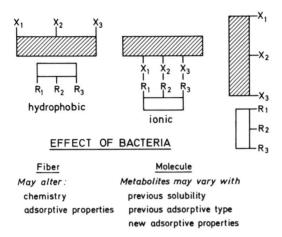

Figure 4. The nature of the adsorption between a substrate and fiber alters the groupings (R) and (X) available for bacterial metabolism, and hence the end product will vary. Bacteria may modify both the fiber chemistry and adsorptive properties, and the substrate availability will be affected by previous solubility and adsorption properties.

in hydration and solvent ionic strength. Similarly, the environment will be changed by bacterial metabolites such as organic acid and sugar phosphates. The degree of cross-linking determines the mean pore size of the fibrous material, and this will determine the rate of ion exchange and the adherence of materials to the fiber. Water-soluble compounds in the cecum are metabolized to less water-soluble products; exposure of parent compounds to bacteria and adsorption to fiber of bacterial products will be affected by the mesh of fiber matrix.

IX. Effect of Bacteria on Solutes

The distribution of bacteria along the gastrointestinal tract is such that the heaviest concentration is in the large intestine. Compounds passing along the gastrointestinal tract remain essentially unaltered by bacteria until they reach the cecum. Bacteria profoundly metabolize solutes in the colon. Some of the end-products of bacterial metabolism are reabsorbed and return to the liver via the portal vein, completing an enterohepatic circulation. Other metabolites are excreted in the feces. Bacteria have an important role in the fate of compounds excreted by the liver. The liver converts lipid-soluble compounds to water-soluble derivatives, which are excreted in the bile. These compounds are not absorbed from the small intestine. It is only after they have been metabolized in the colon by bacteria and become less water-soluble that it is possible for them to be reabsorbed from the intestine.[21]

It is possible to identify different types of compounds that are dealt with in this way. Bile acids are conjugated with glycine and taurine and are, in fact, reabsorbed principally from the ileum, although an unknown but significant amount of bile acids pass into the colon, where deconjugation and 7α-dehydroxylation take place. Other compounds such as drugs, thyroxine, and bilirubin are principally excreted in the feces after metabolic transformation by bacteria. It is clear that bacteria modify the fate of a compound either by converting it into a form that is readily reabsorbed and conserving it via the enterohepatic circulation or by the formation of a bacterial derivative that is more readily excreted in the feces.[9]

A. Solutes of Endogenous Origin

The most common compounds of endogenous origin which are extensively metabolized are bile acids, bilirubin, and hormone derivatives such as thyroxine and steroid hormones. The effect of bacteria on bile acids is to convert a water-soluble compound into one much less water soluble. Thus taurocholic, glycocholic, glycochenodeoxycholic, and taurochenodeoxycholic acids become free bile acids, principally deoxycholic acid and lithocholic acid, respectively. Similarly, bilirubin glucuronide is converted to a rather large series of breakdown products, of which urobilinogen is perhaps the best known.

B. Solutes of Exogenous Origin

Gut microflora and particularly anaerobic organisms are important in the metabolism of exogenous compounds.[53] The bacterial reactions on drugs and other compounds include hydrolysis of glucuronides, glucosides, hydroxylation of certain catechols, decarboxylation, dealkylation, aromatization, dehalogenation, deamination, reduction of various kinds, and ring fission. The metabolic pathway followed by any compound will be determined in part by the gut flora. It is almost certain that the toxicity of many compounds, e.g., azo compounds, is due to a metabolite that arises from bacterial metabolism of the parent compound.[48] The water-soluble neoprontosil is poorly absorbed from the gut but is reduced by gut flora, and the resultant lipid-soluble prontosil is absorbed to a significant extent.[19]

Differences in the bacterial population of the gut depend on environment, species, and diet. One of the environmental variables will be the contents of the cecum. Very little is known of the effect fiber has on the bacterial metabolism of the compounds present in the cecal contents. There has been some work on the effect of diet and intestinal lumenal contents on drug metabolism, but this is largely in the complex area of starvation experiments.[21]

It has been suggested that cancer of the bowel can be due to the production of carcinogens by the action of colonic bacteria on compounds present in the colon, e.g., bile acids and steroids or nitrate being metabolized to carcinogenic dialkylnitrosamines. It has also been shown that the gut bacteria can produce a carcinogen from cicadin, which is found in cycad nuts. This compound is carcinogenic when given orally to rats but not when given by injection, demonstrating that the carcinogenic effect[51] develops only after passage through the gut.

X. Matrix Bacterial–Solute Interaction

A. Possible Modes of Action for Fiber

The matrix will adsorb water, thus exerting a diluting effect on soluble intestinal contents. This in turn will influence the interaction between bacteria and the solutes both in the fiber matrix and in the surrounding area.

It is possible that vegetable dietary fiber acts in the colon by forming a supporting matrix which provides surfaces or a loculating microenvironment in which bacteria and intestinal contents interact. The system will have many properties of a chromatography system exposed to a gradient elution and infected by bacteria. The supporting medium will be altered in its physical characteristics by changes in the vegetable fiber component of the diet. When the predominant physical property of the matrix is water absorption, then biologically active materials may be diluted in the colonic contents. Also, cation-exchange capacity properties may influence mineral metabolism.

Vegetable dietary fiber may also act as a gel with partitioning characteristics for both bacteria and other substances in the intestinal contents. The metabolism of substances in the colon in its turn is influenced by the physical characteristics of the matrix, and this supporting medium may be metabolized and its character changed by the activity of bacteria. Fiber can thus influence the metabolism of material passing into the colon both quantitatively and qualitatively. Metabolites that are reabsorbed from the colon and pass back to the liver for reextraction in the bile thus influence the production of the parent compound. There are also qualitative and quantitative effects on the excretion of the metabolized material.[12]

The contents of the large intestine may, conversely, be regarded as a system where there is a thick suspension of solid in an aqueous phase. There are two boundaries of the aqueous phase, the intestinal mucosa and the surface of a suspended solid. There are, of course, many other boundaries, because the interphase between suspended solid and aqueous phase acts in several ways. There is a passive barrier to diffusion through the aqueous phase. The suspension acts as a site for segregation and coalescence of particles or droplets of insoluble compounds generated in the aqueous phase. Enzymes are absorbed from the aqueous phase in the intestine, and the solid suspension will also act as a site of metabolic activity because of microorganisms passively concentrating at interphases. There is thus a mosaic of internal surfaces separated by an aqueous phase. Each biological process leads to a segregation of metabolism of a substance in solution or suspension in the aqueous phase, altering the chance of its absorption from the colon or transformation by the intestinal contents.[18]

Fiber may also be important nutritionally following the hydrolysis of polysaccharides with changes in pH and redox potential of the cecal contents and attendant changes in the metabolic activity within the colon. It is possible that liberated acidic sugars and sugars may be relevant to the nutrition of the structure of the colon.

The colon consists of a series of chambers, or haustra, in which water and fiber are loculated. The haustral segmentation aids water absorption and the absorption of other material, e.g., bile acids, and also influences the movement of intestinal contents along the colon. We have shown in other experiments that in the normal colon, the intestinal solid phase and the liquid phase pass along the colon at the same rate. However, in diarrhea, the water phase passes at a faster rate than the solid phase, and in diverticular disease the solid phase moves ahead of the liquid phase. This heterogenous system is a highly complex one.[17]

B. Interaction with a Solute of Endogenous Origin

The metabolism of bilirubin glucuronide to urobilinogen, or at least the excretion of the urobilinogen in the feces, appears to be affected by the diet. Malhotra[37] has shown that the urobilinogen content in feces from individuals on a rice diet is considerably greater than in the feces of North Indians who have a wheat and vegetable diet. Whether this is due to an increased secretion of bilirubin glucuronide in the bile or changes in colonic metabolism is not clear. Similarly, there are effects of dietary fiber on bile acid metabolism. Norman[40] has shown that ^{14}C-labeled cholic acid given orally appears in the feces in a form that is only in part dialysable. This suggested to him that there was chemical transformation by intestinal microorganisms and that the physical state of the bile acids in the intestinal contents was changed by the formation of derivatives poorly soluble and easily adsorbed to the residue. Eastwood and Hamilton have shown that vegetable fiber can adsorb bile acids.[10] Bile acids appear to be adsorbed onto dietary fiber with an adsorption that is greatest for those bile acids that are found in the colon, i.e., free bile acids and in particular deoxycholic acid, the principal microbial derivative from cholic acid. The adsorption is pH-dependent and is probably by hydrophobic bonding. Bile acid conjugates found in the bile and duodenum are least strongly adsorbed. However, in studies on the physical state of bile acids and the intestinal contents of germ-free and conventional animals, Gustafsson and Norman[20] were led to believe that deoxycholic acid was adsorbed to the fecal residue but that lithocholic acid, which is the principal derivative of chenodeoxycholic acid, is adsorbed to bacteria.

The effect of adsorption on bacterial metabolism of bile acids is illustrated by the somewhat rare situation that exists after resection of the ileum, for example, in diseased states such as Crohn's disease. Here the concentrations of bile acids that pass into the colon[27] are sufficiently high to inhibit bacterial modification of bile acids. No secondary bile acids appear in the feces, because the effect of the inhibition is to prevent 7α-dehydroxylation. The concentration at which inhibition of 7α-dehydroxylation occurs is approximately 10 mM; the concentration of bile acid in normal stool is 2–6 mM.

In the treatment of ileal resection, an exchange resin such as cholestyramine or lignin may be used to adsorb bile acids and thus prevent inhibition of sodium and water reabsorption, alleviating the diarrhea. The result of successful treatment is a reduction in stool volume and, in parallel, an increase in fecal bile acid concentration, as the fecal bile acid output remains constant despite the symptomatic improvement due to the treatment. In some patients, coincidental with relief of symptoms, secondary bile acids appear in the stool; i.e., there is a reappearance of 7α-dehydroxylation at concentrations of bile acids normally inhibitory, suggesting that adsorption and the presence of surfaces has an effect of bile acid metabolism.[38]

C. Interaction with a Solute of Exogenous Origin

Ershoff[13] has shown that the growing potential of immature rats fed a highly purified low-fiber diet containing massive doses of nonionic surfactant agents, (Myrj 45, Myrj 52, Tween 20) was variably affected. Myrj 45 at a 15–20% level in the diet resulted in slight retardation of growth. However, the other nonionic detergents at 15–20% in the diet caused a significant growth retardation, diarrhea, and death. The deleterious effect of the compounds was completely counteracted by the concurrent feeding of alfalfa meal at 20% level in the diet. The protective effect of alfalfa was retained in the alfalfa residue fraction even after extensive water washing of the pulp. Alfalfa, however, had no such protective effect on animals fed a diet containing another poisonous detergent, Span 20.

Purified cellulose, when incorporated into the diet, had a significant effect in counteracting the toxic effect of Myrj 52, Tween 20, and Tween 60. Rye grass, fescue grass, oat grass, orchard grass, and wheat grass, as well as carrageenan and sodium alginate, each had a significant action in counteracting the toxic effect of Tween 60.

Similar effects have been shown in the study of cyclamate metabolism. It is known that cyclamate is a highly polar compound not readily absorbed when taken by mouth. It passes unchanged to the cecum where it is metabolized by bacteria to cyclohexylamine and other metabolites. The organism that splits cyclamate in man appears to be an enterococcus, but the extent of the effect varies with individuals.[42] A clue to the mechanism of this is gained from experiments in which immature rats fed different concentrations of sodium cyclamate in conjunction with either a highly purified diet or a natural diet showed differing effects.[14] On the stock diet, sodium cyclamate at levels of 2.5–10% of the diet had little adverse effect on weight increase. When, however, comparable levels were fed in a purified diet, sodium cyclamate caused a significant retardation in growth which was proportional to the amount ingested. The animals developed alopecia and diarrhea and, at the 10% level, they died. The deleterious effects of cyclamate were manifest after only 3 days of feeding, which suggests that the introduction of fiber into the diet might affect the metabolism of these compounds. It is not known whether the action is due to the water-holding properties of the fiber, thus diluting the cyclamate, or to pure adsorption of the parent compound.

Work on the role of colonic bacteria and colonic cancer has been extended to the investigation of compounds that might prevent the development of cancer by known carcinogens. Rats fed a low-residue diet and azoxymethane develop tumors in the intestinal tract. The presence of cellulose in the diet is protective against the development of carcinoma in the small intestine. This suggests that azoxymethane-induced tumors in the small intestine are sensitive to bulk and fiber content in the diet.[49] On the other hand, if rats are fed carcinogenic materials such azoxymethane, the incidence of carcinoma of the colon can be materially increased by adding the anion-exchange resin cholestyramine to the diet. The increasing incidence of carcinoma of the distal colon suggests that the carcinogenic material is carried to the distal colon with resulting carcinogenesis.[39] However, a similar carcinogen used with magnesium sulfate, with the production of dilute diarrhea stool, still causes carcinoma of the colon, suggesting no protection by dilution of the stools by this method.[6]

During recent years, however, studies have shown that several antioxidants will inhibit the carcinogenic effects of chemical carcinogens. The most extensive work on this subject is with phenolic antioxidants, in particular butylated hydroxyanisone and butylated hydroxytoluene. Inhibition of this sort occurs when the route of administration results in direct contact of the carcinogen with the target organ, e.g., carcinoma of the forestomach in mice, although it is possible that carcinogenesis at a site remote from administration may be inhibited. Butylated hydroxyanisone and butylated hydroxytoluene are of interest because of their use as food additives. In the United States the daily human consumption of these phenolic antioxidants is on the order of several milligrams per day.

It is possible to protect against the carcinogenic effect of chemical carcinogens by inducing increased mixed-function oxidase activity. The level of mixed-function oxidase activity in the small intestine is related to the amount of fiber in the diet. Thus, one way of increasing the mixed-function oxidase activity is by alterations in the diet. In experiments with mice, hamsters, and rabbits fed Purina Rat Chow® or purified diet, it was shown that loss of mixed-function oxidase activity occurred in animals which had been starved or fed a purified diet, whereas those fed purina Rat Chow® had substantial levels of activity. It was suggested that the inducer or inducers are present in the crude fiber. The precise inducing agent has not been identified, however, and there is no reason to believe that this is necessarily a component of fiber. The active principle may be quite a small molecule.[50] However, if it is a component of fiber, there is an intriguing possibility that it is lignin, an omnipresent component of vegetable fiber, which acts as an antioxidant in the digestive tract. It is possible that lignin in the digestive tract can act as a free-radical scavenger, reducing agent, oxygen scavenger, or peroxide decomposer, thus affecting the formation or function of the actual carcinogenic material.[15]

XI. Summary

Vegetable dietary fiber is a physical complex that acts principally in the colon. Fiber has water-holding, cation-exchange, and adsorptive properties, which vary with the variety of fruits and vegetables studied. There are important effects of fiber on solute and bacterial activity, which influence the fate of bacterial metabolites of solute of endogenous and exogenous origins. Such effects may decide whether the metabolites are excreted in the feces or are conserved by being returned in the portal vein. Fiber itself may be metabolized by bacteria. The matrix provided by fiber has a profound effect on cecal metabolism, an effect which has many facets not yet examined in detail.

ACKNOWLEDGMENTS

We are grateful for discussions and help from Dr. Ingemar Falkehag, Dr. B. DelliColli, Professor R. B. Fisher, Dr. A. A. McConnell, Mr. A. N. Smith, and Dr. R. MacRae, without whose ideas and thoughts much of this work would not have been possible.

References

1. ALLISON, J. B. Biological evaluation of proteins *Physiol. Rev.* **35**:664, 1955.
2. ASPINALL, G. O., and C. T. GREENWOOD. Aspects of Chemistry of Cereal Polysaccharides. *J. Inst. Brew.* **68**:167, 1962.

3. Berg, B., B. v. Hofsten and G. Pettersson. Growth formation by cellvibrio fulvus. *J. Appl. Bacteriol.* **35**:201, 1972.
4. Birkner, H. J., and F. Kern, Jr., Adsorption of bile salts to food residues and drugs. *Clin. Res.* **20**:729, 1972.
5. Branch, W. J., D. A. T. Southgate and W. P. T. James. Binding of calcium by dietary fibre and its relationship to unsubstituted uronic acids. *Proc. Nutr. Soc.* **34**:120A, 1975.
6. Cleveland, J. W., and J. W. Cole. Relationship of experimentally induced intestinal tumour to laxative ingestion. *Cancer,* **23**:1200, 1969.
7. Cummings, J. H. Dietary Fibre. *Gut* **14**:69, 1973.
8. Dellicolli, H. T., M. A. Eastwood, A. A. McConnell and W. D. Mitchell. Unpublished results.
9. Drasar, B. S., and M. J. Hill. *Human Intestinal Flora.* London, Academic Press, 1974.
10. Eastwood, M. A., and D. Hamilton. Studies on the adsorption of bile salts to nonabsorbed components of diet. *Biochem. Biophys. Acta* **152**:165, 1968.
11. Eastwood, M. A. Vegetable fibre: its physical properties. *Proc. Nutr. Soc.* **32**:137, 1973.
12. Eastwood, M. A. The role of vegetable dietary fibre in human nutrition—a hypothesis. *Med. Hypothesis* **1**:46, 1975.
13. Ershoff, B. H. Beneficial effects of alfalfa meal and other bulk-containing or bulk-forming materials on the toxicity of nonionic surfactant agents in the rat. *J. Nutr.* **70**:484, 1960.
14. Ershoff, B. H. Comparative effects of a purified diet and stock ration on sodium cyclamate toxicity in rats. *Proc. Soc. Exp. Biol. Med.* **141**:857, 1972.
15. Falkehag, I. Personal communication.
16. Findlay, J. M., W. D. Mitchell and M. A. Eastwood. The physical state of bile acids in the diarrhoeal stool of ileal resection, *Gut* **14**:319, 1973.
17. Findlay, J. M., W. D. Mitchell, M. A. Eastwood, A. J. B. Anderson, and A. N. Smith. Intestinal streaming patterns in cholerrheic enteropathy and diverticular disease. *Lancet* **1**:146, 1974.
18. Fisher, R. B. Personal communication, 1975.
19. Gingell, R., J. W. Bridges and R. T. Williams. Gut flora and the metabolism of prontosils in the rat. *Biochem. J.* **114**:5P, 1969.
20. Gustafsson, B. E., and A. Norman. Physical state of bile acids in intestinal contents of germ-free and conventional rats. *Scand. J. Gastroenterol.* **3**:625, 1968.
21. Hartiala, K. Metabolism of hormones, drugs and other substances by the gut. *Physiol. Rev.* **53**:496, 1973.
22. Hartley, R. D. Carbohydrates esters of ferulic acid as a component of cell walls of *Lolium multiflorum. Phytochemistry* **12**:661, 1973.
23. Hattori, R., and T. Hattori. Effect of a liquid–solid interface on the life of micro-organisms. *Ecol. Rev.* **16**:63, 1963.
24. Hattori, R. Growth of *Escherichia coli* on the surface of an anion exchange resin in continuous flow system. *J. Gen. Appl. Microbiol.* **18**:319, 1972.
25. Haug, A., and O. Smidsrød. The effect of divalent metals on the properties of alginate solutions. *Acta Chem. Scand.* **19**:341, 1965.
26. Helfferich, F. *In: Ion Exchange.* New York, McGraw-Hill, 1962.
27. Hofmann, A. F. The syndrome of ileal disease and the broken enterohepatic circulation: Cholerrheic enteropathy. *Gastroenterology* **52**:752, 1967.
28. Jirgensons, B., and M. E. Straumanis. *A Short Textbook of Colloid Chemistry*, 2d rev. ed. Oxford, Pergammon, 1962.
29. Knight, A. H., W. M. Crooke and H. Shephard. Chemical composition of pollen with particular reference to cation-exchange capacity and uronic acid content. *J. Sci. Food Agric.* **23**:263, 1972.
30. Kodicek, E. The story of Vitamin D. *Lancet* **1**:325, 1974.

31. KRITCHEVSKY, D., and J. A. STORY. Binding of bile salts in vitro by nonnutritive fiber. *J. Nutr.* **104**:458, 1974.

32. KUDO, M., and T. KONDO. Studies on the hydrophilic properties of lignin. *TAPPI* **54**:2046, 1971.

33. LIN, S. Y. Accessibility of cellulose, a critical review. *Fiber Sci. Technol.* **5**:303, 1972.

34. LUTHER, L., R. SANTINI, C. BREWSTER, E. PEREZ-SANTIAGO and C. E. BUTTERWORTH. Folate binding by insoluble components of American and Puerto Rican diets. *Ala. J. Med. Sci.* **2**:389, 1965.

35. MCCONNELL, A. A., and M. A. EASTWOOD. A comparison of measuring 'fibre' in vegetable material. *J. Sci. Food Agric.* **25**:1451, 1974.

36. MCCONNELL, A. A., M. A. EASTWOOD and W. D. MITCHELL. Physical characteristics of vegetable foodstuffs that could influence bowel function. *J. Sci. Food Agric.* **25**:1457, 1974.

37. MALHOTRA, S. L. Fecal urobilinogen content in populations with different peptic ulcer incidence in India and its possible role in the etiology of ulceration. *Scand. J. Gastroenterol.* **2**:337, 1967.

38. MITCHELL, W. D., J. M. FINDLAY, R. MACRAE, M. A. EASTWOOD and R. ANDERSON. Factors affecting bile acid metabolism in cholerrheic enteropathy. *Digestion* **11**:135, 1974.

39. NIGRO, N. D., N. BHADRACHARI, C. CHOMCHAI. A rat model for studying colonic cancer. Effect of cholestyramine or induced tumors. *Dis. Colon Rectum* **16**:438, 1973.

40. NORMAN, A. Faecal excretion products of cholic acid in man. *Br. J. Nutr.* **18**:173, 1964.

41. REES, D. A. *The shapes of molecules*. Contemporary Science Paperbacks, No. 14. Edinburgh, Oliver and Boyd 1967, p. 113.

42. RENWICK, A. G., and R. T. WILLIAMS. Metabolites of cyclohexylamine in man and certain animals. *Biochem. J.* **129**:857, 1972.

43. SARKANEN, K. V., and C. H. LUDWIG, (eds.). *Lignins*. New York, Wiley-Interscience, 1971.

44. STONE, J. E., and A. M. SCALLAN. A structural model for the cell wall of water swollen wood pulp fibers based on their accessibility to macromolecules. *Cellul. Chem. Technol.* **2**:343, 1968.

45. STONE, J. E., E. TREIBER and B. ABRAHAMSON. Accessibility of regenerated cellulose to solute molecules of a molecular weight of 180 to 2×10^6. *TAPPI* **52**:108, 1969.

46. VAN SOEST, P. J., and R. W. MCQUEEN. The chemistry and estimation of fibre. *Proc. Nutr. Soc.* **32**:123, 1973.

47. WAITE, R. The chemical composition of grasses in relation to agronomical practice. *Proc. Nutr. Soc.* **24**:38, 1965.

48. WALKER, R. The metabolism of azo compounds. A review of the literature. *Food Cosmet. Toxicol.* **8**:677, 1970.

49. WARD, J. M., R. S. YAMAMOTO and J. H. WEISBURGER. Cellulose, dietary bulk and azoxymethane-induced intestinal cancer. *J. Natl. Cancer Inst.* **51**:713, 1973.

50. WATTENBERG, L. W. Potential inhibitors of colon carcinogenesis. *Am. J. Dig. Dis.* **19**:947, 1974.

51. WEISBURGER, J. H. Colon carcinogens. Their metabolism and mode of action. *Cancer* **28**:60, 1971.

52. WERCH, S. C., and A. C. IVY. A study of the metabolism of ingested pectin. *Am. J. Dis. Child.* **62**:499, 1941.

53. WILLIAMS, R. T. Toxicologic implications of biotransformation by intestinal microflora. *Toxicol. Appl. Pharmacol.* **23**:769, 1972.

54. WORTH, H. G. J. The chemistry and biochemistry of pectic substances. *Chem. Rev.* **67**:465, 1967.

5

Microbial Activities Related to Mammalian Digestion and Absorption of Food

Robert E. Hungate

I. Introduction

The alimentary tract of animals is a device for sequestering food from possible seizure by other animals and keeping it under conditions favorable for digestion while the animal moves from place to place. It does not, however, protect the ingesta against attack by microorganisms. Foods that can nourish animals can also nourish microbes. The microbes' capacity for rapid multiplication makes them effective competitors for the animal's food, even though individually they are so minute.

The moisture, temperature, and substrates in the gut of warm-blooded mammals provide exceptionally favorable conditions for microbial growth. Bacteria taken in with the food, or already in the alimentary tract, can increase 500 times in 3 hr, 250,000 times in 6 hr. In most mammals the residence time for digesta exceeds this, and microbes growing in the gut could consume significant fractions of the nutrients if allowed to increase unchecked.

The contents of the alimentary tract are essentially devoid of oxygen, although a small amount may diffuse in from the epithelium. A platinum electrode placed in contact with the lumen surface of the epithelium shows a redox potential slightly higher than that shown when it is surrounded by gut contents. Apparently the O_2 obtained from the blood by the epithelial cells is incompletely used, leaving a low steady-state concentration which maintains a diffusion gra-

ROBERT E. HUNGATE • Professor Emeritus of Bacteriology, Department of Bacteriology, University of California, Davis, California.

dient into the gut. Any O_2 available through this route would increase the growth yield of microbial cells. The amount supplied in this way has not been estimated, but it is probably small, and in most parts of the intestine it is less than the O_2-utilizing capacity of the microbial population. The microbes obtain the major part of their energy anaerobically by fermentation.

II. Fermentation

When O_2 is available, the utility for microorganisms of the various foods in the gut is quite different from their utility under anaerobic conditions. Aerobically, metabolism can release a maximum of 9 kcal (heat of combustion) by the oxidation of a gram of fat, 4.5 from a gram of protein, and 4.1 from carbohydrate. Anaerobically, much less heat can be liberated, and the order of effectiveness of these foods is reversed. The oxygen combined in the food as–OH must be used as the oxidant, and since carbohydrates contain the greater proportion of –OH, they can react anaerobically with greater decrements of free energy. Proteins are intermediate, and fats the least capable of reacting. Reactions in which the elements C, H, and O are rearranged into CO_2 and CH_4 show the greatest decrement of free energy, i.e., the greatest capacity to do chemical work, as in the synthesis of new microbial cells.

What amount of oxygen in a carbon compound enables it to react with the greatest decrement of free energy under anaerobic conditions? The most oxidized or most reduced forms of carbon, CO_2 and CH_4, cannot react with themselves or with each other at moderate temperatures and pressures. If HCHO, HCOOH, and CH_3OH are taken as examples of the oxidation states of carbon which can react anaerobically, the superiority of HCHO emerges. This is evident from the following equations showing the maximum decrement of free energy for each, i.e., with CO_2 and CH_4 as the products:

$$4HCOOH \rightarrow 3CO_2 + CH_4 \qquad \Delta G = -35 \text{ Kcal } (0.3 \text{ kcal/g}) \qquad (1)$$
$$4CH_3OH \rightarrow 3CH_4 + CO_2 \qquad \Delta G = -38 \text{ Kcal } (0.2 \text{ kcal/g}) \qquad (2)$$
$$2HCHO \rightarrow CO_2 + CH_4 \qquad \Delta G = -61 \text{ Kcal } (1.0 \text{ kcal/g}) \qquad (3)$$

In equations (1) and (2), four molecules of substrate participate in an oxidation reduction reaction where six electrons are transferred, i.e., 1½ e/molecule. With carbon at the intermediate (CH_2O) state of oxidation (3) two molecules of substrate are involved in the transfer of 4 e, i.e., two per molecule. Even if the decrements of free energy in each electron transfer were the same in all the reactions, the advantage of 2e over 1½ e/molecule is sufficient for an evolutionary selection of carbohydrate as a preferred substrate for anaerobic growth. The actual advantage of carbohydrate over other states of oxidation of carbon is much greater, as is evident from the decrements of free energy in the three reactions.

Before O_2 accumulated in the atmosphere, organisms used anaerobic oxidation-reduction reactions to accomplish the chemical work involved in the synthesis of cell material. On the basis that carbohydrates provided more reactivity, they must have been a preferred food.

Except in the case of acetate, carbohydrates and their derivatives are not fermented directly to both CO_2 and CH_4 · CO_2 and $-COOH$ are common as oxidized forms of carbon in fermentation products, and $-CH_3$ and $-CH_2-$, together with H_2, are the common reduced products of carbohydrate fermentation. In mixed anaerobic populations, the H_2 is often used by methanogenic bacteria to reduce CO_2 to CH_4, and a fairly complete conversion of carbohydrate to CO_2 and CH_4 can occur, as in the anaerobic stage of sewage disposal plants. Fermentation of carbohydrates always leads to the production of acids or gas or both.

The glycerol moiety of fats can be fermented, as can the long-chain fatty acids, but the latter support a fermentation only in mixed cultures in which CH_4 is produced. Proteins are fermented by many bacteria, but on a weight basis cannot support as much anaerobic growth as carbohydrates. Many of their carbon atoms are already partially reduced or oxidized. Fermentations of protein are common, with many of the products similar to those derived from carbohydrates, but in addition, large quantities of ammonia are formed, making the net reaction alkaline rather than acid.

As the carbohydrate foods initially present on earth were fermented, carbohydrate must have become very limiting until photosynthesis evolved. When a method for photosynthetic conversion of CO_2 and H_2O into carbohydrate and O_2 had developed, carbohydrate became an exceedingly abundant and cheap food. All that was necessary for its production was the cell machinery, carbon dioxide, water, and light.

III. Food Effects on Host-Microbe Relationships

The bonds of carbon also permit it to form chains of carbon atoms, as in sugars, each in an intermediate state of oxidation and thereby possessing a potential for extensive rearrangement. Sugars can rearrange into many of the complex molecules that we know as intermediate products of metabolism. They can also polymerize into very large three-dimensional, semirigid aggregates with low reactivity.

When aquatic plants moved onto land, these large carbohydrate polymers, such as starch and glycogen, were probably already common as storage products. Cellulose, hemicelluloses, and related components of the cell wall presumably composed part of the cell envelope of aquatic ancestral forms, and with the move to land the abundance of these fibrous materials increased. Their greater strength, supporting the photosynthetic organs for optimal exposure to light and air, was advantageous.

As these structural carbohydrates increased, the composition of plants diverged from that of animals. Plant reproductive structures still preserved a composition similar to those of animals, but the vegetative plant body came to contain a much greater proportion of polymerized wall material, with the protein, fat, and storage carbohydrate constituting only a small fraction. The nitrogen content of coniferous woods is only about 0.05%, as compared to 16% in protein. Lignin was presumably a later development and will be omitted from this discussion except to point out that it is widely regarded as resistant to degradation under anaerobic conditions. The natural lignins, celluloses, and hemicelluloses each contain a predominance of certain chemical linkages, but none of them is a single pure substance, and their degree of digestibility can vary according to the nature of the accompanying substances and to the degree of desiccation.

The relationship of animals to their gut microbes depends to a considerable extent on the kind of food consumed. If the food is primarily concentrates, i.e., animal, or fruits of plants (in the broad sense of seeds), rich in protein and easily digested carbohydrates, the relationship is competitive, both animal and microbe seeking the same foods. If the food contains a large proportion of the vegetative parts of plants, i.e., fiber, the relationship is cooperative. The vertebrate has not evolved the enzymes necessary for digesting fiber and depends on microbes for this activity.

Competition and cooperation models will be discussed as a basis for comparing the various relationships of gut microbes to their host.

IV. The Competition Model

The term competition model stems from the fact that the host and microbe can both use the same food. The microbe's capacity for rapid growth is its chief asset in the competition. The animal competes by maintaining a stomach acidity that kills most entering microbes (Figure 1). A high gastric acidity is not characteristic of all animals. It is lacking in gastropod mollusks such as the snail. Snails differ also in having an hepatopancreatic secretion containing an active cellulase synthesized in the digestive glands. But in vertebrates, high acidity in the stomach is the rule. It eliminates or greatly reduces the viability of microbes ingested with the food, or growing in it after ingestion but prior to acidification.

Most vegetative bacteria are killed by hydrochloric acid at pHs characteristic of the stomach. Normally, the human small-intestinal contents are essentially sterile from the duodenum to the ligament of Treitz, and in the next few feet the numbers of bacteria are relatively low.

COMPETITION

COOPERATION

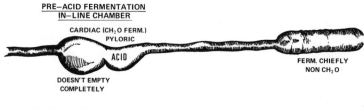

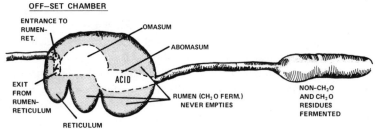

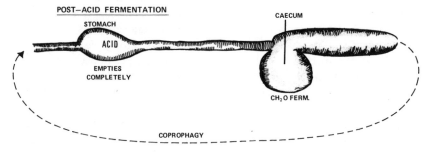

Figure 1. Diagrams illustrating models for mammal–microbe interrelationships.

Coupled with the acidity as protection against microbial competition is the flow of digesta. Peristaltic contractions propel the acid-sterilized food posteriorly, creating a gradient in which flow counteracts the mixing effects of peristalsis and the tendency of the microbes to grow and spread throughout the system. If flow stops, the flora of the entire small intestine soon resembles that of the colon. The ecosystem is in a dynamic state, varying with the influx of digesta and host secretions.

Most carnivorous mammals are good examples of the competition model. The food consumed is readily digested by host enzymes and can be almost completely absorbed. Little plant fiber is ingested, and the alimentary tract has no special chambers for a microbial carbohydrate fermentation. It might seem that the host's competitive success is complete. Yet the feces of these animals contain a large concentration of microbial cells, as is the case also with man and other representatives of the competition model. In all mammals a significant growth of bacteria occurs posterior to the site of acidity, but fermentation of digesta by microbes is small in the region of active digestion and absorption by the host.

What are the substrates for the growth of the microbial cells appearing in the feces? To the extent that substrates are not attacked by host enzymes, they can serve for bacterial growth in the colon. Species of bacteria capable of degrading most organic materials occur on the earth's surface, and pertinent types appear in the gut if their particular substrate is a component of the food. The delicate cell walls of plants may be digested, but the rate is relatively slow as compared to the utilization of soluble substrates. This may be due to the fact that since fiber exists as a separate phase, digestion can occur only at its surface. The residence time for digesta in the large intestine of most carnivores is sufficient for enzymic solubilization of soft plant fibers and cell walls. A cellulolytic bacterium was isolated in pure culture from the human by Khouvine[8] at a time when techniques for growing cellulolytic anaerobes had not been extensively developed. With modern methods it should be possible to demonstrate additional fiber-digesting types in animals consuming a significant quantity of plant cell walls.

The exact site at which bacterial growth occurs during passage of food beyond the acid stomach has not been identified. Some types of bacteria may be able to maintain themselves in the crevices and pits in the lining of the small intestine, and in the mouse certain bacteria have been shown to attach to the epithelium and grow there as a concentrated layer.

In that part of the upper intestine where abundant products of digestion have formed and are not yet absorbed, conditions are ideal for rapid growth of any microbial species adapted to that environment. Some species of streptococci have been demonstrated to grow in this region of the gut, as has *Veillonella*. *Escherichia coli* does not occur here in most humans, although it is a common component of the flora in the colon.

Although the rate of growth of certain bacterial species in the small intestine can be very great, their initial numbers are relatively small. It does not appear that growth on digestion products in the upper small intestine is sufficient to account for the high concentration of microbial cells in the colon and feces of carnivores. Neither are plant residues in the colon sufficient to account for the high numbers of fecal bacteria. Many of these bacteria must grow on substrates other than those released by digestion of food by host enzymes.

The cells sloughed off from the epithelium of the small intestine may account for the substrates supporting the development of many colonic bacteria. The sloughed epithelium presumably contains a fairly high percentage of protein and relatively little carbohydrate. If so, the fermentation in the intestine posterior to the duodenum and jejunum would be expected to differ markedly from that where carbohydrate is the chief substrate.

Presence of soluble sugars in the upper part of the small intestine leads to the postulate that this portion of the gut contains a flora different from that in more posterior regions where soluble carbohydrate is more limiting. Since *Escherichia coli* is characterized by a capacity for a very rapid fermentation of soluble carbohydrates, it is of interest that it does not occur in the sugar-containing region of the small intestine. Its versatility in converting amino acids, as demonstrated by Gale at Oxford, may explain its survival as a minor component of the colonic microflora.

Among nonmammalian vertebrates the competition model is characteristic of most fish, amphibians, reptiles, and birds. Since the food available to these animals contains much concentrate and little plant fiber, a prefermentation would lower, rather than increase, the value of the food to the vertebrate. A few fish such as the carp are reported to consume plant materials, but as yet little information on possible microbial digestion has been reported. Most reptiles are carnivorous, although some of the turtles on the Galapagos Islands may consume plant material.[3]

Birds exhibit essentially the competition model, with exceptions that will be mentioned later. In species of birds that do not consume any plant material, microorganisms harbored in the gastric ceca may be concerned with the synthesis of vitamins.

Among mammals[4] the orders Insectivora (moles and shrews), Tubulidentata (aardvarks), Chiroptera (bats), and Pholidota (anteaters) consume food needing little prefermentation. Similarly, the orders Carnivora and Pinnepedia (seals) depend almost exclusively on animal food.

V. The Cooperation Model, with Preacid Fermentation

The cooperation model for the relationship between alimentary microbes and their host applies to mammals consuming plant materials and is most characteristic when the food contains a large proportion of the vegetative portions of terrestrial plants, containing cell walls composed of pectin, cellulose, hemicellulose, etc. Digestion of these materials requires more time than does the digestion of soluble food components and, accommodating to this requirement, one or more portions of the alimentary tract of herbivores has enlarged into a fermentation chamber allowing a residence time sufficient for digestion of much of the resistant plant fiber. Correlated with this, the stomach does not empty completely but always contains a microbial population capable of rapid attack on fresh ingesta.

A. The In-Line Fermentation Chamber

In the in-line fermentation chamber model (Figure 1), the cardiac portion of the stomach is enlarged and is separated from the pyloric region by a slight constriction. The separation is not as sharp as between certain compartments of the ruminant stomach, but it diminishes the mixing of the cardiac contents with the acid pyloric digesta. The fermentation occurs in the cardiac portion.

The stomach fermentations in the leaf-eating *Colobus* monkeys of Africa and the langur monkeys (*Presbytis*) of India are examples of this model. The langur consumes the leafy parts of plants and has a fermentation similar to that in ruminants,[2] with cellulose substrates being fermented to volatile fatty acids (VFA) and methane. Some of the bacteria accomplishing these functions appear similar to those in the rumen, and the fermentation rate is comparable.

In the colobus monkey, no methane was found in the stomach gas of two specimens collected from their natural habitat,[13] but hydrogen was abundant (26% v/v of the total) in the freshly killed animals. Starchy materials rather than fiber appeared to be the chief food, with very little leafy material consumed. No cellulolytic bacteria could be cultured. Some very large tetracocci were seen by microscopic examination of the stomach contents, particularly in the material bordering the pyloric region. Morphologically they resembled *Sarcina ventriculi*, which typically grows under acid conditions. The stomach contents in the cardiac portion were acid, about pH 5.5, when the pH was tested 30 min after the animal was killed. No reliable measurements of the fermentation rate of the stomach contents could be obtained because of their acidity, but neutralized contents showed a very rapid rate.

The stomach of the quokka (*Setonix brachyurus*), a small rabbit-like marsupial on Rattnest Island near Perth, Australia, is very similar in its anatomy to that observed in the monkeys, with separation into two main compartments. The stomach gas contained both CH_4 and H_2, as well as CO_2, and the fermentation rate and fermentation products were comparable to those of the rumen. The quokkas feed at night, and the animals studied had been consuming green grass.[12] A surprising observation was that the stomach contents could be drawn fairly readily into a fine-tipped pipette. Mastication had apparently been so complete that the slurry in the stomach contained only extremely fine particles. This is in marked contrast to the stomach contents of most herbivores. In the elephant, pieces of twigs 10 cm long and 5 mm in diameter are fairly common, and even in the ruminant, with the additional mastication of rumination, many particles are as large as 10×2 mm.

Other marsupials such as the leaf-eating koala and the grazing wallabies and kangaroos presumably maintain a stomach fermentation comparable to that in the quokka.

The sloth, order Edentata, has leaf-eating food habits that suggest some sort of microbial interrelationship, probably of the in-line preacid type, but it has not been examined with this in view. The food is leaves, buds, and twigs of trees, and the digesta are retained over a lengthy period of 5–6 days,[4] ample time for extensive fiber digestion. The order Sirenia (manatees and dugong) includes large mammals that consume the vegetative parts of aquatic plants. They, too, would be likely sites for an active microbial fermentation.

B. The Offset Fermentation Chamber

The rumen (Figure 1) is the prime example of the offset fermentation chamber. The food is subjected to microbial fermentation before it reaches the acid stomach, and the anatomy is such that there is negligible mixing between the fermentation and acid portions. The ruminant stomach has four compartments; the rumen and reticulum together constitute the fermentation chamber, followed by the omasum and abomasum. The omasum is a unique structure, with leaflike adjacent layers of tissue arranged so that the digesta passing through are close to an absorptive surface. A special esophageal groove in the wall of the reticulum can be closed via a suckling reflex in young ruminants, shunting milk past the fermentation chamber into the abomasum, the acid stomach.

The rumen–reticulum[5] contents contain about 2×10^{10} bacteria/ml and as many as 10^6 protozoa/ml. They carry on an active fermentation, using daily about 150 g of substrate in each liter of contents. Acetic, propionic, and butyric acid, in descending order of abundance, are formed, along with copious quantities of CO_2 and CH_4. An alkaline saliva is produced during feeding and also when the digesta are returned to the mouth for rechewing in the process of rumination. The alkalinity helps to neutralize the fermentation acids. Absorption of the free acids from the rumen also aids in maintaining acidities favorable for microbial growth. Since sodium bicarbonate is the chief base in the saliva, CO_2 is produced from it during neutralization of the acids. Thus all the chief fermentation products are gases or cause gas production. The gas produced is normally voided by eructation through the mouth. Gas production is so rapid that, if abnormal conditions prevent its eructation, a rapidly fatal bloat can result.

The ruminant depends to a large extent on fermentation acids rather than on glucose for many of its metabolic needs. It derives most of its protein from the bodies of the rumen microbes. They pass between the leaves of the omasum into the abomasum where they are killed by the acid. The best explanation for the evolution of the omasum, with its thin sheets of tissue extending into its lumen, is that it is needed to absorb out the VFA before the digesta enter the abomasum, where hydrochloric acid is secreted. If not absorbed, the sodium salts of the acetic, propionic, and butyric acids in the digesta coming from the rumen–reticulum would buffer the abomasal HCl and hamper attainment of the acidities required to kill the rumen microbes. Their killing may promote their digestion,[15] but little evidence on this possibility is available.

In a sense the ruminant's nutrition is comparable to that of a plankton feeder, which strains microorganisms out of large volumes of water to obtain its protein. The ruminant cultures its plankton. The ingesta provide the substrate for the planktonic growth, and because conditions are anaerobic, the waste products from the planktonic metabolism are incompletely oxidized. They are absorbed into host tissues, where O_2 is available for their oxidation to provide ATP and other needed metabolites. The microbes and the ruminant are linked in the biochemical chain of reactions leading to complete oxidation of the carbohydrate substrate. The microbes accomplish their work of growth by anaerobic rearrangement of the carbohydrate into acids and gases; the ruminant oxidizes the acid waste products and meets its nitrogen requirements by digestion of the microbial cells.

The large numbers and kinds of ruminants indicate that this offset model for utilization of the fibrous components of plants is very successful. It has the disadvantage (from the human standpoint of producing ruminants in the shortest possible time) that proteins cannot be fed to increase the rate of ruminant growth. Soluble or readily digestible proteins are rapidly fermented in the rumen and, although microbial cells are produced, the amount of protein in those cells is less than that in the fermented protein. Also, protein is generally more expensive than carbohydrate and, as indicated earlier, its fermentation may not support as much microbial growth as an equal quantity of carbohydrate.

Proteins can be protected against microbial proteinases in the rumen by methods that leave them digestible by the more powerful host proteinases secreted in the abomasum and small intestine. Currently, methods of protecting proteins from digestion in the rumen are of economic interest, but as proteins become increasingly expensive, feeding practices will probably return more to rations making use of microbial fiber digestion.

Offset pregastric fermentation is found in all the ruminants (suborder Ruminantia of the order Artiodactyla). A somewhat similar stomach, lacking the omasum, occurs in the suborder Tylopoda (camels, llamas, vicunas, and alpacas), but the fermentation itself is the same, and tylopods ruminate. The suborder Suiformes (pig, peccary, and hippopotamus) does not ruminate, but the hippopotamus stomach[16] has been shown to contain a fermentative microbiota, including protozoa. The stomach is partially compartmentalized; some of the compartments harbor the microorganisms.

The in-line and offset fermentation chamber models differ in the kinetics of passage of digesta. Passage in the in-line model tends toward plug flow, with the residence time corresponding to the distance the digesta have proceeded along the gut. In the offset model the fresh ingesta are mixed with the digesta already consumed, and passage corresponds to the model for a continuous fermentation, in which the volume of the fermentation chamber remains constant, with the influent and effluent rates equal. Passage of material from this system is proportional to the fraction which the material constitutes of the total, i.e., $dx/dt = \mu x$, with x the fraction of material in the fermentation chamber at any time t, and μ the dilution rate constant in units of reciprocal time.

In the continuous-fermentation model, the fresh ingesta are quickly subjected to a larger concentration of microbes; the average retention time is longer, and the substrate concentration, microbial population, and rate of digestion tend to be constant. In the plug-flow model, the population increases with time, the concentration of substrate diminishes, and the rate of fermentation increases initially and then decreases.

Neither the in-line nor the offset model corresponds exactly to the situation in the actual animal. Whereas the plug-flow model applies to a linear tube of uniform diameter, the parts of the in-line stomach vary in diameter, and peristaltic contractions tend to mix digesta of different residence times. In the offset fermentation chamber, there are diurnal changes in the rates of entrance and exit of digesta, as in the rumen, and the volume is not constant. But these models can be used with advantage in understanding the complex microbe–host interrelationships, particularly when a quantitative analysis is desired.

VI. The Cooperation Model, with Postacid Fermentation

The cooperation model with postacid fermentation (Figure 1) corresponds essentially to the postgastric fermentation in the competition model, but differs in that a large quantity of fibrous plant material is consumed. The posterior parts of the alimentary tract become much enlarged to lengthen the residence time of the increased bulk. This model has the same advantage as the competition model; the proteins, carbohydrates, and fats, digestible by the host enzymes, can be utilized before there is extensive microbial growth, while in addition, there is provision for a later fermentation. The fermentation chamber is usually the combined cecum and large intestine, although in some animals there is more of a separation of these two organs.

Cecal fermentations have not been as completely studied as those of the rumen, but measurements of the fermentation rate in the porcupine[7] indicate that significant quantitites of VFA and methane are produced, at rates comparable to those in the rumen.

The substrate for the postacid fermentation in this model is chiefly the plant fiber indigestible by host enzymes, but, as in the competition model, cells sloughed off from the ends of the villi in the small intestine provide a source of protein and other nutrients. These are important for fiber digestion, since much of the food protein will have been removed by host enzymes. Quantitative figures are not available for the cecal model, but one would expect that microbial assimilation of the sloughed epithelial nitrogen will be greater than in the competition model because of the carbohydrate substrates in the fiber.

The VFA are absorbed from the cecum and large intestine, but no digestion and absorption of microbial cells or sloughed-off host cells have been demonstrated.

Although the cecal fermentation occurs posterior to the host's digestive processes and does not interfere with utilization of foods digestible by the host, it has the disadvantage that the microbial cell bodies produced cannot be utilized before they are voided.

This inefficiency in cecal fermentation has been corrected in many postacid fermentation systems by the host habit of coprophagy. By this means the animal can utilize the proteins in the microbes produced in the cecal and large-intestinal fermentations. For example, hares and rabbits produce two sorts of feces. The night feces are soft and moist and are refected (eaten). They contain relatively few undigested residues of plant fiber. The other feces, containing the residues, are dry and hard and are not refected.

Among mammals the cecal type of fermentation is exemplified in the orders Lagomorpha (hares, rabbits, and pikas), Rodentia (a great many representatives ranging from mice to beaver, porcupines, and capybara), Proboscidea (elephants), and Perissodactyla (horse, tapir, and rhinoceros). The hydraxes (proboscidians) consume primarily plant material and probably use this type of fermentation, but they have not been examined. Among animals with the cecal type of fermentation, the elephants do not practice coprophagy, and the author is not aware of its occurrence to any extent in the Perissodactyla. It is common in the lagomorphs and among the rodents.

Refection serves also to inoculate the stomach with an anaerobic population of microorganisms. These can attack food in the cardiac portion, as is seen in the rabbit. The stomach of rabbits contains a large pocket of gas, presumably derived in part from a stomach fermentation. The stomach of some rodents is compartmentalized, with an active fermentation in certain compartments. The role of refection in such rodents has not been analyzed.

The food of most birds consists chiefly of animals and plant seeds or other easily digested material. Some birds, such as ducks and geese, consume large quantities of grass as well as aquatic plants. They have been examined for microbial fermentation of cellulose, but the results have been completely negative.[10] An important consideration in these birds is the disadvantage of retaining digesta long enough to allow fiber digestion. Retention requires a large fermentation chamber, and the energy cost of supporting such a chamber in flight is considerably greater than that of an animal on the ground. The goose consumes green plant forage in large quantity, but studies of its rate of passage[10] show that the average retention time is only 2 hr. This is insufficient for digestion of plant fiber, which is apparently not utilized in the goose.

The members of the grouse family (Tetraonidae) are not long-distance flyers, in contrast to the migratory geese. In the grouse the gastric ceca are well developed, consisting of two long, blind, and fairly narrow tubes opening into the large intestine. The food consists of plant buds and twigs. These are ground in the gizzard. The softer parts are very finely comminuted, but the woody parts of the twigs are little affected and are voided as small cylinders. The finely ground material is fermented in the gastric ceca at a rate lower than that in the rumen, with formation of VFA.[9] The authors do not mention coprophagy among these birds, but noted that droppings containing cecal material had a distinctive appearance and odor. From the fact that bird droppings show visible separation of the uric acid from the food residue portion of the feces, it seems possible that some birds may obtain microbial protein through coprophagy. It may be noted, however, that coprophagy could be very disadvantageous if the animal has an intestinal parasite capable of resisting the stomach acidity.

VII. Types of Microorganisms

Aside from man, the most-studied microbial ecosystem of the alimentary tract is that in the rumen. Gram-negative bacteria in this habitat are species of *Bacteroides, Butyrivibrio, Veillonella, Selenomonas, Treponema, Succinivibrio, Succinimonas, Anaerovibrio, Megasphaera*, and *Acholeplasma*. Gram-positive genera represented are *Ruminococcus, Eubacterium, Lachnospira, Methanobacterium, Streptococcus, Propionibacterium, Lactobacillus*, and *Clostridium*. The gram-positive microbes often are relatively more numerous when readily digested concentrates are fed, whereas on forages the gram-negatives predominate. An interesting discrepancy has been noted: whereas the microbial culture count for the rumen of concentrate-fed bovines is about the same as the direct microscopic count, in forage-fed animals it is usually not more than 20% of the direct count.

The bacterial floras of nonrumen alimentary microbial ecosystems (except that in human feces) have not been studied as extensively as those of the rumen. Many of the genera found in the rumen occur also in man, although the species differ. The flora in the langur monkey, and probably that in the quokka, resembles the flora of the rumen, but, as mentioned, the colobus flora is quite different. Comparisons of the flora in the posterior portion of various alimentary tracts will probably show more species common to this habitat than will be the case with the preacid fermentations, but much more information is needed before accurate conclusions can be drawn.

Certain protozoa have become modified from their aerobic ancestral types and live successfully in the alimentary tracts of a number of mammals. Vestigial mitochondria have been identified in some of them, and limited utilization of oxygen via a peroxidase has been demonstrated for one genus,[13] but even this genus is killed by long exposure to O_2. Both the protozoa and the bacteria are obligately anaerobic. Most of the protozoa are ciliates. There are a few small flagellates, usually present in low numbers, but these may become abundant when the ciliates are eliminated.

Some of the ciliates (holotrichs) are uniformly ciliated over the cell surface. They readily absorb and ferment soluble sugars, in contrast to the more complex rumen ciliates, the entodiniomorphs (family Ophryoscolecidae). These use chiefly starch and cellulose, although some appear also to use the chlorophyllous portions of plant cells. In the ophryoscolicids the ciliation is restricted to a band of dorsal cilia around the buccal cavity in the genus *Entodinium*, with an additional dorsal band in other genera. Most rumen ciliates ingest bacteria as a source of nitrogen.

Entodiniomorphs of the families Cycloposthiidae and Ditoxidae are abundant in the cecum of the horse, tapir, and rhinoceros, and many species of holotrichs are also common in these habitats. The hippopotamus contains a large variety of protozoa, some of which have characteristics relating them to entodiniomorphs in the rumen and in the horse cecum.[16] The quokka was discovered[11] to contain a number of species of ciliates, all of them entirely different from those in eutherian mammals. The elephant cecum contains large entodiniomorphs, with many ciliary bands, and a related type has been found in the feces of the chimpanzee. Roundworms were also observed in the elephant. No protozoa were seen in the stomachs of the langur and colobus monkeys, possibly because leaf-eating is a recent evolutionary development in this group, or perhaps because the stomach is at times too acid.

Comparisons of the nutritional value of the bacteria and the protozoa from the rumen indicate that in the rat a greater fraction of the protozoal nitrogen is assimilated than is the case with bacterial protein. Much of the bacterial nitrogen is in the nucleic acids and in the cell envelope in forms not directly assimilable into protein by the mammal. Mammals deriving nitrogen from the microbial cells formed in alimentary fermentations secrete significantly larger amounts of pancreatic ribonuclease[1] than do mammals exhibiting the competition model of interaction.

VIII. Abnormalities in Alimentary Fermentations

Availability of substrate almost always limits the rate of fermentation in an alimentary tract, except in the duodenum and jejunum, where the soluble products formed by host digestive enzymes accumulate. But few viable microbes are in this habitat. Efficiency in achieving maximal growth per unit of carbohydrate fermented has been postulated[6] to explain the relative success of the bacteria competing with each other for limiting concentrations of food. As long as food is limiting, alimentary ecosystems remain stable under a wide variety of feeding conditions, with VFA, CO_2, and H_2 or CH_4 as the principal waste products of the microbial fermentation. In the rumen, fluctuations in the proportions of acetic, propionic, and butyric acids occur with concomitant changes in the proportion of methane. The total rate of fermentation can vary with the amount and quality of the feed, but essentially the same products are formed as long as substrate is limiting.

This ecosystem is destroyed if an excess of substrate is provided. In nature, where concentrated and readily digestible foods are not commonly available to ruminants, there is no problem. With the feeding to domestic ruminants of large quantities of carbohydrates in the form of grains or sugars, an excess is sometimes consumed. As little as 200 g of sugar or grain, administered to a sheep not adapted to this diet, can cause death within 24 hr. The excess carbohydrate, coupled with nutrients already in the rumen, support maximum growth of lactic acid bacteria, chiefly *Streptococcus bovis*. So much lactic acid is produced that it exceeds the buffering capacity of the rumen, the pH dropping to 4.0–4.5. At this concentration of lactic acid, the quantities absorbed cause a severe and often fatal acidosis.

Both L(+) and D(−) lactate are formed and absorbed through the rumen wall into the blood stream. Ruminant tissues can oxidize much of the L(+)-lactate but the host D(−) lactic dehydrogenase is entirely inadequate to cope with the excess of the D(−) acid. Usually, following the initial acidification by *S. bovis*, its growth is slowed by the acid, but there is then a growth of acidoduric lactobacilli, which continue production of lactic acid until the substrate is exhausted.

Ruminants can be adapted gradually to consume quantities of grain and sugar that initially would have been fatal, and a high productivity can be attained, as in feedlots. But this rumen microbial ecosystem is less stable than when the ration contains forage. Many of the animals suffer damage due to chronic acidosis, some go off feed, and some occasionally succumb to more extreme acidities.

In feedlot animals adapted to a high grain ration there can be a high incidence of bloat. It resembles the bloat encountered when fresh legumes (or more rarely grasses) are consumed in quantity. The digesta become a viscous mass in which the gas released by the fermentation is entrapped in the form of small gas bubbles. These do not coalesce in the top of the rumen into a large continuous gas phase which can be voided by eructation through the mouth. The rumen contents increase in volume like rising bread dough and can within a fairly short time exert a pressure that cuts off the animal's respiration and circulation, causing death. The underlying causes of bloat are not understood, although microbial growth, type and amount of saliva, and the genetic constitution of the ruminant have all been shown to influence its incidence. Bloat can be prevented by ensuring that foam-breaking chemicals are ingested along with the bloat-provoking feed or are introduced artificially.

Some bases for understanding the shift of the normal rumen ecosystem to the one causing acidosis have been proposed.[6] The rate of growth of the various microbial components of the rumen population determines their survival, in competition with the other microbes. When carbohydrate is limiting, a premium is placed on efficiency in using it for growth. If growth is a function of the rate at which adenosine triphosphate (ATP) can be derived from the carbohydrate, those fermentations yielding the maximum (at least 4) ATP per molecule of sugar have the potential to form more ATP and thus more cells per unit time, other factors being equivalent. All microbial components must adapt to this situation in order to maintain themselves in the ecosystem.

Administration of excess carbohydrate introduces a new factor. Efficiency in obtaining ATP per unit substrate no longer determines the ATP per unit time. Organisms deriving less ATP per sugar molecule, but taking in sugar at a faster rate, may be able to produce more ATP per unit time. The lactic acid bacteria have this capability. With an excess of starch or sugar, they grow with a generation time as short as 20 min, increasing 1000 times in about 3 hr. This is a general characteristic of the lactic acid bacteria, but most species attain this rapid growth only in an environment containing abundant peptones and amino acids. *Streptococcus bovis* does not require split products of the proteins (these are limiting in the rumen), and it differs also from most lactic acid streptococci in being able to utilize starch. It can process excess carbohydrate, producing more ATP per unit time than can the normal rumen microbiota, even though it produces only two ATP per hexose, as compared to four obtained by the normal saccharoclastic flora when carbohydrate is limiting.

There are few examples in nature where the concentration of soluble carbohydrates is in excess, with an abundance of other required nutrients. As mentioned, this does occur in the duodenum and jejunum of mammals having no preacid fermentation. Acidity in the stomach has killed off most of the bacterial competitors, and the host enzymes have released soluble sugars, amino acids, and peptides. Conditions are ideal for an extremely rapid development of lactic acid bacteria, and some have been shown to grow in the jejunum. But their small numbers, following the acidification, prevents their becoming serious competitors for the products of digestion. Rapidly growing streptococci have been isolated from the small intestine. If physiological disturbances in the host prevent secretion of acid in the stomach, if more carbohydrate is consumed than can be digested and absorbed, or if flow of digesta stops, conditions leading to a lactic fermentation and high acidity can develop also in animals exhibiting the competition model of interrelationship with the microbes.

IX. Other Effects of the Microbial Ecosystem on the Host

Microbial fermentation has been shown to exert effects leading to structural changes in host tissues. When rodents or lambs are reared under axenic conditions, the wall of the fermentation chamber does not attain the strength and bulk developed when the normal microbial ecosystem is present. The greater development is not due directly to the increased bulk of the food in the alimentary tract of conventional animals. Nylon sponges, stuffed into the rumen of a suckling lamb as a substitute for forage bulk, cause a great increase in the size of the rumen, but the weight of the organ does not increase.

The wall of the rumen is normally covered with myriads of papillae which greatly increase the absorptive areas. These are missing or greatly reduced in axenic animals or in young ones prior to forage consumption and fermentation. The normal development of both rumen and cecum depends on the microbiota. The fermentation acids are partially metabolized during passage through a wall of these organs. Relatively little of the acetic acid is oxidized, but a large fraction of the butyric and propionic acids are utilized by the host cells. Feeding of the acids to axenic animals has been shown to elicit normal development of the gut wall.

X. Summary and Conclusion

The vertebrate alimentary tract provides an almost ideal environment for the growth of anaerobic microbes. Their activities, if unchecked, would seriously impair the nutritive value of foods digestible by the host enzymes. This potential competition for food has been met in many vertebrates by the evolution of a high stomach acidity, effective in killing all or a great majority of alimentary microbes. Coupled with flow of digesta, microbial numbers are kept low in the small intestine, and host digestive and absorptive activities can garner soluble amino acids and sugars without serious losses to microbial fermentation. This type of host–microbe relationship, found in carnivores, is interpreted on the basis of a competition model.

In this model the host advantage is brief. There is a dense microbial population in the large intestine, using as substrates the intestinal epithelial cells sloughed off from the intestinal wall, as well as any food residues not digested by host enzymes. These residues are minor in carnivores. The significance to the host of the colonic microbial activity in carnivores is not clear. The acids formed in the posterior gut by fermentation of protein may have value, and vitamin synthesis may be important, but many nitrogenous products of dubious or negative value are also formed, in contrast to the fermentation products from carbohydrates.

Most herbivorous vertebrates have adapted their physiology to cooperate with the microbes, making use of the microbial abilities to digest resistant plant components for which the animal has no effective enzymes. An essential feature of this cooperation model is the development of a fermentation chamber in which these plant components reside long enough to allow microbial digestion and fermentation. The chamber may be anterior (prefermentation) to the site of acidification by host gastric juice or posterior (postfermentation).

Among animals patterned on the prefermentation model, primitive types show a partial separation of the stomach into a cardiac portion, accommodating the fermentation, and a succeeding pyloric (acid) portion.

The rumen is the most advanced expression of the preacid fermentation, with a complete separation of the fermentation from the acid compartment. With the supply and removal services provided by the ruminant, this is probably the most efficient host–microbe system for utilization of the less-digestible carbohydrates of plants. The volatile fatty acids produced as waste products of the microbial carbohydrate fermentation are absorbed by the host and oxidized to meet its energy requirements. The microbes formed in the fermentation are killed in the acid stomach, then digested as a source of protein.

But the preacid fermentation has a disadvantage if plant foods of fairly high quality, digestible by host enzymes, are consumed in quantity. These are all fermented rapidly by the bacteria, with resultant loss of food value for the host, particularly in the case of the proteins. The amino acids are destroyed, and the microbial protein produced is lower in quantity and (usually) biological value than the protein in the food.

This disadvantage has been met in the postacid model. In the cecal–colonic type of fermentation, animals can utilize their own enzymes to break down the easily digestible parts of the food. The resistant carbohydrates are then fermented by the cecal microbiota, and the host absorbs the acids produced. In the simple cecal model, the microbial bodies are voided and lost, but in many animals with postacid fermentation, the feces are eaten as a means of recovering the microbial protein.

These models indicate some patterns of host–microbe interrelationship but cannot describe adequately the great diversity of expression found in indivdual animals and groups. Many kinds of microbes are common to more than one type of mammalian digestive system, but others appear to be peculiar to particular types. Among the protozoa, where morphology alone can indicate differences, specificities for particular hosts or groups of hosts are especially evident. Much work remains before many of the host–microbe interrelationships can be reliably interpreted, but the new viewpoints and unexpected advantages resulting from an understanding of microbial activity are likely to reward such efforts. Particularly, quantitative information on the rates of various processes, both for the whole system and for individual components, will be illuminating.

References

1. BARNARD, E. A. Biological function of pancreatic ribonuclease. *Nature* **221**:340, 1969.
2. BAUCHOP, T., and R. W. MARTUCCI. Ruminant-like digestion of the langur monkey. *Science* **161**:698, 1968.
3. CLARKE, R. T. J. Personal communication.
4. CRANDALL, L. S. *The Management of Wild Animals in Captivity*. Chicago, University of Chicago Press, 1964.
5. HUNGATE, R. E. *The Rumen and Its Microbes*. New York, Academic Press, 1966.
6. HUNGATE, R. E. The rumen microbial ecosystem. *Annu. Rev. Ecol. Syst.* **6**:39, 1975.
7. JOHNSON, J. L., and R. H. MC BEE. The porcupine caecal fermentation *J. Nutr.* **91**:540, 1967.
8. KHOUVINE, Y. Digestion de la cellulose par le flore intestinale de l'homme. *B. cellulosae dissolvens, n. sp.* Ph.D. thesis, Faculté des Sciences de Paris, Imprimerie de la Cour D'Appel, Paris, 1923.
9. MC BEE. R. H., and G. C. WEST. Caecal fermentation in the willow ptarmigan. *The Condor* **71**:54, 1969.
10. MATTOCKS, J. G. Goose feeding and cellulose digestion. *Wildfowl* **22**:107, 1971.
11. MOIR, R. J. Personal communication.
12. MOIR, R. J., and R. E. HUNGATE. Unpublished experiments.
13. OHWAKI, K., R. E. HUNGATE, L. LOTTER, R. R. HOFMANN, and G. MALOIY. Stomach fermentation in East African *Colobus* monkeys in their natural state. *Appl. Microbiol.* **27**:713, 1974.
14. PRINS, R. A. and E. R. PRAST. Oxidation of NADH in a coupled oxidase-peroxidase reaction and its significance for fermentation in rumen protozoa of the genus *Isotricha*. *J. Protozool.* **20**:471, 1973.
15. ROBINSON, J. P., and R. E. HUNGATE. *Acholeplasma bactoclasticum* sp. n., an anaerobic mycoplasma from the bovine rumen. *Int. J. Syst. Bacteriol.* **23**:171, 1973.
16. THURSTON, J. P., and C. NOIROT-TIMOTHÉE. Entodiniomorph ciliates from the stomach of *Hippopotamus amphibius*, with descriptions of two new genera and three new species. *J. Protozool.* **20**:562, 1973.

6

The Use and Function
of Fiber in Diets
# of Monogastric Animals

John A. Lang and George M. Briggs

I. Introduction

During the past century, nutritional scientists have successfully identified many nutrients essential to laboratory animals. However, as each essential dietary component was identified and added to diets in pure form, the fiber content of these rations was dramatically reduced. Fibrous materials, especially those of plant origin, are often chemically heterogeneous. In the formulation of chemically defined diets, such complex mixtures were avoided whenever possible. Thus the reduction of fiber in experimental diets has been a natural consequence of the search for chemically defined diets.

Many nutritionists have been concerned about the removal of fiber from these semipurified diets and believe that an essential dietary ingredient has been omitted. We shall review and evaluate the evidence for the necessity of fiber diets for laboratory animals.

We shall begin this chapter by discussing what types of materials serve as sources of fiber in natural-ingredient rations and then proceed to describe the physical and chemical nature of some purified fibers. Following a brief outline of the theories of how fiber might promote better health, we shall discuss representative experiments in fiber research with experimental animals.

JOHN A. LANG • Postdoctoral Fellow, Department of Nutritional Sciences, University of California, Berkeley, California.   GEORGE M. BRIGGS • Professor of Nutrition, Department of Nutritional Sciences, University of California, Berkeley, California.

II. Fiber in Natural-Ingredient Diets

There is a great deal of confusion over the use of the word "fiber" because there is no universally accepted definition of the concept. In reference to diets, scientists are not necessarily using the word according to its strict dictionary definition.* In common usage, the word "fiber" denotes a physical state of being threadlike. Nutritionists use the term not only for substances that are physically fibrous, such as cellulose, but also for those which, while not physically fibrous, have similar chemical and biological properties. Most animals cannot digest fibrous material such as cellulose.[52] Thus, in reference to diets, most definitions of fiber refer to a material that is neither digested nor absorbed and provides bulk or physical substance to a diet. The problem of misleading or vague terminology in fiber research is an important topic and is discussed in detail in this work.

The tendency for nutritionists to use the term "fiber" instead of "roughage" or "bulk" probably arises from the extensive use of the traditional analytical method for crude fiber. This system of analysis was developed by Henneberg and Stohmann at Weende Experiment Station in Germany between 1858 and 1863[28] (for details see chapter 3). The acid and base treatments in the chemical analysis are intended to simulate the natural digestive processes in the stomach and intestines of monogastric animals. Thus, those materials that are chemically analyzed as crude fiber (cellulose, lignin, etc.) are considered to be indigestible or unavailable polysaccharides.

This analytical method is still widely used. The reference book, *Compositions of Foods* (Agricultural Handbook Number Eight), uses crude-fiber determinations to estimate "unavailable carbohydrate."[65]

Crude fiber is an essential component of commercial natural-ingredient (stock) diets. Natural-ingredient diets were empirically and then scientifically formulated to yield maximal growth rates and optimal maintenance conditions. The levels of fiber in these commercial diets reflect current optimal usage and can be used as a starting point in determining fiber requirements. Commercial companies carefully monitor the nutrient content of each dietary component and, by means of a computer program, adjust the level of each natural-ingredient component so that the final mix has the published nutrient composition.[56] Therefore, the level of a specific dietary component, such as alfalfa, may change somewhat, but the overall composition of the essential nutrients is maintained constant. Variation in the composition of commercial natural diets is an often-overlooked variable in experimental research.

*"(1) A slender, threadlike structure that combines with others to form animal or vegetable tissue.
(2) Any substance that can be separated into threads or threadlike structures . . ."[66]

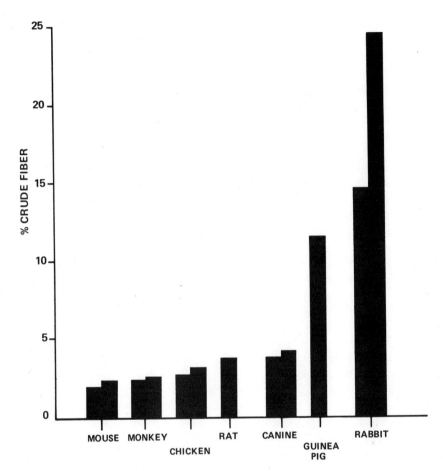

Figure 1. Crude-fiber content (percentage) of Ralston Purina Laboratory Chows.® Multiple listings represent specialized diets, such as weaning rations and high-protein formulas for lactation.

Figure 1 shows the published fiber content of some Ralston Purina Chow® diets.[56] The widespread use of the abbreviation "Chow"* by many nutritionists for any natural-ingredient diet indicates that the Ralston Purina diets are representative of those used in the animal feed industry.

The presence of significant amounts of crude fiber in natural-ingredient diets is not evidence, per se, that crude fiber is an essential dietary requirement.

*The term "Chow" is a registered trademark of Ralston Purina Company.

The crude-fiber content of dietary components is one of the variables that commercial companies carefully monitor and control.[56] This parameter is important in terms of economics as well as nutrition. It has been pointed out that crude-fiber analysis has been used as an estimate of the unavailable portion of a food. Those commodities that have a low crude-fiber content, such as fish meal or sugar, generally have a high economic value. In contrast, those commodities that have high crude-fiber content and are indigestible, such as rice straw or rice hulls, have such a minimal economic value that they constitute an agricultural waste-disposal problem. If they are fed to animals at high levels, the commodities will be transformed into a fecal disposal problem. Thus, because of low cost, it would benefit commercial companies to maintain as high a crude-fiber content as is acceptable.

There is a large variation in the crude-fiber content of commercial natural-ingredient diets (Figure 1). This reflects the ability of some animals to accept and digest large quantities of fiber. Most organisms, including monogastric mammals and their microflora, cannot digest fiber because they do not possess the enzymes that catalyze the hydrolysis of cellulose and other fibrous constituents. Associated with high levels of cellulose in their environment, a few organisms such as termites, brown rot fungi, and snails have developed the necessary enzymes to utilize large amounts of fiber. Cows, sheep, and other ruminants have multiple stomachs and associated microflora that can ferment high-fiber diets. Herbivores such as rabbits and guinea pigs possess modified stomachs, elongated intestinal tracts, or enlarged ceca, which contain specific microflora that can digest fibrous materials.

Mammals that do not harbor symbiotic cellulolytic bacteria (rats, mice, monkeys, dogs, and chickens) cannot tolerate high levels of cellulose. The commercial natural-ingredient diets for these animals have a low and narrow range of crude fiber (2–4%). The fiber content of diets for herbivores ranges from 12 to 25% crude fiber (see Figure 1).

III. Fiber in Semipurified Diets

Typical examples of materials that serve as fiber in semipurified diets are listed in Table I. The most popular types of commercially purified fiber are Alphacel®, Solka Floc®, and Cellophane Spangles®. These products are distributed by the two major suppliers of dietary ingredients, Nutritional Biochemical Company and the Teklad Test Diets, Sprague-Dawley Division of the Mogul Corporation (formerly General Biochemicals). It is interesting that cellulose-type compounds are the only products these distributors have promoted as a nonnutritive bulk or nonnutritive fiber.

TABLE I. Crude Fiber Content of Purified Sources of Fiber

Source (trade name)	Crude fiber (%)[a]	Distributor and/or manufacturer	Chemical and physical properties
Avicel®	61.7	FMC Corp. Marcus Hook Pennsylvania 19061	Primarily crystalline cellulose; microcrystalline cellulose. Food and pharmaceutical uses.
Whatman No. 1 Cellulose Powder®	71.3	Reeve Angel 9 Bridewell Pl. Clifton, New Jersey 07014	Finely ground filter paper.
Alphacel®	61.8	Distributed by ICN Nutritional Biochemicals 26201 Miles Rd. Cleveland, Ohio 44128	Finely ground cellulose; nonnutritive bulk.
Celluflour®	68.4	Chicago Dietetic Supply, Inc. La Grange, Illinois 60525	Finely ground cellulose.
Solka Floc®	65.3	Distributed by Catalog No. 160390 Teklad Test Diets, Sprague-Dawley Div. of the Mogul Corp. 2826 Latham Dr. Madison, Wisconsin 53711 (Formerly General Biochemicals) Manufactured by Brown Company Berlin, New Hampshire	85% wood alpha cellulose with 15% nonglucose hemicellulose; nonnutritive fiber; pet food and pharmaceuticals.
Cellophane Spangles	55.5	Distributed by Catalog No. 160110 Teklad Tests Diets, Sprague-Dawley Div. of the Mogul Corp. 2826 Latham Dr. Madison, Wisconsin 53711 (Formerly General Biochemicals) Manufactured by E.I. DuPont Wilmington, Delaware	90% Cellophane®, 10% chemically undefined. One can consider Cellophane® to be a modified, regenerated, amorphous form of cellulose; nonnutritive fiber. Used for decorating envelopes.

[a] The crude fiber analysis was provided by Dr. Robin Saunders of the Western Regional Research Laboratory (WRRL) of the United States Department of Agriculture.

A material that provides no nutritive value is an important tool for the nutritional scientist. For example, if one wants to study the effect of adding some component to a complete diet, one must realize that by simply adding that component to the diet, one is diluting all other dietary ingredients.[50] In a well-controlled experiment, one could vary the concentration of the test substance at the expense of a nonnutritive component of the diet and consequently eliminate the dilution effect. With animals that do not produce cellulase enzymes, it has been common practice to use some form of cellulose as the controlled nonnutritive variable.

Since cellulose has been the most popular form of nonnutritive fiber, we shall examine its chemical and physical structure for clues as to how these properties relate to the role of cellulose as a nonnutritive fiber.

Cellulose is the earth's most abundant organic compound and forms the fibrous tissue of plant materials. Chemically, cellulose is a linear polymer of glucose monomers connected by $\beta(1\text{-}4)$-glycosidic linkages. The degree of polymerization (DP), is defined as the number of glucose monomer units in a molecule. $\beta(1\text{-}4)$-glucans with a DP greater than seven are insoluble. Cellulose molecules have a DP between 100 and 1,000,000 and are consequently very insoluble in aqueous solutions.[8]

The long chains of cellulose molecules aggregate and form structures called microfibrils.[42] Within the microfibril, some of the cellulose molecules pack together very tightly to form "crystalline cellulose," which produces X-ray diffraction patterns typical of crystals. In other regions of the microfibril, the molecules are loosely packed with random orientation. This region of random chains is called "amorphous cellulose." On a gross morphological scale, many microfibrils bunch together to form a cellulose fiber. The close-packed crystalline region of cellulose is exceptionally resistant to enzymatic hydrolysis. Relatively few organisms can effectively hydrolyze and utilize crystalline cellulose, while a number of organisms can degrade amorphous cellulose (random chains, noncrystalline). Reese et al.[52] proposed that those organisms that can degrade crystalline cellulose must produce a unique kind of cellulase, the hypothetical C_1 enzyme. Although many organisms have an ability to hydrolyze the beta-glycosidic linkage of soluble forms of cellulose, most fail to degrade native cellulose because they do not produce the C_1 enzyme. The unique crystalline cellulase enzyme that permits the complete hydrolysis of cellulose has been recently purified and characterized.[15,36]

The enzymatic degradation of cellulose by those few organisms that produce the crystalline cellulase enzyme is a very slow process. The slow rate of hydrolysis of cellulose is presumably due to the fact that cellulose is a difficult substrate. The large size, insolubility, and density of the cellulose particle presents the enzyme with a considerable resistance to hydrolytic action. The kinetics of hydrolysis of crystalline cellulose particles does not follow classical Michaelis–Menten kinetics but instead appears to be restricted and follows Shutz kinetics.[51] McLaren has derived the equations for this type of kinetics assuming that only the surface of an insoluble substrate is available to the enzyme.[43] McLaren's concept was confirmed empirically when it was shown that the enzyme reaction rate is proportional to the surface of the substrate rather than the total amount of substrate.[35,37] The observation that the velocity of cellulose hydrolysis is a function of particle size[5] supports the hypothesis that the enzyme reaction is restricted to the surface of cellulose particles. One means of counteracting a slow reaction rate is to lengthen the enzyme–substrate incubation period. Many of the mammals that harbor cellulolytic bacteria have evolved specialized anatomy (rumens, elongated intestinal tract, and enlarged cecum) which permits prolonged periods of contact with the substrate. These physiological changes appear to be an evolutionary response to the inherently slow rate of the hydrolysis of cellulose.

The types of fiber commonly used in experimental diets have a number of distinguishing properties. In Table I, the dietary fiber components are arranged in decreasing order of crystalline cellulose content.

Alphacel® is probably the most widely used fiber component for semipurified diets. It is a refined form of wood cellulose and is primarily alpha cellulose. Defined in terms of the method of analysis, alpha cellulose is cellulose that is not soluble (average DP greater than 200) in 17.5% sodium hydroxide solutions.[21]

Solka Floc® is the trade name for a family of purified cellulose products manufactured from wood by Brown Company. Solka Floc BW 200® is distributed by Teklad Test Diets (catalog No. 160390, nonnutritive fiber, cellulose). It differs from Alphacel® in that it contains 15% hemicellulose in addition to 85% alpha cellulose. Hemicellulose is a poorly defined class of compounds[68] and represents, in this case, the fraction of cellulose that is not alpha cellulose.[45] The relatively soluble, low-molecular-weight hemicellulose includes approximately 4% pentosans.[45]

Cellophane Spangles® is also distributed by Teklad Test Diets (catalog No. 160110) for use as a nonnutritive fiber. The Cellophane® is manufactured by DuPont and sold to Virkotype Division of Lawter Chemicals, which in turn cuts the film into "precision-chopped glitter." This product is used in industry to decorate greeting cards. Cellophane® is regenerated cellulose and is basically amorphous or noncrystalline cellulose. Cellophane Spangles® may have an advantage over the other sources of dietary fiber because it glitters and potentially attracts animals. This property has been used as an argument for its use in guinea pig diets.

DuPont manufactures many different kinds of Cellophane® for commercial packaging purposes. One class, the "K" Cellophanes®, is "basically vinylidene chloride copolymer coated cellulose film."[61] The monomer, 1,1 dichloroethylene used to manufacture this polymer is toxic.[44] Another class of Cellophane® in Cellophane Spangles® is the MSD type which are "nitrocellulose/wax coated cellulose films.[61] The coatings and their various constituents make up roughly 10% of these Cellophane® products.[61] In addition, the relative amounts of these two classes of Cellophane® in Cellophane Spangles® may change according to the availability of Cellophane® material from DuPont.[17] Because up to 10% of its composition is chemically poorly defined, potentially toxic, and varies considerably from lot to lot, we do not recommend the use of this source of fiber.

The remaining fiber products listed in Table I are less commonly used than those previously described. Whatman Cellulose Powder® is a highly refined and processed form of alpha cellulose and is noted for its very low ash content (0.015%). Avicel® is wood cellulose that has been treated with dilute acid to remove the amorphous cellulose. Avicel® is primarily crystalline cellulose and is therefore the most enzyme-resistant form of cellulose. Celluflour® was originally processed by Mr. Hoelzel of Chicago Dietetic Supply, Inc. This company no longer manufactures Celluflour® but now distributes the product of Brown Company.[19] Thus, over the years, the cellulose nature of Celluflour®, as well as its susceptibility to enzyme attack, has probably changed. This change should be kept in mind when comparing and evaluating experiments using Celluflour®.

All of the purified forms of fiber listed in Table I are derived from wood pulp. Although the manufacturers try to control the quality and uniformity of their products as carefully as possible, there are unavoidable variations in chemical structure. Variations can arise from changes in the nature of the raw materials (species of tree, seasonal changes) as well as from the wide variety of pulping methods (alkaline-kraft, acid sulfite, or the neutral sulfite process, to mention only a few of the basic options) and variations in the extent of bleaching. The chemical and physical variations in the semipurified sources of fiber listed in Table I are of less magnitude than the uncontrolled variations in raw, natural sources of fiber; nevertheless, the possibility of these variations affecting biological activity should not be ignored.

With the exception of Cellophane Spangles®, all the fibers in Table I are finely milled white powders. The particle size of the substance has been shown to influence the rate of enzymatic hydrolysis.[5,35,37] In the process of purifying these substances, most of the physically fibrous nature of these materials has been destroyed. This is especially true of regenerated cellulose such as Cellophane Spangles®, and of Avicel®. In theory, the latter is composed only of microcrystallites. Although the fibrous nature of these materials has been lost, their beta-glycosidic linkages still make these substances very resistant to enzymatic hydrolysis. Thus, these "fiberless fibers" can still function as bulking agents.

Carbohydrate gums are the ultimate in fiberless fibers. They have none of the physical characteristics of fiber because they are soluble in aqueous solutions and form transparent solutions. Nevertheless, they should be considered a fiber because their chemical structure makes them resistant to enzymatic hydrolysis or digestion.

A few representative samples of natural and synthetic gums are listed in Table II. Polysaccharide gums are used extensively by the food industry. They

TABLE II. Crude-Fiber Content of Carbohydrate Gums

Gum	Crude fiber (%)[a]	Distributor or manufacturer	Chemical and physical properties
Agar	0.49		Polymer of D-galactose residues and 3,6 anhydro-L-galactose residues with a half-ester sulfate on approximately every 10th galactose residue.
Carrageenan	N.D.[b]		Irish moss, galactose monomers, each sulfonated.
CMC (carboxymethyl-cellulose)	0.29	Manufactured by Hercules Inc. Wilmington, Delaware 19899 and Dow Chemical Co. Midland, Michigan 48640	Gum cellulose, substituted cellulose, water soluble, produces viscous solutions, cationic. Food additive and pharmaceuticals (binder).
Guar gum	1.5-2.0%	Distributed by Stein Hall Co. 605 Third Ave., New York, New York	Galactomannose polymer, nonionic; substituted mannose polymer linked by $\beta(1\text{-}4)$-glycosidic linkages, galactose branching by $\alpha(1\text{-}6)$ linkages.
Gum arabic	N.D.		Polymer or arabinose, galactose, rhamnose, and glucuronic acid; molecular weight 24,000
Methylcellulose	89.7	Manufactured by Dow Chemical Co. Midland, Michigan 48640	Soluble in cold water, insoluble in hot water

[a]The crude fiber analysis was provided by Dr. Robin Saunders of the Western Regional Research Laboratory (WRRL) of the United States Department of Agriculture.
[b]N.D., not determined.

serve as thickening agents, gelling agents, foam stabilizers, emulsifiers, etc. The use of gums in fabricated foods was reviewed by Glicksman in 1975.[22]

Carboxymethylcellulose and methylcellulose are two examples of a whole family of substituted cellulose gums. The chemical nature and degree of substitution of the side groups can be controlled to optimize such properties as viscosity, ionic charge, solubility, and resistance to enzymatic degradation. The principle advantages to those working with experimental diets of these semisynthetic polysaccharides are: (1) chemical uniformity of the fiber molecule relative to natural sources of fiber, (2) freedom from impurities, and (3) chemical variety. These semisynthetic gums, with their unique properties, are ideal model compounds to test possible theories of the mechanism by which fiber promotes better health.

Because they are soluble in aqueous solutions, most carbohydrate gums have levels of crude fiber that are insignificant (less than 1%; see Table II). Since the gums are indigestible carbohydrates, the crude-fiber method of analysis cannot be used as a measure of all the unavailable or nonnutritive carbohydrates. The deficiency of the crude-fiber analysis has led to the search for new detergent methods for chemical analysis[62,63] and new terminology, such as "dietary fiber."[60] The expanded concept of dietary fiber includes a larger quantity of substances associated with plant cell-wall materials, such as hemicellulose, pectins, lignins, and tannins, in addition to cellulose.

The limitations of the crude-fiber analysis and the complexity of the problem involved in developing a single method for fiber analysis are most evident in the case of methylcellulose. While gums such as carboxymethylcellulose contain little crude fiber (0.29%), methylcellulose appears to contain an extremely high amount of crude fiber (89.7%). The unusually high crude-fiber determination for methylcellulose is due to the fact that, while this gum is soluble in cold aqueous solutions, it is insoluble in hot solutions[44] and therefore precipitates during analysis. This is an anomaly of the analytical method; methylcellulose is quite soluble under normal physiological conditions.

Table III lists a few of the natural sources of fiber. The primary sources of crude fiber in Ralston Purina® natural-ingredient diets are sun-cured alfalfa, dehydrated alfalfa, beet pulp, wheat middlings, soybean hulls, and oat hulls.[56] The principal disadvantage of evaluating experiments using these substances is that they are complex mixtures. They contain many other nutrients in addition to their fiber content, and much of this may be digested and utilized by the test animal. These experiments must be carefully designed so that the potential benefits of nutrients other than fiber are recognized.

Natural sources of fiber are tested and ranked according to feed quality. The grading of these natural products is an attempt to account for unavoidable variations that are due to differences in time of harvest, soil conditions, weather conditions, plant species, processing methods, and storage conditions. For example, dehydrated alfalfa is graded and sold on commercial markets according to

TABLE III. Crude Fiber Content of Unrefined Sources of Fiber

Source	Comment	Fiber, dry matter basis (%)
Alfalfa:		
Dehydrated	Minimum 15% protein content	28.4[a]
Dehydrated	Minimum 30% protein content	21.7[a]
Sun cured		28.0[a]
Wheat (Triticum):		
Bran	Feed grade	11.2[a]
Bran	Human grade	8.0[b]
Mill run		8.9[a]
Middlings	Wheat standard middlings	7.3[c]
Flour	Human grade	0.3[c]
Beet pulp	Dried	20.9[a]
Corncobs	Ground	34.6[c]
Soybean hulls	Soybean flakes	36.1[c]
Oat hulls		32.2[c]
Rice straw		35.1[c]
Rice hulls		44.5[c]

[a]Nutrient requirements of laboratory animals.[48]
[b]Bing.[2]
[c]Atlas of Nutritional Data of U.S. and Canadian Feeds.[47]

its protein content. The crude-fiber content of dehydrated alfalfa varies inversely with the protein level and ranges from 19.9 to 28.4%.[48] Significant differences in grades of wheat bran are also often overlooked. Feed-quality wheat bran has an 11.2% crude-fiber content, while human-food-grade wheat bran has an 8% crude-fiber content.[2] The exact grade of an agricultural commodity should be noted in any publication so that the investigation can be correctly interpreted and repeated. Obviously, there are many variations within grades; thus, natural plant products should be carefully characterized in chemical and physical terms.

IV. Theories on the Health-Benefiting Role of Fiber in Diets

There are many theories as to how dietary fiber could promote better health. We shall discuss the experimental evidence in terms of three basic mechanisms by which dietary fiber could provide better health.

A. Fiber Functions as a Bulking Agent

Undigested, high-molecular-weight compounds form fecal bulk. Bulk in the intestinal tract induces a faster transit time and reduces the opportunity for microbial production of toxins or carcinogens. In addition, this faster transit time shortens the length of exposure to any toxin that may have been ingested.

B. Fiber Functions as a Toxin Antagonist

Large insoluble compounds are ideal absorbants. In addition to their many hydroxyl groups, charged side groups greatly enhance the binding strength of ionized molecules. Cellulose and its many derivatives and analogs are routinely used to purify compounds by selective adsorption. Dietary fiber could bind potential toxins, prevent their digestion or absorption, and facilitate their excretion.

C. Fiber Functions as a Nutrient

Fiber sources may be a source of unknown nutrients or compounds that have medicinal properties. Natural plant fibers are a heterogeneous mixture of complex compounds. The chemical structure and biological activity of the individual molecule is often unknown. Even undigested molecules can cause an indirect health benefit by stimulating the growth of beneficial microflora or by retarding the growth of pathogenic organisms.

V. Experimental Studies with Dietary Fiber

Within this theoretical framework, we shall now examine the experimental evidence for the need of dietary fiber in semipurified diets. For the sake of clarity, this material has been organized according to animal species.

A. Rats

Presumably functioning as a bulking agent, cellulose has been shown to stimulate the rate of gastric emptying time,[58] intestinal motility, and histochemical enzyme changes in the small intestine.[54] Cecum enlargement and enhanced sodium transport could be produced by the addition of polyethylene glycol to the drinking water of rats.[38] The cecum changes were found to be reversible. The latter study clearly demonstrates that fiber does not have to be solid or insoluble in order to function as a bulking agent. The belief that intestinal physiology requires bulk for normal function is supported by the work of Lubbock et al.[41] In 1937, they reported that rats develop diverticular disease when on a low residue diet.

The antitoxic properties of fiber in rat and mouse rations were reviewed by Ershoff in 1974.[16] Fiber reduced the toxic effects of nonionic detergents (Tween 60 and Tween 20), glucoascorbic acid, sodium cyclamate, and the red dye amaranth.[16] The antitoxic benefits of sources of fiber were attributed, at least in part, to some factor or factors other than cellulose content per se.[16] For example, pectin has been reported to prevent hypercholesterolemia and elevated levels of liver cholesterol.[67] There are numerous reports on the ability of fibrous materials from cereals or other plants to facilitate the excretion of cholesterol and bile salts. This research has direct implications to cardiovascular disease and cancer.

Fiber is not always an antagonist to toxins; it may even enhance a toxin's potency. For example, an undesirable side effect of the antiinflammatory drug indomethacin is the production of intestinal ulcers. When fed with fiber, the drug produced numerous ulcers in rats. A low-fiber diet reduced the apparent toxicity of indomethacin.[14]

Most studies indicate that cellulose is not digested by the rat to any appreciable extent. These investigations justify the description of cellulose as nonnutritive fiber and nonnutritive bulk. However, Conrad et al. isolated a ^{14}C-labeled substance from soybeans.[7] When this material was fed to rats, only 50% of the radioactive label was recovered in the feces.[7] Yang et al. could account for only about half the amount of Alphacel® fed to rats.[70] These investigators measured increased volatile fatty acids when cellulose was fed to rats and attributed the apparent digestion of cellulose to microbial fermentation in the cecum. The methodology and nutritional design of the latter report is weak. Part of the results may be explained by the fact that apparent digestibility of fiber is often a function of the level of fiber fed[33] and the limits of analytical methods. One must also eliminate both chemical and microbial destruction of cellulose in the feces.

The cellulose derivatives carboxymethylcellulose[20,71] and methylcellulose[4] have been shown by radioactive-tracer studies to undergo no significant digestion. The rat's inability to digest soluble forms of cellulose strongly supports the many studies that contend that cellulose is not digested in the rat to any significant extent.

Chemically defined liquid diets have been developed for use in gnotobiotic research. These diets appear to be nutritionally adequate for the rat and produce no serious health problems.[24] The adequacy of these diets, which obviously contain no fiber, indicates that fiber is not an essential dietary ingredient.

B. Rabbits

In their investigations with radioactive carboxymethylcellulose, Ziegelmeyer et al. found no significant digestion in rats or humans, but rabbits digested 50% of this soluble derivative of cellulose.[71] Using the acid-detergent method of fiber analysis, Hoover and Heitmann reported a 34% digestibility of the fiber.[32] These authors estimated that the rabbit derived 12% of its daily basal caloric requirement from volatile fatty acids produced in the cecum. Crude-fiber digestibility by rabbits on an alfalfa diet has been reported to be 16.2%.[59] The wide variations in these reports on the digestibility of fiber by rabbits emphasizes the important influence of the type of fiber fed and the method of fiber analysis.

C. Guinea Pigs

Slade and Hintz reported that rabbits were only half as efficient as guinea pigs in digesting crude fiber.[59] Guinea pigs digested 38.2% of the fiber in their diets. Loosli et al. also report that the guinea pig is approximately equivalent to horses and ponies in its ability to digest fiber and about twice as efficient as rabbits in digesting alfalfa crude fiber.[40] The guinea pig is an efficient herbivore, whose natural diet includes green vegetation with a low caloric density.[34] One would expect that an animal that feeds on a high-fiber, low-caloric-density diet would evolve the capacity to digest and utilize cellulose. The guinea pig's capability has been demonstrated to be a result of cellulolytic activity in the cecum.[27]

In 1945, Woolley and Sprince reported that cellulose (Celluflour®) stimulates guinea pig growth.[69] Numerous investigations have confirmed the health benefits of cellulose roughage in guinea pigs.[3,9,53] The studies by Heinicke and Elvehjem with guinea pigs indicate that wood cellulose is superior to Cellophane Spangles® or gum arabic.[29]

The numerous studies with the guinea pig provide the strongest argument for those who contend that fiber is an essential dietary ingredient for man. A requirement for fiber per se by the guinea pig has not been definitively proven, because not all of the nutrients essential to the guinea pig have been identified. In addition to those nutrients required by rats or man, guinea pigs appear to require an unidentified factor of plant origin. A water-soluble, unidentified growth factor has been prepared from alfalfa.[39] The solubility of this growth factor demonstrates that it cannot be cellulose. The potential role of fiber can be evaluated only when all other essential nutrients are included in the test diet. It will be interesting to determine whether guinea pigs are attracted to plants because they require a substance found only in plants or because of their ability to utilize plant materials efficiently.

D. Chickens

In 1947, it was reported that the addition of 15% wood cellulose to a purified chicken diet causes a significant increase in growth.[11] In the next year, it was reported that sawdust in amounts up to 20% of the purified ration could be tolerated with no harmful effects.[12] It is evident that cellulose was not being used as a nutrient, since it had been reported in 1939 that the digestibility of sawdust cellulose by chickens was only 2.33%.[26]

In an elegant experiment, Forbes' laboratory demonstrated how crude fiber apparently improved an imperfect diet.[25] The experiment was based upon the theory that animals would consume feed until their energy needs were met. The replacement of available carbohydrate by nonnutritive fiber would cause an increase in voluntary feed intake. If the diets were deficient in a nutrient, the increased food intake would support growth by supplying more of the limiting nutrient. Forbes reported a linear relationship between growth and intake of tryptophan and arginine. This relationship was evident either by inducing the chickens to eat more by dilution of the diet with nonnutritive fiber or by increasing the concentration of tryptophan and arginine in the diet.[25] The nonessentiality of fiber in the chick diet is supported by the adequacy of a chemically defined, liquid diet for the chick.[64]

The history of the controversy over the need for fiber in the chick diet emphasizes the complexity and importance of apparent nutrient interactions and the subtlety of nutritional imbalances. The critical importance of good nutritional design in fiber research cannot be overemphasized.[23]

There are many reports of other monogastric animals requiring and receiving health benefits from fiber in their diet. The mouse,[1,10] meadow vole,[57] hamster,[55] grouse,[46] beaver,[31] and pig[13,18] are just a few examples of other animal species reportedly showing beneficial effects of fiber in their diets. Many of these studies are difficult to evaluate because the source and purity of the fiber used has not been adequately documented and the nutritional adequacy of the diet has not been demonstrated.

VI. Conclusion

Fiber can be an important tool in experimental nutrition. For example, when testing the potential biological activity of a substance, the test material is substituted for a purified, nonnutritive fiber instead of the whole diet. Fiber serves as a controlled variable and eliminates artifacts arising from the dilution of essential nutrients present in the whole diet.

The debate over the need for dietary fiber has a long history and will doubtlessly continue for many decades. At this point in time, we must conclude that we have not seen convincing evidence that dietary fiber is an essential, irreplaceable dietary component. It is difficult to prove a negative, but the development of chemically defined liquid diets for rats, chickens, and man suggests that there is no essential, unique function for dietary fiber.

Although dietary fiber may not be an essential dietary component, many studies indicate that dietary fiber may have health-benefiting properties, such as reducing the effects of toxins, reducing cholesterol levels, and preventing diverticular disease. Unfortunately many of these theories involve slow processes and therefore require long-term tests which must be carefully controlled.

We have tried to point out some of the pitfalls of previous research so that future experiments with fiber products might be better designed. As we have seen with the studies with chickens, artifacts arising from the use of nutritionally inadequate diets are very subtle and difficult to identify. There is a great need for definitive experiments demonstrating the unique benefits of fiber. In general, definitive experiments require (1) careful nutritional design, (2) purification of the active component in dietary fiber, (3) detailed chemical and physical characterization of the compound, (4) quantitative measurement of the mechanism by which dietary fiber promotes improved health, and (5) use of nutritionally adequate diets.

In addition to feeding a substance and measuring a biological response, a more complete understanding of the role of fiber may result from elucidating the relationship between biological structure and function. The investigations on cellulose revealing the relationship between its chemical and physical properties and its resistance to digestion may well serve as a model for this approach.

Scientists in fiber research should also be aware of the nutritionally harmful effects of unrestricted additions of fiber to diets. The addition of large amounts of indigestible material will dilute the essential nutrients in a diet (the dilution effect[50]) and may cause the animal to reject the food or become nutritionally deficient. In addition to its potential detoxifying properties, fiber absorbs and makes unavailable many well-established essential nutrients. For example, phytic acid, oxalic acid, and phenolic compounds are associated with plant fibers and have been shown to restrict the availability of essential elements such as zinc.[49]

The implications of fiber research may even extend to the field of experimental medicine. Pharmaceutical companies frequently use Avicel®, carboxymethylcellulose, or Solka Floc® as carriers or binders in medicine tablets. As we have seen, these cellulose products are the most common sources of purified fiber in semipurified diets. A placebo is a tablet composed of a carrier or binder that contains no medication and therefore serves as a control in the testing of new drugs. A consistent proportion of patients have been cured by placebo treatments, and this curative rate for placebo has usually been attributed to psychological factors. In light of the antitoxic and health-benefiting potentials of cellulose fiber, removal of fiber from placebo tablets may help them to serve as a better control in pharmacological testing.

References

1. BELL, J. M. A comparison of fibrous feedstuffs in nonruminant rations: effects on growth responses, digestibility, rate of passage and ingesta volume. *Can. J. Anim. Sci.* **40**:71, 1960.
2. BING, F. C. Indigestible carbohydrate. Letter to the editor. *Am. J. Clin. Nutr.* **27**(11):1201, 1974.

3. BOOTH, A. N., C. A. ELVEHJEM and E. B. HART. The importance of bulk in the nutrition of the guinea pig. *J. Nutr.* **37**:263, 1949.

4. BRAUN, W. H., J. C. RAMSEY and P. J. GEHRING. The lack of significant absorption of methylcellulose, viscosity 3300 cp from the gastrointestinal tract following single and multiple oral doses to the rat. *Food Cosmet. Toxicol.* **12**:373, 1974.

5. CAULFIELD, D. F., and W. E. MOORE. The effect of varying crystallinity of cellulose on the enzymatic hydrolysis. *Wood Sci.* **6**(4):375, 1974.

6. CHATFIELD, C., and G. ADAMS, (eds.). *Proximate Composition of American Food Materials*, Circular 549. Washington, D.C., USDA, 1940.

7. CONRAD, H. E., W. R. WATTS, J. M. IACONO, H. F. KRAYBILL and T. E. FRIEDEMANN. Digestibility of uniformly labeled carbon 14 Soybean cellulose in the rat. *Science* **127**:1293, 1958.

8. COWLING, E. B. Structural features of cellulose that influence its susceptibility to enzyme hydrolysis. *In: Advances in Enzymic Hydrolysis of Cellulose and Related Materials*, E. T. Reese (ed.). New York, Pergamon Press, 1963.

9. CRAMPTON, E. W., and J. M. BELL. Studies on the dietary requirements of guinea pigs, II. Effects of roughage. *Sci. Agric.* **27**:57, 1949.

10. DALTON, D. C. Effect of dilution of the diet with an indigestible filler on feed intake in the mouse. *Nature* **205**:807, 1965.

11. DAVIS, F., and G. M. BRIGGS. The growth-promoting action of cellulose in purified diets for chicks. *J. Nutr.* **34**(3):295, 1947.

12. DAVIS, F., and G. M. BRIGGS. Sawdust in purified chick rations. *Poultry Sci.* **27**(1):117, 1948.

13. DeGOEY, L. W., and R. C. EWAN. Effect of level of intake and diet dilution on energy metabolism in the young pig. *J. Anim. Sci.* **40**(6):1045, 1975.

14. DREES, D. T., T. L. ROBBIN and L. CRAGO. Effect of low residue foods on Indomethacin-induced intestinal lesions in rats. *Toxicol. Appl. Pharmacol.* **27**:194, 1974.

15. EMERT, G. H., E. K. GUM, JR., J. A. LANG, T. H. LIU and R. D. BROWN, JR. Cellulases. *In: Food-related Enzymes*, J. R. Whitaker (ed.). Advances in Chemistry Series. Washington, D.C., American Chemistry Society, 1974.

16. ERSHOFF, B. H. Antitoxic effects of plant fiber. *Am. J. Clin. Nutr.* **27**:1395, 1974.

17. FALLAT, A. Personal communication: Virkotype Division, Lawter Chemicals, 111 Rock Ave., Plainfield, New Jersey 07060, 1975.

18. FORBES, R. M., and T. S. HAMILTON. The utilization of certain cellulosic materials by swine. *J. Anim. Sci.* **11**:480, 1952.

19. FOWLER, R. S. Personal communication: Chicago Dietetic Supply, Inc., 405 E. Shawmut Ave., La Grange, Illinois 60525, 1975.

20. FRAWLEY, J. P., A. K. WEIBE and E. D. KLUG. Studies on the gastrointestinal absorption of purified sodium carboxymethylcellulose. *Food Cosmet. Toxicol.* **2**:539, 1964.

21. GASCOIGNE, J. A., and M. M. GASCOIGNE. *Biological Degradation of Cellulose*. London, Butterworths, 1960, p. 9.

22. GLICKSMAN, M. Carbohydrates for fabricated foods. *In: Fabricated Foods*, E. E. Inglett (ed.). Westport, Connecticut, AVI Publishing Co., 1975.

23. GREENFIELD, H., and G. M. BRIGGS. Nutritional methodology in metabolic research with rats. *Annu. Rev. Biochem.* **40**:549, 1971.

24. GREENSTEIN, J. P., S. M. BIRNBAUM, M. WINITZ and M. C. OTEY. Quantitative nutritional studies with water soluble, chemically defined diets. I. Growth, reproduction and lactation in rats. *Arch. Biochem. Biophys.* **72**:396, 1957.

25. GRIMINGER, P., H. M. SCOTT and R. M. FORBES. Dietary bulk and amino acid requirements. *J. Nutr.* **62**(1):61, 1957.

26. GUISTI, G. The digestibility by chickens of the cellulose of sawdust. *Clin Vet.* **62**:283, 1939.

27. HAGEN, P., and K. W. ROBINSON. The production and absorption of volatile fatty acids in the intestine of the guinea pig. *Aust. J. Exp. Biol. Med. Sci.* **31**:99, 1953.

28. HANSEN, R. G., R. M. FORBES and D. M. CARLSON. *A Review of the Carbohydrate Constituents of Roughages*. Bulletin 634 of the Agricultural Experiment Station of the University of Illinois; and Publication No. 88 of the North Central Region USDA, Urbana, Illinois, 1958.
29. HEINICKE, H. R., and C. A. ELVEHJEM. Effect of high levels of fat, lactose and type of bulk in guinea pig diets. *Proc. Soc. Exp. Biol. Med.* **90**:70, 1955.
30. HEININKEL, H., S. LINDVALL and P. REIZENSTEIN. Gastrointestinal activity of cellulose in man after oral administration. *Gastroenterologia* **93**(2):69, 1960.
31. HOOVER, W. H., and S. D. CLARKE. Fiber digestion in the beaver. *J. Nutr.* **102**:9, 1972.
32. HOOVER, W. H., and R. N. HEITMANN. Effect of dietary fiber level on weight gain, cecal volume and volatile fatty acid production in rabbits. *J. Nutr.* **102**:375, 1972.
33. KEY, J. E., JR., P. J. VAN SOEST and E. P. YOUNG. Effect of increasing dietary cell wall content on the digestibility of cellulose and hemicellulose in swine and rats. *J. Anim. Sci.* **31**:1172, 1970.
34. KING, J. A. Social relations of the domestic guinea pig living under seminatural conditions. *Ecology* **37**:221, 1956.
35. KING, K. W. Fragmentation during enzymic degradation of cellulose. *Biochem. Biophys. Res. Commun.* **24**:295, 1966.
36. LANG, J. A. The purification of C_1 components from the cellulase complex of *Trichoderma viride*. Ph.D. dissertation, Virginia Polytechnic Institute and State University, 1970.
37. LIU, T. H., and K. W. KING. Fragmentation during enzymic degradation of cellulose. *Arch. Biochem. Biophys.* **120**:462, 1967.
38. LOESCHKE, K., E. UHLICH and R. HALBACH. Cecal enlargement combined with sodium transport stimulation in rats fed polyethylene glycol (36966). *Proc. Soc. Exp. Biol. Med.* **102**:96, 1973.
39. LOFGREN, P. A., P. S. REYES, K. W. MCNUTT, G. M. BRIGGS and G. O. KOHLER. Unidentified growth factor(s) present in alfalfa water-solubles for young guinea pigs. *Proc. Soc. Exp. Biol. Med.* **147**:331, 1974.
40. LOOSLI, J. K., B. L. RICHARDS, JR., L. A. MAYNARD and L. M. MASSEY. The effect of sulfur dioxide on the nutritive value of alfalfa hay. *Cornell U. Agr. Exp. Sta. Memoir* **227**:3, 1939.
41. LUBBOCK, D. M., W. THOMSON and R. C. GARRY. Epithelial overgrowths and diverticula in the gut of rats fed on a human diets. *Br. Med. J.* 1252, 1937.
42. MANLEY, R. ST. J. Fine structure of native cellulose microfibrils. *Nature* **204**:1155, 1964.
43. MCLAREN, A. D. Enzyme reactions in structurally restricted systems. *Enzymologia* **26**(1):1, 1963.
44. *Merck Index*, 8th ed., P. G. Stecher, (ed.). Rahway, New Jersey, Merck and Co., 1968.
45. MORSE, E. E. Personal communication: Berlin-Gorham Division, Brown Co., 650 Main St., Berlin, New Hampshire 03570, 1975.
46. MOSS, R., and J. A. PARKINSON. The digestion of heather (*Calluna vulgaris*) by red grouse (*Lagopus lagopus scoticus*). *Br. J. Nutr.* **27**:285, 1972.
47. NATIONAL ACADEMY OF SCIENCES. *Atlas of Nutritional Data on United States and Canadian Feeds*, Publication 1919. Washington, D.C., 1971.
48. NATIONAL RESEARCH COUNCIL. *Nutrient Requirements of Laboratory Animals*, 2nd rev. ed. Washington, D.C., National Academy of Sciences, 1972.
49. NATIONAL RESEARCH COUNCIL. *Toxicants Occurring Naturally in Foods*. Washington, D.C., National Academy of Sciences, 1973.
50. PETERSON, A. D., and B. R. BAUMGARDT. Food and energy intake of rats fed diets varying in energy concentration and density. *J. Nutr.* **101**:1057, 1971.
51. RAUTELA, G. S., and K. W. KING. Significance of the crystal structure of cellulose in the production and action of cellulase. *Arch. Biochem. Biophys.* **123**:589, 1968.
52. REESE, E. T., R. G. H. SIU and H. S. LEVINSON. The biological degradation of soluble cellulose derivatives and its relationship to the mechanism of cellulose hydrolysis. *J. Bacteriol.* **59**:485, 1950.

53. REID, M. E., and G. M. BRIGGS. Development of semisynthetic diet for young guinea pigs. *J. Nutr.* **51**:341, 1953.
54. RIECKEN, E. O., R. H. DOWLING, C. C. BOOTH and A. G. E. PEARSE. Histochemical changes in the rat small intestine associated with enhanced absorption after high-bulk feeding. *Enzyme* **5**(4):231, 1965.
55. SALLEY, J. J., and W. F. BRYSON. Vitamin A deficiency in the hamster. *J. Dent. Res.* **36**:935, 1957.
56. SHELTON, D. C. Personal communication: Ralston Purina Co., Checkerboard Square, St. Louis, Missouri 63188, 1975.
57. SHENK, J. S., F. C. ELLIOTT and J. W. THOMAS. Meadow vole nutrition studies with semisynthetic diets. *J. Nutr.* **100**:1437, 1970.
58. SHURPALEKAR, K. S., O. E. SUNDARALLI and M. NARAYANARA. Effect of inclusion of extraneous cellulose on the gastric emptying time in adult albino rats. *Can. J. Physiol. Pharmacol.* **47**(4):387, 1969.
59. SLADE, L. M., and H. F. HINTZ. Comparison of digestion in horses, ponies, rabbits and guinea pigs. *J. Anim. Sci.* **28**:842, 1969.
60. SPILLER, G. A., and R. J. AMEN. Plant fibers in nutrition: Need for better nomenclature. Letter to the editor. *Am. J. Clin. Nutr.* **28**(7):674, 1975.
61. VANDERHORST, P. J. Personal communication: Patents and contracts, E. I. DuPont, Wilminton, Delaware 19898, 1975.
62. VAN SOEST, P. J. Nonnutritive residues: A system of analysis for the replacement of crude fiber. *J. Assoc. Off. Agric. Chem.* **49**(3):546, 1966.
63. VAN SOEST, P. J., and R. H. WINE. Use of detergents in the analysis of fibrous feeds. IV. Determination of plant cell-wall constituents. *J. Assoc. Off. Agric. Chem.* **50**(1):50, 1967.
64. WAIBEL, P. E., B. S. RAO, K. E. DUNKELGOD, F. J. SICCARDI and B. S. POMEROY. A chemically defined diet for the chick. *J. Nutr.* **88**:131, 1966.
65. WATT, B. K., and A. L. MERRILL, (eds.). *Composition of Foods*. Agriculture Handbook No. Eight, rev., 1963. Washington, D.C., USDA, 1963.
66. WEBSTER'S NEW WORLD DICTIONARY of the American Language, College Edition, J. H. Friend and D. B. Guralnik (eds.). Cleveland, Ohio, World Publishing Co., 1953.
67. WELLS, A. F., and B. H. ERSHOFF. Beneficial effect of pectin in prevention of hypercholesterolemia and increased liver cholesterol in cholesterol-fed rats. *J. Nutr.* **74**:87, 1961.
68. WISE, L. E. Hemicelluloses. *In: Wood Chemistry*, 2d ed., L. E. Wise and E. C. Jahn (eds.). New York, Reinhold, 1952, Vol. I, Chap. 10, p. 369.
69. WOOLLEY, D. W., and H. SPRINCE. The nature of some new dietary factors required by guinea pigs. *J. Biol. Chem.* **157**:447, 1945.
70. YANG, M. G., K. MANOHARAN and A. K. YOUNG. Influence and degradation of dietary cellulose in cecum of rats. *J. Nutr.* **97**:260, 1969.
71. ZIEGELMEYER, W., A. COLUMBUS, W. KLAUSCH and R. WIESKE. The influences of carboxymethylcellulose on digestion in rats, dogs and humans. *Arch Tierernahr* **2**:35, 1951.

7

Dietary Fiber and Lipid Metabolism

Jon A. Story and David Kritchevsky

I. Introduction

Extensive research on the effects of various nutrients on health and disease in man have included numerous investigations into the influence of various types of fat on atherosclerotic heart disease. Although most studies have centered on dietary fat, the effects of carbohydrates, proteins, and trace minerals have also been investigated. In addition, dietary fat and carbohydrate have been studied in connection with gallbladder disease, and protein malnutrition in relation to mental development. There appears to be a rationale for studying these dietary ingredients since they are metabolized and further utilized and could, therefore, play an active role in the etiology of disease.

Evidence is now accumulating that a relatively nonmetabolized dietary component, fiber, may also be of importance. Fiber is an all-inclusive name for a variety of nonnutritive substances. Eventually these substances will have to be analyzed and definitively classified, but for the purposes of this presentation, coming, as it were, at the dawn of this era in nutrition research, we will refer to them all by the generic name, fiber.

Since this exposition deals with the effects of fiber on lipid metabolism, a brief overview of the subject of cholesterol metabolism may be in order. Cholesterol is a monohydric sterol found in every cell in the body. It is present in large amounts in nervous tissue and brain. It is also present in fair quantity in skin,

JON A. STORY • Research Scientist, Wistar Institute of Anatomy and Biology, Philadelphia, Pennsylvania.   DAVID KRITCHEVSKY • Associate Director, Wistar Institute of Anatomy and Biology; Wistar Professor of Biochemistry, University of Pennsylvania, School of Veterinary Medicine, Philadelphia, Pennsylvania.

adrenals, and liver. In the blood, the amounts of cholesterol and its esters are about the same as in liver, but they exist in different ratios. Cholesterol is transported in the blood as part of a lipoprotein complex.

The blood lipoproteins may be subfractionated by density or electrophoretic mobility. Cholesterol and its esters occur in all the lipoprotein fractions but are found primarily in the low-density lipoproteins. Esterified cholesterol is normally found in the blood and in tissues that convert cholesterol to other biologically active substances. Those tissues are liver, adrenals, gonads, and corpus luteum.

The synthesis of cholesterol takes place primarily in liver and intestine. All 27 carbon atoms of cholesterol arise from the two carbons of acetate (15 from the methyl group and 12 from the carboxyl). The obligatory steps in the biosynthetic pathway include the reduction of hydroxymethylglutaryl-CoA to mevalonic acid and the cyclization of the $C_{30}H_{50}$ hydrocarbon, squalene.

The major catabolic products of cholesterol are the bile acids. The conversion of cholesterol to bile acids (which is summarized in Figure 1) involves inversion of hydroxyl group at C-3, insertion of hydroxyl groups at position 7 or 12 or both, and scission of the side chain. The rate-limiting step in bile acid formation appears to be the 7α-hydroxylation of cholesterol. The primary bile acids are $3\alpha,7\alpha,12\alpha$-trihydroxycoprostanoic acid (cholic) and $3\alpha,7\alpha$-dihydroxycoprostanoic acid (chenodeoxycholic); intestinal bacteria dehydroxylate these bile acids at the 7 position to yield the secondary bile acids, deoxycholic acid and lithocholic acid, respectively. In the bile and blood, the bile acids circulate as amides (called conjugates for historical reasons) of taurine (the taurocholanoic acids) or glycine (glycocholanic acids).

The bile contains cholesterol, bile salts, and phospholipids (primarily lecithin) in a relatively delicate balance—any substantial change in the proportions of these three components may lead to gallstone formation.

This brief summary has been treated fully in many biochemistry and physiology texts and published symposia.[13,28,30,41,44]

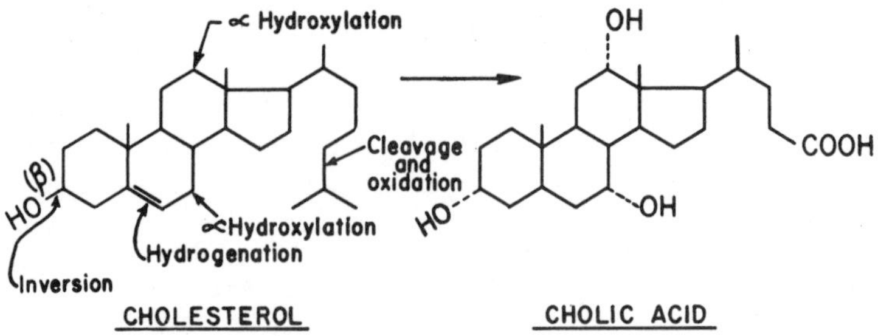

Figure 1. Steps involved in the conversion of cholesterol to cholic acid.

II. Early Studies

Recent epidemiological observations[7,8,12,53,55] that indicated that the poor black in Africa who subsists on a high-fiber diet is free of many of the diseases of the Western world, including heart disease and gallstones, stimulated interest in the relation between fiber and lipid metabolism. However, interest in the effects of various nonnutritive substances is not new; in the ongoing search for the dietary clue to susceptibility or resistance to coronary disease, many workers have touched on this subject.

One reflection of lipid metabolism is the level of circulating cholesterol. It is relatively simple to obtain a blood sample and analyze it for cholesterol and other lipids. In man, it is the only way, short of biopsy. There have been many studies of the effects of various polysaccharides (primarily pectin) on blood lipid levels.

The effects of pectin in animals are summarized in Table I. In general,

TABLE I. Effects of Pectin and Other Polysaccharides
on Serum and Liver Cholesterol Levels in Animals

Species	Substance	%	Duration (weeks)	Cholesterol[a] Serum	Liver	Comment[b]	References
Rat	Pectin	5	4	D	D	–	56
Rat	Pectin	5–10	4	D	D	–	17
Rat	Vegetable gums	10	4	D	D	–	17
Pig	Pectin	5	12	I	I	1	19
Rat	Pectin	5	4	D	D	–	40
Chick	Vegetable gums	1/4–10	4	D	–	–	18
Chick	Pectin	3	3	D	–	–	20
Chick	Cereals	43	3	D	–	–	20
Rat	Pectin	5–10	4	N–D	D	–	48
Rat	Guar gum	5–10	4	N–D	D	–	48
Rat	Pectin	18	4	N	–	–	25
Rat	Pectin	5	4	D	D	2	1
Rat	Polysaccharides	5	2	V	V	3	27
Rat	Pectin	3	3	D	–	4	45
Rat	Starches	56	16	D	D	5	54
Rabbit	Pectin	5	5	D	–	6	6

[a]D = decrease; I = increase; N = no effect; V = variable.
[b]1 Serum triglycerides rose but liver triglycerides fell.
2 Tomato pectin and citrus pectin had similar effects.
3 Pectin and carboxymethylcellulose caused decreases, alginic acid and cellulose caused increases.
4 Absorption of vitamin A was impaired.
5 Aortic total cholesterol decreased.
6 Cholesterol levels decreased in absence of dietary cholesterol, increased in its presence.

pectin and other polysaccharides are hypocholesterolemic, but a few of these substances do not show this effect,[27,48] and even pectin is not uniformly active in all species.[6,19] Leveille and Sauberlich[40] suggested that the mechanism of action of pectin was inhibition of cholesterol absorption and increase of fecal bile acid excretion. Beher and Casazza[4] have reported that psyllium hydrocolloid (4%) increases bile acid excretion in rats. This observation has also been made in humans.[21,49]

The comparisons of serum lipid levels in man is one of the few methods available for testing the effects of dietary fiber on lipid metabolism. Experiments of this type are becoming numerous and many are summarized in Table II. Although bran has been the most popular material tested, experiments reported to date seem to indicate it has little effect on serum cholesterol.[9,14,22,24] Pectin and guar gum, on the other hand, seem to be quite effective in lowering serum cholesterol.[24,26]

III. Effect of Fiber on Lipids and Atherosclerosis

A. Commercial Laboratory Rations vs. Semipurified Diets

Research on the influence of dietary fiber on the lipid metabolism of laboratory animals has provided a great deal of data relevant to the problem of atherosclerosis. Kritchevsky[29] called attention to the lack of agreement of results concerning the atherogenicity in rabbits of cholesterol-free diets containing saturated fat. A review of available data revealed that feeding saturated fat with a semipurified diet resulted in significant atherosclerosis, whereas the same fat fed

TABLE II. Effect of Bran or Other Fibers on Human Lipid Levels

Type of fiber	Level (g)	Number of subjects	Duration (weeks)	Cholesterol	Triglycerides	Comment[b]	Reference
Pectin	15	6	3	D	–	–	26
Cellulose	15	6	3	N	–	–	26
Bran	37	14	12	N	N	–	14
Bran	38	14	4–9	N	D	S	22
Bran	6–8	22	11	N	N	N	9
Wheat fiber	36	5	2	N	–	–	24
Pectin	36	7	2	D	–	–	24
Guar gum	36	7	2	D	–	–	24

Serum lipids[a] spans Cholesterol and Triglycerides.

[a]D = decrease; N = no effect.
[b]S = Significant drop in serum calcium levels. N = No effect on serum calcium levels.

TABLE III. Influence of Components of Commercial Diets
on Atherosclerosis in Rabbits[a]

Diet	Cholesterol ± SEM		Average atheromata		Aortic cholesterol (mg/g)	
	Serum (mg/dl)	Liver (mg/100g)	Arch	Thoracic	Free	Ester
Semipurified (SP)[b]	207 ± 36	1111 ± 137	1.2	0.5	4.68	2.28
Commercial rabbit feed (LR)	40 ± 9	318 ± 22	0.2	0.1	2.84	0.14
SP + LR fat (2%)[c]	249 ± 41	829 ± 77	1.1	0.7	5.06	1.53
Extracted LR + 14% HCNO	64 ± 9	599 ± 37	0.5	0.3	2.82	0.43
LR + 12% HCNO	35 ± 2	466 ± 24	0.3	0.2	2.06	0.03

[a] Source: Kritchevsky and Tepper.[35,36]
[b] Semipurified diet (25% casein, 40% dextrose, 15% cellulose, 14% hydrogenated coconut oil (HCNO), 5% salt mix and 1% vitamin mix).
[c] Fat extracted from LR (2%) added to SP at expense of HCNO (12%).

with commercial rabbit diet resulted in little or no atherosclerosis. He surmised that some other dietary component, e.g., carbohydrate, roughage, or trace minerals, present in the commercial diet, must be modifying the effects of the fat. To ascertain the possibility that the small amount of unsaturated fat present in the commercial rabbit diet (iodine value, 115–120) was exerting a protective effect, Kritchevsky and Tepper[35,36] fed saturated fat with the residue remaining after ether extraction of commercial rabbit diet. These rabbits were compared with groups fed the same fat with normal commercial rabbit diet and a semipurified diet. The residue remaining after ether extraction of commercial rabbit diet, augumented with saturated fat and vitamins, exerted a protective effect, as compared to the semipurified diet (Table III). These experiments proved that non-lipid components of the commercial rabbit ration were responsible for its beneficial effect vis-à-vis saturated-fat-induced atherosclerosis.

B. Specific Types of Fiber (Wheat Straw and Alfalfa)

Moore[43] compared the effects of type of fiber in a semipurified, cholesterol-free diet on atherogenesis. Serum cholesterol levels and degree of atherosis were lower when wheat straw or peat were substituted for cellulose or Cellophane Spangles® in the diets (Table IV).

Cookson et al.[10] observed a similar hypocholesteremic effect with alfalfa (Table V). Rabbits given large oral doses of cholesterol (600 mg/day) maintained normal serum cholesterol levels and remained relatively free from atherosclerotic lesions when large amounts of alfalfa (90%) were included in the diet. They

TABLE IV. Effects of Wheat Straw and Peat on Plasma
Cholesterol and Aortic Atherosis[a]

Source of fiber[b]	Plasma cholesterol (mg/dl)	Degree of aortic atherosis[c]
Wheat straw	114 ± 12	12.7 ± 2.9[e]
Solka floc[®d]	133 ± 10	20.8 ± 2.9[f,g]
Cellophane[®]	216 ± 14	37.5 ± 6.8[e,f,h]
Cellophane[®] (14%) + peat (5%)	141 ± 12	10.7 ± 2.0[g,h]

[a]Source: Moore.[43]
[b]Basal semipurified diet containing 16.3% wheat starch, 10% sucrose, 25% casein, 1% methyl cellulose, 9.7% salt and vitamin mixtures, 20% butter fat, and 19% fiber.
[c]Arbitrary units.
[d]Cellulose.
[e–h]Any two means with same superscript are significantly different (P<0.05).

TABLE V. Inhibition of Cholesteremia by Alfalfa[a]

Diet	Duration (weeks)	Serum cholesterol (mg/dl)
Basal diet (B)[b] + cholesterol (C)[c]	10	1345 ± 212
BC + alfalfa[d]	10	45 ± 6
BC + alfalfa[e]	1–8	57 ± 9
BC[e]	9–20	954 ± 257
BC + alfalfa[e]	21–30	111 ± 35

[a]Source: Cookson et al.[10]
[b]35% oats, 45% wheat, 10% soybean meal, 3% vitamin and salt mix, 5% beet molasses, 2% brewers' yeast.
[c]0.6 g cholesterol/day orally.
[d]Fed as 9 parts alfalfa to 1 part basal diet.
[e]Same animals fed diets as before.

theorized that alfalfa interfered with cholesterol absorption. In a subsequent experiment, Horlick et al.[23] observed large increases in the excretion of neutral steroids in rabbits fed large amounts of alfalfa and 300–600 mg/day cholesterol.

Cookson and Fedoroff[11] have shown a quantitative relationship between cholesterol fed and the amount of alfalfa needed to prevent hypercholesteremia. Barichello and Fedoroff[3] found further evidence that alfalfa was altering cholesterol absorption by comparing the quantity of alfalfa needed to prevent hypercholesteremia in cholesterol-fed (300 mg/day) rabbits with ileal bypass with the quantity required by intact animals. The quantity of alfalfa required was less in the animals that had the ileal bypass, thus supporting the theory of decreased cholesterol absorption in alfalfa-fed animals.

C. Isocaloric, Isogravic Diets

Similar evidence has been reported for rats. Kritchevsky et al.[31] fed rats isocaloric, isogravic diets in which 50% of the calories were derived from carbohydrate, fat, or protein; cellulose was the fiber source. These animals were compared with a group fed a commercial rat diet. Rats on the commercial rat diet had consistently lower serum and liver cholesterol levels. They also excreted more of a radioactive dose of cholesterol given orally 48 hr before termination of the experiment. More labeled cholesterol was recovered in the serum and liver of the rats fed the semipurified diets. In a subsequent experiment[38] similar isocaloric, isogravic diets were prepared with cellulose or alfalfa used as a source of fiber. In every case, the alfalfa-fed animals had a lower serum-plus-liver-cholesterol pool than the corresponding animals fed the same diet with cellulose as fiber. They also excreted more radioactively labeled neutral and acidic steroids than their cellulose-fed counterparts after being fed labeled cholesterol as in the earlier experiment (Table VI). Apparently the fiber, alfalfa, interferes with the absorption of cholesterol as evidenced by the increased excretion of cholesterol. The increase in acidic steroid excretion would suggest an increase in bile acid synthesis, which would be an important factor contributing to lowered cholesterol absorption.

TABLE VI. Effect of Isocaloric, Isogravic Diets on Serum plus Liver Cholesterol and Fecal Steroid Excretion[a]

Diets[b]		Serum plus liver cholesterol (mg)	Fecal steroids (dpm $\times 10^3$)	
50% of calories	Fiber source		Neutral	Acidic
Dextrose	Cellulose	58 ± 3	2.01	0.76
	Alfalfa	54 ± 1	3.64	0.65
Sucrose	Cellulose	43 ± 3	2.50	0.44
	Alfalfa	40 ± 1	4.17	0.83
Fat	Cellulose	41 ± 2	2.05	0.58
	Alfalfa	40 ± 1	3.74	0.78
Protein	Cellulose	46 ± 1	1.45	0.66
	Alfalfa	39 ± 1	2.49	1.37

[a]Source: Kritchevsky et al.[38]
[b]Isocaloric, isogravic diets (3.6 cal/g). All diets contained 5% salt mix and 1% vitamin mix with quantities of dextrose, sucrose, corn oil, casein, and cellulose or alfalfa adjusted to equal the percentage contribution mentioned.

D. Semipurified Diets in Nonhuman Primates

Kritchevsky et al.[32] fed baboons a semipurified diet in which the nonnutritive component was cellulose. The purpose of their experiment was to compare the effects of various carbohydrates on cholesteremia and atherosclerosis. They found that all the test animals had elevated serum lipids and increased aortic sudanophilia compared to the controls who were fed bread, vegetables, and fruit. Analysis of the biliary bile acid spectra of these baboons showed that the ratio of primary to secondary bile acids was lower in all the animals on the semipurified regimen. The results suggested a decreased synthesis of bile acids in baboons fed a semipurified diet and were borne out by data from biosynthetic studies. Similar results were obtained in rabbits fed semipurified or commercial diets.[37]

Portman and Murphy[47] previously found that feeding rats a semipurified diet containing no fiber greatly reduced pool size and increased the half-life of cholic acid when compared to chow-fed animals. Addition of Celluflour® returned these variables to the range of the chow-fed rats. Portman[46] has reviewed these and similar experiments. Kyd and Bouchier,[39] in experiments on gallstone production in rabbits, also concluded that bile acid synthesis was reduced in animals fed semipurified diets.

IV. Possible Mechanisms of Action

A. Binding of Bile Acids and Bile Salts

These animal experiments suggest many possible mechanisms of action to explain the influence that fiber exerts on lipid metabolism. Most prominent is that of the influence on cholesterol absorption and on bile acid synthesis and excretion. The two may be related.

Eastwood and Boyd[15] found that significant quantities of bile acids were bound to nonabsorbable materials in the small intestine. In further experiments[16] it was found that this binding of bile salts by dietary material could be duplicated *in vitro* and that lignin, a constituent of most types of fiber, actively bound bile salts from phosphate buffer solutions.

This information, added to earlier experimental evidence, suggested the possibility that fiber inhibited cholesterol absorption by binding bile salts. Such binding could result in a failure in micellar formation, essential for cholesterol absorption, and this, in turn, would increase bile acid excretion, effecting an increase in bile acid synthesis to replace the lost bile salts. Both events would drain cholesterol pools.

TABLE VII. Binding of Sodium Taurocholate by Various Types of Fiber[a]

| Binding substance | Sodium taurocholate bound (μmoles) | |
	Expt. 1[b]	Expt. 2[c]
Alfalfa	16.9 ± 0.7	–
Wheat straw	1.8 ± 0.8	
Sugar cane pulp	1.2 ± 0.5	8.9 ± 1.7[d]
Sugar beet pulp	1.0 ± 0.4	6.6 ± 1.2[d]
Bran	0.7 ± 0.7	7.3 ± 1.0[d]
Oat hulls	0.4 ± 0.6	8.2 ± 2.0[d]
Cellophane Spangles®	0.4 ± 0.7	1.4 ± 1.0
Cellulose	0.5 ± 0.5	2.4 ± 1.8

[a]Source: Kritchevsky and Story.[33]
[b]40 mg binding substance; 100 μmoles sodium taurocholate; incubated in 0.15 M NaCl for 1 hr at 37°C.
[c]100 mg binding substance; 240 μmoles sodium taurocholate; incubated as above.
[d]Significantly more bound in Experiment 2 as compared to Experiment 1 (P <0.01).

Kritchevsky and Story[33,50] verified the binding of bile salts from the saline solutions by several types of fiber commonly found in animal diets (Table VII). Purified types of fiber seemed unable to bind bile salts. Many of the substances tested showed increased binding of taurocholate when their concentration was increased, but cellulose and Cellophane Spangles® did not exhibit this property. This behavior may explain the differences observed between their hypocholesteremic properties. Alfalfa was found to bind approximately 20% of the amount of sodium taurocholate bound by a commercially available bile acid sequestrant, cholestyramine. Many types of foodstuffs were found to bind sodium taurocholate as seen in Table VIII.[51]

Others have also noted the ability of various dietary substances to bind bile salts. Balmer and Zilversmit[2] demonstrated the ability of several types of fiber, as well as cholestyramine, to bind sodium taurocholate and cholesterol from micellar solutions. Birkner and Kern[5] found significant binding of sodium glycocholate and sodium chenodeoxycholate by hemicellulose and residues of various foodstuffs, such as apple, celery, lettuce, potato, and string bean.

Kritchevsky and Story[34] have pointed out the shortcomings of equating the binding capacity of a certain type of fiber for sodium taurocholate with its ability to bind all bile acids. Tables IX and X compare the binding capacity of several types of fiber for the major bile acids and their conjugates. Lignin seems to have a greater overall capacity for binding than any of the other types of fiber. Cellulose seems unable to bind any type of bile acid, again indicating its lack of effect on lipid metabolism in the intact animal. Bran, which binds relatively little taurocholate, binds appreciably greater amounts of the other bile salts. Alfalfa also binds less taurocholate than any of the other bile salts tested. These findings may be of importance in the light of the fact that the taurine/glycine ratio in

TABLE VIII. Binding of Sodium Taurocholate by Various Foodstuffs[a]

Foodstuff	Sodium taurocholate bound[b]
Curry powder	1.96
Cloves	1.48
Lyophilized lettuce	1.40
Dried parsley	1.30
Oregano	1.30
Chili powder	1.10
Alfalfa	1.00
Thyme	0.90
Paprika	0.90
Cinnamon	0.74
Lyophilized cabbage	0.63
Dried celery	0.32

[a]Source: Story and Kritchevsky.[51]
[b]100 mg foodstuff; 100 μmoles sodium taurocholate; incubated in 0.15 M NaCl for 1 hr at 37°C. Alfalfa = 1.00.

TABLE IX. Comparison of *in Vitro*
Binding of Bile Acids by Several Types of Fiber

Acid	Alfalfa	Bran	Cellulose	Lignin
Cholic acid	1.00[a]	0.51	0.15	2.20
Chenodeoxycholic acid	1.25	0.92	0.10	1.17
Deoxycholic acid	0.52	0.27	0.01	0.87

[a]50 mg binding substance; 50 μmoles bile acid (sodium salt); incubated in phosphate buffer (pH 7.0) for 2 hr at 37°C. Alfalfa bound 19.9% of the sodium cholate and has been taken as 1.00.

TABLE X. Comparison of *in Vitro*
Binding of Bile Acid Conjugates by Several Types of Fiber

Bile acid conjugate	Alfalfa	Bran	Cellulose	Lignin
Taurocholate	1.00[a]	0.20	0.14	3.20
Glycocholate	1.67	0.55	0.17	3.26
Taurochenodeoxycholate	2.19	1.42	0.00	3.68
Glycochenodeoxycholate	2.16	3.10	0.03	3.65
Taurodeoxycholate	1.65	0.49	0.10	4.48
Glycodeoxycholate	4.03	1.13	0.68	7.62

[a]50 mg binding substance; 50 μmoles bile salt; incubated in phosphate buffer (pH 7.0) for 2 hr at 37°C. Alfalfa bound 6.9% of the sodium taurocholate and has been taken as 1.00.

human bile acid conjugates is about 1:3 or 1:4. The differences in binding of bile conjugates have been observed by other investigators.[2,4]

B. Water-Holding and Cation Exchange

Many fiber-containing materials possess, in addition to their bile salt-binding capacity, physical characteristics that could influence lipid metabolism. The water-holding capacity of fiber, which results in a great increase in total fecal output, could influence the absorption of cholesterol and bile acids. McConnell et al.[42] have measured the water-holding and ion-exchange capacities of many fiber-containing materials, including turnip, celery, carrot, and apple (Table XI). They theorize that the materials in the large intestine may act as a physiological chromatography column, whose gel-filtration and ion-exchange properties could influence bile acid absorption and, hence, metabolism.

It is apparent that the mechanisms involved in dietary effects of fiber on lipid metabolism are complex. All of the factors mentioned may play a part in the total effect any particular type of fiber exerts. The effects discussed above have been largely physical. The influence of amount and type of fiber on the cellular metabolism within the intestinal tract has not been investigated to any extent but could also contribute to the effects observed in animals and man.

V. Summary and Conclusions

Investigation into the effects of dietary fiber on lipid metabolism is relatively recent. The accumulated data indicate a general hypolipidemic and antiatherogenic effect. However, the type of fiber used is the determining factor; hence, fiber cannot be considered to be a uniform, generic, dietary material. To sort out its effects, more must be learned about the composition of each fiber, instituting a system of subclassification with more precise data on the mode of action of each.

Translation of present laboratory research into recommendations for the clinical application of nonnutritive fibers, however, must await thorough testing. Some adverse effects of excessive consumption of fiber have been noted, for instance, overuse of fiber has been indicated as a causitive agent in sigmoid volvulus,[52] a twisting of the sigmoid colon. This condition is much more prevalent in populations that consume large quantities of dietary fiber.

Thus fiber, like all dietary constituents, must be consumed moderately. A well-balanced diet including an adequate supply of fiber-containing foods would

TABLE XI. Water-Holding Capacity and Cation-Exchange
Capacity of Several Vegetables[a]

Vegetable	Water-holding capacity[b]	Cation-exchange capacity[c]
Apple	12.1	1.9
Bran	3.0	–
Carrot	23.4	2.4
Celery	19.2	1.5
Green bean	8.1	1.4
Lettuce	23.7	3.1
Potato	2.0	0.4
Tomato	10.8	1.0
Turnip	9.0	2.3

[a]Source: McConnel et al.[42]
[b]g water/g acetone dried powder.
[c]Meq/g acetone dried powder.

appear to be of some benefit in controlling hyperlipidemia and its sequelae. Any pharmacological use of fiber still needs thorough testing and evaluation of its effect on the overall physiological condition of humans.

ACKNOWLEDGMENTS

This research is supported, in part, by U.S. Public Health Service Research grants HL-03299 and HL-05209 and Research Career Award HL-0734 from the National Heart and Lung Institute and RR-05540 from the Division of Research Resources, and by grants-in-aid from the National Dairy Council and the National Live Stock and Meat Board.

References

1. ANDERSON, T. A., and R. D. BOWMAN. Comparative cholesterol-lowering activity of citrus and tomato pectin. Proc. Soc. Exp. Biol. Med. 130:665, 1969.
2. BALMER, J., and D. B. ZILVERSMIT. Effects of dietary roughage on cholesterol absorption, cholesterol turnover and steroid excretion in the rat. J. Nutr. 104:1319, 1974.
3. BARICHELLO, A. W., and S. FEDOROFF. Effect of ileal bypass and alfalfa on hypercholesterolaemia. Br. J. Exp. Pathol. 52:81, 1971.
4. BEHER, W. T., and K. K. CASAZZA. Effects of psyllium hydrocolloid on bile acid metabolism in normal and hypophysectomized rats. Proc. Soc. Exp. Biol. Med. 136:253, 1971.
5. BIRKNER, H. J., and F. KERN, JR. In vitro adsorption of bile salts to food residues, salicylazosulfapyridine, and hemicellulose. Gastroenterology 67:237, 1974.

6. BORGMAN, R. F., and F. B. WARDLAW. Serum cholesterol and cholelithiasis in rabbits treated with pectin and cholestyramine. *Am. J. Vet. Res.* **35**:1445, 1974.

7. BURKITT, D. P. Epidemiology of large bowel disease: The role of fiber. *Proc. Nutr. Soc.* **32**:145, 1973.

8. CLEAVE, T. L. *The Saccharine Disease*. Bristol, John Wright and Sons, Ltd., 1974.

9. CONNELL, A. M., C. L. SMITH and M. SOMSEL. Absence of effect of bran on blood-lipids. *Lancet* **1**:496, 1975.

10. COOKSON, F. B., R. ALTSCHUL and S. FEDOROFF. The effects of alfalfa on serum cholesterol and in modifying or preventing cholesterol-induced atherosclerosis in rabbits. *J. Atheroscler. Res.* **7**:69, 1967.

11. COOKSON, F. B., and S. FEDOROFF. Quantitative relationships between administrated cholesterol and alfalfa required to prevent hypercholesterolaemia in rabbits. *Br. J. Exp. Pathol.* **49**:348, 1968.

12. CUMMINGS, J. H. Progress report: Dietary fiber. *Gut* **14**:69, 1973.

13. DANIELSSON, H. Mechanism of bile acid biosynthesis. *In: The Bile Acids*, Vol. 2, P. P. Nair and D. Kritchevsky (eds.). New York, Plenum, 1973, p. 1.

14. EASTWOOD, M. A. Dietary fiber and serum lipids. *Lancet* **2**:1222, 1969.

15. EASTWOOD, M. A., and G. S. BOYD. The distribution of bile salts along the small intestine of rats. *Biochim. Biophys. Acta* **137**:393, 1967.

16. EASTWOOD, M. A., and D. HAMILTON. Studies on the adsorption of bile salts to nonabsorbed components of diet. *Biochim. Biophys. Acta* **152**:165, 1968.

17. ERSHOFF, B. H., and A. F. WELLS. Effects of gum guar, locust bean gum and carrageenan on liver cholesterol of cholesterol-fed rats. *Proc. Soc. Exp. Biol. Med.* **110**:580, 1962.

18. FAHRENBACH, M. J., B. A. RICCARDI and W. C. GRANT. Hypocholesterolemic activity of mucilaginous polysaccharides in white leghorn cockerels. *Proc. Soc. Exp. Biol. Med.* **123**:321, 1966.

19. FAUSCH, H. D., and T. A. ANDERSON. Influence of citrus pectin feeding on lipid metabolism and body composition of swine. *J. Nutr.* **85**:145, 1965.

20. FISHER, H., and P. GRIMINGER. Cholesterol-lowering effects of certain grains and of oat fractions in the chick. *Proc. Soc. Exp. Biol. Med.* **126**:108, 1967.

21. FORMAN, D. T., J. E. GARVIN, J. E. FORESTNER and C. B. TAYLOR. Increased excretion of fecal bile acids by an oral hydrophilic colloid. *Proc. Soc. Exp. Biol. Med.* **127**:1060, 1968.

22. HEATON, K. W., and E. W. POMARE. Effect of bran on blood lipids and calcium. *Lancet* **1**:49, 1974.

23. HORLICK, L., F. B. COOKSON and S. FEDOROFF. Effect of alfalfa feeding on the excretion of fecal neutral sterols in the rabbit. *Circulation (Suppl. II)* **36**:18, 1967.

24. JENKINS, D. J. A., A. R. LEEDS, C. NEWTON and J. H. CUMMINGS. Effect of pectin, guar gum, and wheat fiber on serum cholesterol. *Lancet* **1**:1116, 1975.

25. KARVINEN, E., and M. MIETTINEN. Effect of apple and pectin diets on serum and liver cholesterol in rats. *Acta Physiol. Scand.* **72**:62, 1969.

26. KEYS, A., F. GRANDE and J. T. ANDERSON. Fiber and pectin in the diet and serum cholesterol concentration in man. *Proc. Soc. Exp. Biol. Med.* **106**:555, 1961.

27. KIRIYAMA, S., Y. OKAZAKI and A. YOSHIDA. Hypocholesterolemic effect of polysaccharides and polysaccharide-rich foodstuffs in cholesterol-fed rats. *J. Nutr.* **97**:382, 1969.

28. KRITCHEVSKY, D. *Cholesterol*. New York, Wiley, 1958.

29. KRITCHEVSKY, D. Experimental atherosclerosis in rabbits fed cholesterol-free diets. *J. Atheroscler. Res.* **4**:103, 1964.

30. KRITCHEVSKY, D. Biochemistry of steroids. *In: Lipids and Lipidoses*, G. Schettler (ed.). Berlin, Springer-Verlag, 1967, p. 66.

31. KRITCHEVSKY, D., R. P. CASEY and S. A. TEPPER. Isocaloric, isogravic diets in rats. II. Effect on cholesterol absorption and excretion. *Nutr. Rep. Int.* **7**:61, 1973.

32. KRITCHEVSKY, D., L. M. DAVIDSON, I. L. SHAPIRO, H. K. KIM, M. KITAGAWA, S. MALHOTRA, P. P. NAIR, T. B. CLARKSON, I. BERSOHN and P. A. D. WINTER. Lipid metabolism and experimental atherosclerosis in baboons: Influence of cholesterol-free, semisynthetic diets. *Am. J. Clin. Nutr.* **27**:29, 1974.

33. KRITCHEVSKY, D., and J. A. STORY. Binding of bile salts *in vitro* by nonnutritive fiber. *J. Nutr.* **104**:458, 1974.

34. KRITCHEVSKY, D., and J. A. STORY. *In vitro* binding of bile acids and bile salts. *Am. J. Clin. Nutr.* **28**:305, 1975.

35. KRITCHEVSKY, D., and S. A. TEPPER. Factors affecting atherosclerosis in rabbits fed cholesterol-free diets. *Life Sci.* **4**:1467, 1965.

36. KRITCHEVSKY, D., and S. A. TEPPER. Experimental atherosclerosis in rabbits fed cholesterol-free diets: Influence of chow components. *J. Atheroscler. Res.* **8**:357, 1968.

37. KRITCHEVSKY, D., S. A. TEPPER, H. K. KIM, D. E. MOSES and J. A. STORY. Experimental atherosclerosis in rabbits fed cholesterol-free diets. IV. Investigation into the source of cholesteremia. *Exp. Mol. Pathol.* **22**:11, 1975.

38. KRITCHEVSKY, D., S. A. TEPPER and J. A. STORY. Isocaloric, isogravic diets in rats. III. Effects of nonnutritive fiber (alfalfa and cellulose) on cholesterol metabolism. *Nutr. Rep. Int.* **9**:301, 1974.

39. KYD, P. A., and I. A. D. BOUCHIER. Cholesterol metabolism in rabbits with oleic acid-induced cholelithiasis. *Proc. Soc. Exp. Biol. Med.* **141**:846, 1972.

40. LEVEILLE, G. A., and H. E. SAUBERLICH. Mechanism of the cholesterol-depressing effect of pectin in the cholesterol-fed rat. *J. Nutr.* **88**:209, 1966.

41. MASORO, E. J. *Physiological Chemistry of Lipids in Mammals.* Philadelphia, Saunders, 1968.

42. MCCONNELL, A. A., M. A. EASTWOOD and W. D. MITCHELL. Physical characteristics of vegetable foodstuffs that could influence bowel function. *J. Sci. Food Agric.* **25**:1457, 1974.

43. MOORE, J. H. The effect of the type of roughage in the diet on plasma cholesterol levels and aortic atherosis in rabbits. *Br. J. Nutr.* **21**:207, 1967.

44. MOSBACH, E. H. Hepatic synthesis of bile acids. *Arch. Intern. Med.* **130**:478, 1972.

45. PHILLIPS, W. E. J., and R. L. BRIEN. Effect of pectin, a hypocholesterolemic polysaccharide, on vitamin A utilization in the rat. *J. Nutr.* **100**:289, 1970.

46. PORTMAN, O. W. Nutritional influences on the metabolism of bile acids. *Am. J. Clin. Nutr.* **8**:462, 1960.

47. PORTMAN, O. W., and P. MURPHY. Excretion of bile acids and β-hydroxysterols by rats. *Arch. Biochem. Biophys.* **76**:367, 1958.

48. RICCARDI, B. A., and M. J. FAHRENBACH. Effect of guar gum and pectin N.F. on serum and liver lipids of cholesterol-fed rats. *Proc. Soc. Exp. Biol. Med.* **124**:749, 1967.

49. STANELY, M., D. PAUL, D. GACKE and J. MURPHY. Comparative effects of cholestyramine, metamucil and cellulose on bile salt excretion in man. *Gastroenterology* **62**:816, 1972.

50. STORY, J. A., and D. KRITCHEVSKY. Binding of bile salts by nonnutritive fiber. *Fed. Proc.* **33**:663, 1974.

51. STORY, J. A., and D. KRITCHEVSKY. Binding of sodium taurocholate by various foodstuffs. *Nutr. Rep. Int.* **11**:161, 1975.

52. SUTCLIFFE, M. M. L. Volvulus of the sigmoid colon. *Br. J. Surg.* **55**:903, 1968.

53. TROWELL, H. Dietary fiber and coronary heart disease. *Eur. J. Clin. Biol. Res.* **17**:345, 1972.

54. VIJAYAGOPALAN, P., and P. A. KURUP. Effect of dietary starches on the serum, aorta and hepatic lipid levels in cholesterol-fed rats. *Atherosclerosis* **11**:257, 1970.

55. WALKER, A. R. P., B. F. WALKER, B. D. RICHARDSON and A. WOOLFORD. Appendicitis, fiber intake and bowel behavior in ethnic groups in South Africa. *Postgrad. Med. J.* **49**:243, 1973.

56. WELLS, A. F., and B. H. ERSHOFF. Beneficial effects of pectin in prevention of hypercholesterolemia and increase in liver cholesterol in cholesterol-fed rats. *J. Nutr.* **74**:87, 1961.

8

Dietary Fiber and Colon Function

W. D. Mitchell and Martin A. Eastwood

I. Introduction

In man the colon has several functions, two of the more important being its role
in conserving important intestinal contents and its role as a reservoir.

In addition to electrolytes, feces contain bacteria, unabsorbed dietary residues, shed epithelial cells, and organic and inorganic materials. An important
group of organic compounds found in feces are bile acids, and it is well established that the colon contributes to the enterohepatic circulation of bile acids.
The efficiency of the enterohepatic circulation of bile acids is such that approximately 3–5 g can be cycled through the intestine 6–10 times per day with only a
small amount (100–500 mg) lost in the feces. However, in situations where
there is terminal ileum dysfunction, and excessive amounts of bile acids pass into
the colon, then bile acids and bile acid conjugates affect colon function, especially the colonic absorption of electrolytes and water.[30,52,51,53] This disturbance
of electrolyte and water absorption has been observed especially in patients with
ileal resection.[38,78] These patients have profuse diarrhea, which has been shown
to be associated with the large amount of chenodeoxycholic acid, a primary bile
acid, which is present in their stool but which is not normally found in feces.[54] It
has been postulated that chenodeoxycholic acid inhibits colonic absorption of
Na^+ and thus inhibits the absorption of water.[55]

The absorption of water by the colon is necessary in the formation of a solid
concentrated stool from the liquid effluent which is passed from the ileum. This
conservation is made possible by the circular muscle of the large bowel contract-

W. D. MITCHELL ● Wolfson Gastrointestinal Laboratories, Western General Hospital, Edinburgh,
Scotland.   MARTIN A. EASTWOOD ● Consultant Physician, Wolfson Gastrointestinal
Laboratories, Department of Medicine, Western General Hospital, Edinburgh, Scotland.

ing and slowing the onward movement of stool. In doing this the bowel, particularly the descending colon, becomes segmented.

The activity of the colon may be regarded as being of two types: propulsive contractions or mass movements and nonpropulsive contractions which segment. The first activity will dictate the onward movement of intestinal contents, and the second will form part of the braking or controlling mechanism whereby continence is achieved. The colonic mass movements are involved in the transfer of intestinal contents from the right to the left colon. The pharmacology and physiology of the muscle masses involved in colonic motility are poorly understood, as are abnormalities in the colonic motor activities. However, it is possible that many of the problems in colonic disease are associated with excessive contractions of the muscles in the distal colon with a distal intestinal stasis and enhanced motor activity or intestinal pressure.

Colonic movement or pressure can be recorded in a number of ways: (1) radiologically, (2) by intraluminal pressure measurements using open-ended tubes, and (3) by following the movement of a bolus along the bowel. It is recognized that the constrictions that give barium enemas their typical haustrated appearance have three origins: (1) anatomical areas of narrowing; (2) bunching of the mucosa, possibly due to the contraction of the longitudinal muscle; and (3) contraction of the circular muscle of the colon.[11]

The motility of the colon is influenced by many factors both humoral and nervous. The thoracolumbar and pelvic nerves are known to influence distal colonic motility. Lesions of the central nervous system disrupt colonic motor function, and if the lumbosacral cord in man is destroyed, the segmenting motor activity in the lower colon is increased. The normal colon is also affected by humoral stimuli; e.g., food acts by increasing its segmented motor activity, and it is probable that this response is mediated by gastrointestinal hormones.[12] It is also postulated that some humoral material derived from the duodenum or upper small intestine may affect the sigmoid colon; e.g., cholecystokinin stimulates motor activity of the sigmoid in man whereas secretin inhibits it.[18]

It is evident, therefore, that the colon is not the simple organ that it was initially thought to be, and although man can live satisfactorily without a colon, it does not necessarily mean he is better off without it. The question is, how does diet, and more especially vegetable fiber, affect colonic function?

II. Effect of Fiber

A. Effect on Stool Weight

There has been considerable discussion recently on whether or not vegetable fiber is a necessary constituent of the diet.[16] A claim has been made for a causal relationship between a lack of fiber in the diet and a whole range of diseases, particularly diseases of the colon.[6] These diseases are found quite commonly in European and North American communities but are apparently very rare in Africans. It has been shown that the rural African eats a high-fiber-containing diet compared to modern Europeans and North Americans. A consequence of this is that the rural African has a well-filled colon and passes approximately 500 g of stool per day. In contrast, an individual who eats a low-fiber-containing diet passes a small stool of approximately 100 g/day.[8] If the colonic diseases that are suggested to be peculiar to the modern European and North American can be prevented by having a well-filled colon, then diets that can cause increased stool weight may be beneficial. However, the problem of eating a rural-African-type diet in Britain is considerable. Various substances, however, have been shown to cause increase in stool weight. One of these, cereal bran, has been investigated fairly extensively.

Bran has been used as a laxative from early times. Indeed, in 430 B.C. Hippocrates stated that "Wholemeal bread clears out the gut and passes through as excrement."[49] More recently in 1941 Fantus and Frankl[24] published a paper on "The Mode of Action of Bran." Their opening sentence, "The extensive use of bran as a laxative and in diets justifies an enquiry into its various properties," shows that even then various workers were aware that the laxative property of bran, although important, was not the only property that should be investigated. These early experiments produced findings very similar to those reported recently, especially with respect to the effect of bran on stool weight, i.e., "the most significant change in the stool following the addition of bran to an uncontrolled diet is a softening and an increase in stool weight." They also showed an increase in volatile fatty acids, but this increase was not proportional to the increase in stool weight; in addition, the proportion of acetic to butyric acid showed no consistent correlation with the laxative action as measured by an increase in stool weight.

In 1943 Streicher and Quirk[70] studied the use of bran for constipation. They found that of 135 patients they studied, 109 improved clinically and 26 remained unchanged when given bran. Of the 135 patients studied radiologically, 31 demonstrated delayed cecal emptying time before bran, and 16 of these patients showed delayed motility during bran, while the remaining 15 improved on bran. In addition, the transit time, or total emptying time of the gastrointestinal tract, as it was described, decreased in 27 out of 90 patients, while 63 patients showed a delayed total emptying time on bran.

It was thought that, although bran relieved constipation, it might cause increased irritation of the colon. Such irritation is implied by the term roughage commonly used to describe high-fiber-containing diets. However, work by Fantus et al. in 1941[26] showed that irritation as indicated by an increased quantity of mucus was no greater during the "bran" and "after-bran" periods than during the "pre-bran" period. Indeed, in these cases where bran therapy was effective, manifestations of increased mucus disappeared. In 1942 Hoppert and Clark[40] investigated the continued consumption of bran by six normal young men. Two ounces of bran daily over a 30-week period had a pronounced and beneficial effect of laxation, and the improved laxation during the 30-week period resulted in no loss in body weight in these subjects. They also found that bran given at the rate of 4 ounces per day showed no functional habituation and also no apparent diminution in its laxative properties. In fact, the 4 ounces of bran per day was well tolerated as a natural laxative. In an attempt to make bran more palatable, Hoppert and Clark in 1942[41] prepared muffins containing 2 ounces of bran. These were then consumed by eight normal young men on a basal diet low in crude fiber for a period of 4 days. This was followed by a period of 3 days during which each subject received daily a known number of bran muffins in addition to the basal diet. It was found that three or more muffins even for such short periods as this particular study produced a marked effect on the number and consistency of stools produced. The stools became soft and were usually passed without any discomfort. When only two muffins were taken there was no increase in the number of stools passed, but the stool became softer. More recently Eastwood et al.[23] showed that in a group of normal subjects the ingestion of approximately 16 g of cereal bran per day produced softer and considerably heavier stools. This increase in weight appears to be due to an increase in water content of the stool.

An important property of bran is that it can absorb three times its weight of water. From data on the water content of the stool in the subjects who have been fed bran, it appears that the increased fecal weight is due to the increase in water content of the stool. Therefore, it appears that the bulking action of bran acts by absorbing water present in the colon and produces a much softer and more bulky stool.

Various other compounds such as cellulose, hemicelluloses, and various mucilaginous gums are efficient bulking agents, and all have the common property with bran of being able to adsorb water. More importantly they are present to some extent in all fruit and vegetable fiber. The dietary approach to increase stool weight by altering vegetable and fruit intake is difficult because of the limited knowledge of which fruits and vegetables to prescribe. However, in 1930 the bulking action of the crude fiber and cellulose content in certain common foods was also under consideration and investigation. In 1936 Williams and Olmsted[77] produced a table of the ability of various plant sources to increase stool weight. They found that cotton seed wholes have the least effect, then (in increasing efficacy) cellulose, alfalfa-leaf meal, wheat, bran, canned peas, corn-germ meal, sugar-beet pulp, cabbage, carrots, and finally agar-agar. There was a 20-fold difference in capacity to bind water between cotton-seed wholes and agar-agar.

In 1945 Hoppert and Clark[42] carried out a preliminary study to determine the relative effects on laxation of bran and also certain fruits and vegetables, taken as supplements to a diet low in fiber. They showed that bran and cabbage had the most consistent and favorable laxative effect, whereas certain vegetables, e.g., lettuce, gave less pronounced effects. An attempt was made to obtain the coefficients of digestion, as they called it, with respect to crude fiber and cellulose in each of the fruits and vegetables examined. However, because of the limitations in the available analytical methods, the data obtained were inconsistent and no emphasis could be placed on the results. The study serves to show, however, that a dietary approach to alter bowel function was under investigation nearly 40 years ago.

The investigations of Hoppert and Clark[42] showed that there are probably several factors responsible for the laxative action of fiber-containing foods. The bulk-forming properties of food products were not always proportional to the crude fiber or their cellulose content. The investigators suggested that there was an optimum intake of bulk-forming foods of the bran class beyond which no added benefit could be obtained by increasing the intake. In the case of prepared bran, they suggested that the optimum intake was approximately 1 ounce per day, equivalent to 40 mg of crude fiber/kg of body weight. Stools of appreciable size and consistency were produced by certain natural regulators such as bran and cabbage, and there was no direct correlation between the moisture content and the ease of elimination under normal conditions. They found that diets deficient in fiber usually caused constipation as characterized by hard compact stool and rectal irritation. This was frequently accompanied by the appearance of blood flecks. They also give coefficients of digestibility of crude fiber and cellulose in foods from various sources. These coefficients range from low values for cereal products through intermediate values for vegetables and high values for fruits.

It was suggested by Hoppert and Clark that the laxative properties of foods should be evaluated on the basis of the physiological results produced by their ingestion and not only on the basis of their crude-fiber values. They suggest that until this has been done with at least common fiber-containing foods, there can be no logical basis for establishing a definite minimal daily intake during normal laxation.

The work of McConnell et al.[50] adds much to this very early article. It is now possible to tabulate fruits and vegetables as water adsorbers, e.g., turnip < potato < banana < cauliflower . . . < pea < winter cabbage < lettuce < . . . < apple < carrot < mango < bran.[50] Thus we are now in the position to suggest which fruits and vegetables should cause increased fecal excretion.

Another important characteristic in the water-holding-capacity of dietary fibers is particle size. The recent report by Kirwan et al.[45] on the effect of various types of bran on bowel function suggested that the particle size of the bran was important. They found that coarse bran was able to hold more water than the fine bran. Milling of the coarse bran to produce particles of 1 mm considerably lowered its water-holding capacity, whereas milling of the fine bran to 1 mm had little effect on its original water-holding capacity. It would appear that in the coarse bran the fiber matrix is arranged in such a way as to produce interstices that enable more water to penetrate them, whereas the milling of the bran appears to destroy this property. An analogy with sponges of different sizes and with different pore sizes is appropriate. In 1941 Fantus et al.[25] investigated the effect of particle size and produced a paper on the mode of action of bran with a title "The Influence of Size and Shape of Bran Particles." Unlike Kirwan et al.,[45] Fantus et al.[25] found no correlation between particle size of bran and the laxative action when measured by stool weight, and they suggested that the crude fiber prepared from bran was the active principle of bran.

Although vegetable and cereal fibers are known to adsorb water, the constituents of fiber responsible for this have not been conclusively indentified. It is probable that the water-holding property of fiber is due to the polysaccharide content. These compounds also have a cation-exchange capacity and will therefore adsorb sodium, potassium, calcium, and all positively charged elements. It is possible, therefore, that water will be adsorbed directly to the fiber and also indirectly as a result of the adsorption of sodium to the various vegetable dietary fibers.

The bulking action of dietary fiber will have the added effect of diluting colonic contents, e.g., bile acids, which according to Burkitt[7] should lower the risk of colonic cancer. In 1973 Eastwood et al.[23] fed 16 g cereal bran daily for 4 weeks to eight subjects and observed that the concentration of bile acids in the feces decreased. This was later confirmed by Findlay et al.[29] in normal subjects and in patients with diverticular disease.

B. Effect on Transit Time

Transit time is the time taken for the passage of material from the mouth to anus. It varies considerably from person to person, and even in an individual the transit time is not constant. There are many reports of people swallowing articles of all shapes and sizes. From the early nineteenth century[9] there is the report of a man who swallowed 4 inches of flute, which took 3 days to pass. In another case a man swallowed a fork, but the transit time was several weeks.

A variety of methods have been devised in an attempt to standardize transit time measurements; these have ranged from colored glass beads[1] to the more sophisticated radioisotope methods.[44] Various substances can influence colonic transit time. One of these substances is dietary fiber. Rural Africans who have a high fiber intake not only produce large bulky stools but are reported to have very fast transit times, of the order of 30 hr, in contrast to North Europeans and North

Americans, who are reported to have transit times in excess of 48 hr. Burkitt et al.[8] postulated a relationship between daily stool weight and transit time and showed that there was a logarithmic relationship, with short transit time being associated with high stool weight. Cereal bran, when given with the normal diet, not only causes an increase in stool weight but also causes a decrease in transit times. This has been shown in normal subjects and also in patients with prolonged transit time as in diverticular disease.

However, not all subjects shorten transit time when given bran. It has been shown[32] that when subjects with a rapid transit time (< 24 hr) are given bran, their transit time increases. This would suggest that there is some regulating mechanism or that there is an ideal ''bran'' transit time that will vary depending on the source of vegetable fiber in the diet. The mechanism whereby cereal bran or dietary fiber causes a decrease in the transit time through the colon is not adequately explained. However, it has been suggested that this decrease is probably due to filling of the colon with a more bulky, watery stool, which can pass through the colon much more easily with less straining, enabling the muscles to expel the feces much more readily.

The usual method of measuring transit time is the use of radiopaque markers, described by Hinton et al.[36] These have the disadvantage that they are particulate matter and are not dispersed homogeneously through the fecal mass. The precision with which Hinton's markers measure transit time has been criticized by Eastwood et al.[21] They suggested that differences in transit time of the order of 46 hr would be required to show a significant difference between individuals in the same trial, or 19 hr for comparison of the means of groups of six people or replicates, as compared with the statement of Hinton et al.[36] that ''two studies could give identical results or results differing by only 1 hour.''

Also, as the radiopaque markers are solid substances, they will behave like solid particles in the stool and not as liquids. Thus a misleading estimation of transit time will be obtained, as what is actually being measured is the transit time of solid material of the same specific gravity as the markers.

Specific gravity is quite important in the measurement of transit time, as was shown by Kirwan and Smith[44] when pellets differing from each other very slightly with respect to specific gravity (0.93, 1.10, and 1.35) passed at different rates through the gastrointestinal tract. In an earlier study Hoelzel[37] found that the transit of solid particles slowed down when the specific gravity increased from 1 to 2.6. Kirwan and Smith,[44] however, found that the transit of capsules of specific gravity 0.93 and 1.35 was faster than the capsules of specific gravity 1.1. They suggested that specific gravity of 1.1 was similar to that of gut contents and a capsule of this specific gravity would therefore be propelled at the same rate, whereas a lighter or heavier capsule would separate from the gut contents and take up a position close to the bowel wall and thus be more effectively and more rapidly excreted.

These findings raise the possibility that vegetable dietary fiber may affect transit time by altering the specific gravity of the colonic contents. The studies by Kirwan and Smith[44] and Waller[73] suggest that small-bowel transit time is relatively constant in normal subjects, patients with constipation, and patients with diarrhea, and that differences in gastrointestinal transit are due to changes in colonic transit. It may be that because of the large volume of water in the small intestine, changes in specific gravity that are due to unabsorbed solids will be minimal, but in the colon where the water content is markedly reduced, the presence of solid dietary fiber will influence the specific gravity of the colonic contents and hence the colonic transit time.

These observations raise the possibility of a differential rate of transit of solid particles of differing specific gravity as well as a differential rate of transit of solid and liquid phases in the gastrointestinal tract. Kirwan and Smith[44] suggest that this differential rate of particles through the gastrointestinal tract may be part of the reason why bran, which is known to increase stool weight, can alter intestinal transit time. In order to measure transit of both solid and liquid material, it is necessary to use solid and liquid markers. Two suitable markers are chromium sesquioxide, which is insoluble and marks the solid phase, and polyethylene glycol 4000, which is water-soluble and marks the liquid phase. Both these markers have the desirable property of being unabsorbed from the gastrointestinal tract. Chromium sesquioxide, however, has a disadvantage in that it is a compound of fairly high specific gravity, and in some cases the chromium sesquioxide tends to settle out and is therefore not homogeneously mixed throughout the fecal mass as would be desired from an ideal marker.

C. Effect on Solid and Liquid Phases in the Colon

It has been assumed that all phases present in the complex mixture of solids and liquids that consitute colonic contents are treated in a uniform manner.

In the proximal part of the colon there is a high proportion of water, which decreases as the contents move distally. The water in the proximal part can be considered to be in three phases (Figure 1, Chapter 4).

1. Molecular or gel water inherently part of the residue.[48]

2. Interstitial water is distributed throughout the interstices of the dietary residue and cannot be expressed from formed stool by centrifuging at 14,000 g.[27]

3. Free water lies freely in the bowel and is not particularly associated with the dietary residue. It is this free water that makes a stool diarrheal. This free water can be expressed on centrifugation at 14,000 g, but no free water can be expressed from a formed stool on centrifugation.[27]

Polyethylene glycol (PEG) 4000 is a nontoxic, water-soluble, unabsorbable polymer and is therefore suitable as a marker for studying the transit of the liquid phase in the gastrointestinal tract.[4,43,67,75] It will align itself with interstitial water

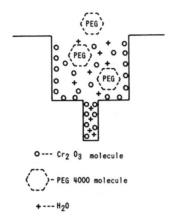

O --- Cr_2O_3 molecule

( ) - PEG 4000 molecule

+ --- H_2O

Figure 1. A diagramatic representation of pores in vegetable fiber showing the accessibility of small molecules, e.g., H_2O and Cr_2O_3, and the inaccessibility of large molecules, e.g., PEG 4000, into pores of different diameter.

and free water in the colon, but because of its molecular weight, it will not mark all of the gel water. Thus the pore sizes of the fiber matrix in the colon will determine the efficiency with which PEG 4000 will mark the gel water (Figure 1).

Chromium sesquioxide has been well used as a marker in studies of colonic function, and it is nontoxic, nonabsorbable, and insoluble in water.[46,74,17]

If it is assumed that in the gastrointestinal tract of the normal person the solid and liquid phases are processed in such a way that they become a gel prior to passing out as feces, it is reasonable to assume that the solid and liquid phases will pass through the gastrointestinal tract at the same rate. To test this hypothesis, it is necessary to give both markers simultaneously in equal doses and collect the stool specimens. This was investigated recently[28] in normal subjects, patients with ileal resection diarrhea, and diverticular disease patients. The results suggest that an equal rate of throughput of solid and liquid markers exists in the normals, whereas in the ileal resection diarrhea the liquid phase travels faster than the solid phase, and in the diverticular disease patients the solid phase travels faster than the liquid phase (Figure 2). Typical excretion patterns of these three groups are shown in Figure 3 a–c, from which it can be seen that the normals appear to reach equilibrium very quickly, whereas the other two groups would appear to take much longer.

The ability of cereal bran to adsorb an appreciable amount of water probably enables it to cause an increase in fecal weight, and it is possible that it could alter the rate of throughput of liquid and solid phases in the gastrointestinal tract. The recent work by Findlay et al.[29] shows that this occurs in an unexpected way in normal subjects. When 20 g of bran was given daily to normal subjects, the solid phase came through faster than the liquid phase, i.e., a flow pattern similar to untreated diverticular disease patients was observed. However, when the fecal

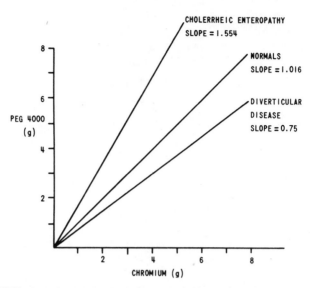

Figure 2. PEG plotted against chromium: regression slopes for normals and patients with cholerrheic enteropathy and diverticular disease. Source: Findlay et al.[28] Reproduced by permission of *Gut*.

concentrations of chromium sesquioxide and PEG 4000 were compared before and after bran, it was found that the chromium sesquioxide concentration was unchanged, but the PEG 4000 concentration had fallen by almost 50%. They suggested that the component causing the increase in stool weight was marked by chromium sesquioxide but not by PEG 4000. They concluded that the increase in stool weight in the bran-fed normal subjects was associated gel water, but not an increase in interstitial water, which would be marked with PEG 4000. Thus the solids and associated gel water travel faster through the gut than liquids, indicating streaming of solid and liquid phases.

D. Effect on Bacterial Flora

The nature of the flora of the large intestine has been thought to be controlled to a large extent by diet,[19,62] but studies on fecal flora of man have produced inconsistent results. In 1971 Hill et al.[35] demonstrated that feces of people from a range of countries showed differences in the relative numbers of various types of fecal bacteria. It was thought that this reflected the effect of diet, although other differences existed between these populations. The intergroup variations were reflected not only in the counts of the constituent bacterial groups present in

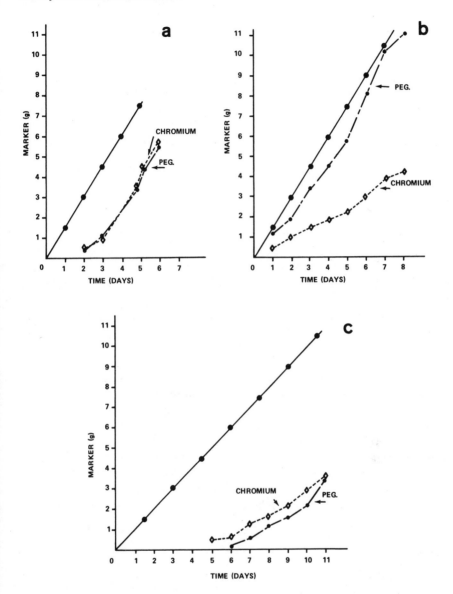

Figure 3. Cumulative inputs and outputs of PEG 4000 and Cr_2O_3 of three subjects: (a) normal, (b) cholerrheic enteropathy, and (c) diverticular disease. Vertical axis, input and output in grams (g) of markers, Cr_2O_3 and PEG 4000. Horizontal axis—duration of experiments in days. Solid line—synchronous input of both markers (cumulative) Interrupted lines—outputs of the two markers (cumulative). Source: Findlay et al.[28] Reproduced by permission of *Gut*.

the feces but also in the relative proportions of species comprising these groups. The most striking differences were that people in Britain and America eating their normal diets had more bacteroids and fewer enterococci in their feces than people in Uganda or South India who were taking a largely vegetarian diet. One particular bacterial species, *Sarcina ventriculi*, has been found virtually confined to vegetarian populations; in some vegetarians the fecal count of these microorganisms has been found to be as high as $10^8/g$.[13] Although differences have been shown between groups of people living on different diets, attempts to change the composition of the bacterial flora by variations in the diet have not been totally successful. In 1969 Moore et al.[56] showed that the fecal flora in 12 teenage girls remained unaltered after changing to a vegetarian diet. Although it has not been possible to alter completely the fecal flora by dietary manipulation, changing the composition of the diet can alter the proportion of the existing microorganism; e.g., changes in the carbohydrate content of the diet have been shown to cause an increase in the relative numbers of the microorganism bifido bacteria, whereas increasing the fat content of the diet increased the number of bacteroids.[39]

Although it is difficult to demonstrate changes in the bacterial species and genera comprising the flora, changes in dietary components have been shown to effect changes in the metabolic properties of the flora. This is particularly well demonstrated with regard to cyclamates. In 1972 Renwick and Williams[65] and Drasar et al.[20] showed that the ingestion of cyclamate by man resulted in the urinary excretion of cyclohexylamine, which is produced by the breakdown of cyclamate by the gut flora. Before exposure to cyclamate, the gut flora was unable to perform this reaction. Thus compounds present in the diet may well induce important changes in the bacterial flora which may or may not be beneficial to the host.

Although marked changes in the bacterial species and genera in the fecal flora have not been demonstrated with dietary alteration, this may be due to the inadequacy of the existing methods used to identify the bacteria in the laboratory. Thus many of the substrates, the metabolism of which is important in bacterial identification, may be without relevance in the context of bacterial growth in the intestine.

III. Dietary Fiber and Diverticular Disease

Diverticular disease is one of the diseases that is thought to develop because of a lack of fiber in the diet.[57,60] It is very common in European and North American communities but is apparently rare in rural Africans. Thus it can be considered as a disease of the affluent society. The hypothesis is that the modern European or North American eats a diet low in vegetable fiber and passes a small, relatively firm stool, which requires appreciable effort. The rural African eats a high-fiber-containing diet, has a well-filled colon, and passes large bulky stools with effortless ease. Thus in individuals on a low-fiber diet, the propulsion of the small, firm stools along the colon will result in increased pressures in the lumen of the bowel. This pressure will cause areas of weakness in the colon to bulge and thus produce small pockets or diverticula.[57] It has been suggested that the majority of people aged over 60 years in the European and North American countries have diverticula in the colon. The symptoms of diverticular disease are often ominous: widespread abdominal pain, nausea, vomiting, disturbed bowel habit, palpable abdominal masses and bleeding per rectum. However, only a small proportion (about 10%) of people with diverticular disease develop these symptoms and probably less than 10% require surgery.[58] In 1971 Painter and Burkitt[60] suggested that an unrefined diet containing adequate fiber may prevent diverticulosis for the following reasons:

1. The colon copes with a large volume of feces, is of a wide diameter, and does not develop a diverticulum. Such a colon has a wide bore, segments less than a narrow colon, and is therefore less prone to diverticulosis.

2. In most instances food residue passes through the African's gut within 48 hr, whereas in an Englishman this may take more than twice as long. Thus the African's colon removes less water from the stool and has to propel a less viscous fecal stream. Therefore the African's colon probably produces less pressure and is less apt to become ''trabeculated'' and to bear diverticula.

3. In Western countries custom often demands the suppression of the call to stool; this favors drying of the feces and increased pressure generation. On the other hand, the black African passes large moist feces without straining.

In summary, the swiftly passed, soft stool subjects the sigmoid to less strain and does not favor the development of diverticula. Painter and Burkitt suggested that if it is the lack of fiber which causes diverticulosis, then the symptoms should be alleviated by replacing the fiber, and they suggested that the fiber replaced should be in the form of bran.

Advocating a high-residue diet for patients with diverticular disease was a controversial step, as low-residue diets were, at that time, considered to be the treatment of choice. The low-residue-diet treatment arose because it was thought that constipation and fecal stagnation led to infection of the diverticula.[69] Undigested fragments of food and bone had also been associated with diverticulitis[5] and were believed to cause perforation. Therefore low-residue diets were advocated by various authorities such as Slesinger[68] and Willard and Bockus.[76] Thus for approximately 50 years, low-residue diets were the mainstay of medical treatment for diverticular disease despite the lack of any evidence they were of benefit.

It is now believed that this diet is contraindicated in patients who suffer from diverticular disease and that a high-residue diet should be given instead.

That bran is beneficial in the treatment of diverticular disease is unquestionable, and various studies have shown that it is very effective.[29,59] There have, however, been only limited studies on the way in which bran can affect the colon function both in normal subjects and in diverticular disease.[23,29,45,61] These recent publications attempt to explain the effect of bran by the physical characteristics of the bran fiber itself.

Findlay et al.[29] also examined the throughput of solid and liquid phases in the colon using a solid marker, chromium sesquioxide, and a liquid-phase marker, PEG 4000. They examined normal subjects and diverticular disease patients and gave both groups of subjects 20 g bran daily. Before the bran regime, the diverticular disease patients and control subjects were all given radiopaque markers of the Hinton type[36] and also PEG 4000 and chromium sesquioxide capsules. Stools were then collected for 7 days and the chromium sesquioxide and PEG content measured; transit times were also measured. These studies were repeated after 4 weeks on 20 g of bran. The results showed that 20 g of unprocessed bran significantly shortened the intestinal transit time in the diverticular disease group and in normal subjects. Stool weight was increased in normals and to a lesser extent in the patients with diverticular disease.

The bran also appeared to modify fecal flow patterns by acting as a vehicle for the molecular or gel water in the normal people and as a vehicle for interstitial water in diverticular disease. These suggestions on the alteration of fecal flow patterns by the bran were deduced from the results of the PEG and chromium sesquioxide throughputs. In the diverticular disease patients the rate of throughput of the solid phase was faster than that of the liquid phase. Findlay et al.[29] suggested that this might be due to retropropulsion[66] of the liquid phase of the gut contents caused by high distal intraluminal pressure in the diverticular disease. As the same quantity of each marker was taken before and after bran, the fall in concentration of both markers along with the insignificant change in stool weight suggest that dilution of both the solid and liquid phase markers took place within the gastrointestinal tract. Painter,[57] in his hypothesis to explain the cause of diverticular disease, suggested that underfilling of the colon could result in a increase in intraluminal pressure. If this hypothesis is correct, then colonic filling should result in a decrease in intraluminal pressure and probably also in the elimination of streaming of solid and liquid phases. This was found in the group of diverticular disease patients.[29] Streaming of solid and liquid phases did not occur during treatment with bran; in addition, the pressure changes in the colon as measured by motility indices decreased significantly. It is highly likely that vegetable fiber will also have an effect similar to that of bran on altering colon function in diverticular disease. Studies have yet to be carried out to investigate the role of the different vegetables such as carrot, potato, and cabbage on transit time, stool weight, and colonic pressures in this disease state. It is likely that they will be equally as effective as bran. However, even if they are of therapeutic value, this does not suggest an etiological role in the development of the disease.

IV. Dietary Fiber and Colonic Cancer

Because of the intimate relationship between the large bowel and food, cancer of the large bowel is thought to be related to diet. Indeed, several studies have suggested that there is a relationship between colonic cancer and diet,[7,31,79] and although nearly all the major components have been incriminated, no single dietary component has been identified as the cause of colonic cancer. This inability to identify the dietary component that could be responsible is at present confusing the role of diet in the etiology of colonic cancer. What is known is that the colonic bacteria can produce carcinogens or cocarcinogens from dietary components or from intestinal secretions produced in response to the diet.[3,35] Diet is also responsible for the amount of substrate for carcinogen formation, for the nature of the bacterial flora acting on the substrate and, to a certain extent, for the conditions in the colon under which the bacterial action takes place. Because of these interrelationships, the correlation between diet and incidence of colonic cancer may be explained. The question regards the role of dietary fiber in the etiology of colonic cancer.

A suggestion that has received considerable support is that diet, and especially fiber, is important in the etiology of colonic cancer.[6,8,72] In support of his hypothesis, Burkitt and his colleagues[6,8] suggest that the typical diet in the Western world is low in fiber and causes fecal stasis. This leads to prolonged exposure of various compounds such as bile acids to the bacterial flora, which may alter the molecule and perhaps produce carcinogens leading to carcinogensis. However, the correlation between dietary fiber and colonic cancer is poor. Hill in 1971[33] showed that the amount of bile acid in the feces is dependent on the fat content of the diet. This may provoke a dilemma because in the prevention of coronary heart disease the patient is advised to take polyunsaturated fats, which lower serum cholesterol but which cause increased fecal bile acid excretion. Aries et al.[3] found that the bacterial flora of people in the low-risk countries contained fewer bacteroides but more streptococci. Also, they had a lower anaerobic-to-aerobic bacterial ratio than those in the high-risk countries. Hill and Aries[34] also showed that the concentration of total fecal bile acids and neutral steroids was higher in people living in high-risk areas and that the degree of dehydroxylation of acid steroids was much higher in those people in the higher-risk countries. In addition, they found that the proportion of the bacterial products of cholesterol were much higher in feces of people from high-risk countries than in feces of people from low-risk areas. Similar findings have been reported by Reddy and Wynder.[64] They showed that Americans on a normal diet had a greater proportion of coprostanol and coprostanone, i.e., bacterial metabolites of cholesterol in their feces, than did American vegetarians, Seventh Day Adventists, Japanese, and Chinese. They also found that the Americans on the normal diet excreted more fecal bile acids than the other groups.

However, when acidic and neutral steroids are expressed as a concentration, i.e., as mg/g dry feces or mg/g wet weight feces, changing to a diet with more unabsorbed fiber present markedly affects these concentrations. The work by Findlay et al.[29] showed that when the diet was supplemented with 20 g bran/day, the stool weight increased but the concentration of fecal bile acids decreased. Thus we have a situation where the concentration of fecal constituents can be altered by changing the dietary intake of unabsorbed fiber.

The hypothesis of Burkitt suggests that prolonged transit time causes increased degradation of bile acids to potential carcinogens. It is possible that patients with prolonged transit time might have an abnormal amount of potential carcinogenic substances such as deoxycholic acid present in their feces, or alternatively they might have an altered bacterial flora compared to those with a short transit time. There is, however, little evidence that transit time as such grossly affects the composition of the gut flora or their ability to degrade steroids. In a study of obese patients and medical students living on liquid diets containing no fiber,[14,15] it was found that although the transit time was very long—up to 14 days—degradation of fecal steroids by gut bacteria was less than that reported by other workers in previously studied populations. They also suggested that the transit times found with the Africans is sufficient for considerable bile acid metabolism to proceed.

A study carried out by Thwe et al.[71] in Edinburgh compared two groups of normal subjects: one group with short transit times (up to 24 hr) and the other group with longer transit times (greater than 48 hr). They found that there was no difference in the bacterial flora between the two groups and that the daily excretion of fecal bile acids was not significantly different between the two groups (Tables I and II). Fecal bile acid concentration did not differ significantly, and the percentage of deoxycholic acid in the feces remained the same in both the short-transit-time and the fast-transit-time group. If the carcinogen is produced in the gut, and if the important factor is not the daily production of the carcinogen but the concentration of this carcinogen that is in contact with the gut wall, then dilution of the colonic contents may be of therapeutic value. Thus the bulking effect of dietary fiber, be it cereal or vegetable, might cause the carcinogen to be diluted to a level where it is no longer dangerous.

The role of fiber in altering bile acid metabolism is potentially very important. In 1968 Eastwood and Hamilton[22] showed that a component of vegetable fiber, lignin, can bind bile salts, and in 1966 Leveille and Sauberlich[47] showed that pectin, another component of vegetable fiber, can bind bile salt and lower blood cholesterol. Earlier, Coleman and Baumann[10] had shown that the addition of fiber to purified diets increased sterol excretion in the rat, and Portman and Murphy[63] showed that fiber caused increased excretion and alteration in the nature of fecal steroid degradation. In 1962 Antonis and Bersohn[2] reported that the fecal excretion of bile acids and neutral steroid was greater in Africans on a

TABLE I. Main Components of Fecal Bacterial Flora

Subjects	Coliform	Strept. faecalis	Clostridia	Lactobacilli	Bacteroides	Fusiforms
Rapid transit group (N = 5)						
Mean	5.8×10^7	6.7×10^6	5.3×10^7	1.3×10^9	8.6×10^8	5.0×10^7
± SD	$\pm 11.3 \times 10^7$	$\pm 9.2 \times 10^6$	$\pm 85.0 \times 10^7$	$\pm 1.2 \times 10^9$	$\pm 10.8 \times 10^8$	$\pm 5.0 \times 10^7$
Slow transit group (N = 4)						
Mean	9.8×10^7	6.2×10^6	3.0×10^8	1.9×10^9	8.3×10^8	1.2×10^8
± SD	$\pm 12.3 \times 10^7$	$\pm 5.2 \times 10^6$	$\pm 2.0 \times 10^8$	$\pm 1.2 \times 10^9$	$\pm 8.7 \times 10^8$	$\pm 1.3 \times 10^8$
t test	NS	NS	NS	NS	NS	NS

TABLE II. Fecal Weight and Fecal Bile Acids in Normal Subjects
with Rapid and Slow Gastrointestinal Transit Times

Subjects	Transit time (hr)	Wt. of wet feces (g/24 hr)	Wt. of dry feces (g/24 hr)	Lithocholic acid (mg/24 hr)	Deoxycholic acid (mg/24 hr)	Total fecal bile acids (mg/24 hr)	Fecal bile acid concentration (mM)	Deoxycholic acid (%)
Rapid transit group (N = 5)								
Mean	24.7	147.3	31.1	119.4	212.7	332.1	6.52	63.9
± SD	± 1.48	± 54.8	± 7.3	± 71.0	± 128.4	± 199.4	± 4.81	± 0.9
Slow transit group (N = 4)								
Mean	62.8	118.7	30.6	113.9	197.4	311.3	8.08	63.3
± SD	± 22.9	± 38.1	± 12.6	± 5.7	± 43.6	± 48.5	± 4.8	± 5.1
t test	$t = -3.7801$ $P < 0.005$	NS	NS	NS	NS	NS	NS	NS

high-fiber diet with 40% of calories as added butterfat than in the Europeans on a low-fiber diet with comparable fat intake.

Therefore it may be that dietary fiber has at least a dual function: (1) It produces a bulky stool, which dilutes potential carcinogens and therefore prevents colonic cancer. (2) It acts as a bile acid adsorbent, increasing bile acid excretion and thereby lowering serum cholesterol levels, thus acting as a hypolipidemic agent.

V. Summary

Dietary fiber, and in particular cereal bran, has been shown to increase stool weight, reduce gastrointestinal transit time, and modify the throughput of solid and liquid phases in the gastrointestinal tract in normals and in diverticular disease patients. In addition, it has been shown to reduce colonic pressure in diverticular disease. Its bulking action causes dilution of fecal constituents, e.g., bile acids, and it is thought that this might act as a protective mechanism in the prevention of colonic cancer. Although the effects of increased intake of dietary fiber have been well documented, the mechanism by which the fiber causes the effects is not well understood and requires more intensive investigation.

References

1. ALVAREZ, W. C. The rate of passage of food residues through the digestive tract. *In: An Introduction to Gastroenterology*, 4th ed., W. C. Alvarez (ed.). London, Heineman, 1949. Chap. 27, p. 617.
2. ANTONIS, A., and I. BERSOHN. The influence of diet on fecal lipids in South African white and Bantu prisoners. *Am. J. Clin. Nutr.* **11**:142, 1962.
3. ARIES, V., J. S. CROWTHER, B. S. DRASAR, M. J. HILL and R. E. O. WILLIAMS. Bacteria and the aetiology of cancer of the large bowel. *Gut* **10**:334, 1969.
4. BEEKEN, W. L. Clearance of circulating radiochromated albumin and erythrocytes by the gastrointestinal tract of normal subjects. *Gastroenterology* **52**:35, 1967.
5. BLAND-SUTTON, J. On the effect of perforation of the colon by small foreign bodies, especially in relation to abscess of apiploic appendage. *Lancet* **2**:1148, 1903.
6. BURKITT, D. P. Related disease-related cause. *Lancet* **2**:1229, 1969.
7. BURKITT, D. P. Epidemiology of cancer of the colon and rectum.*Cancer* **28**:3, 1971.
8. BURKITT, D. P., A. R. P. WALKER and N. S. PAINTER. Effect of dietary fibre on stools and transit times and its role in the causation of disease. *Lancet* **2**:1408, 1972.
9. BURNE, J. *A Treatise on the Causes and Consequences of Habitual Constipation*. London, Longman, Orme, Brown, Green and Longman, 1840.
10. COLEMAN, D. C., and C. A. BAUMANN. Intestinal sterols. III. Effects of age, sex and diet.*Arch. Biochem. Biophys.* **66**:226, 1957.

11. CONNELL, A. M. Applied physiology of the colon: Factors relevant to diverticular disease. *In: Clinics in Gastroenterology*, A. N. Smith (ed.). London, Saunders, 1975, Chap. 2, p. 23.

12. CONNELL, A. M., F. AVERY-JONES and E. N. ROWLANDS. Motility of the pelvic colon. IV. Abdominal pain associated with colonic hypermotility after meals. *Gut* 6:105, 1965.

13. CROWTHER, J. S. Sarcina ventriculi in human feces. *J. Med. Microbiol.* 4:343, 1971.

14. CROWTHER, J. S., B. S. DRASAR, P. GODDARD, M. J. HILL and K. JOHNSON. The effect of a chemically defined diet on the faecal flora and faecal steroid excretion. *Gut* 14:790, 1973.

15. CROWTHER, J. S., B. S. DRASAR, P. GODDARD, M. J. HILL and K. JOHNSON. The effect of chemically defined diets on the faecal flora and faecal steroids. *Gut* 14:831, 1973.

16. CUMMINGS, J. H. Dietary fibre. *Gut* 14:69, 1973.

17. DAVIGNON, J., W. J. SIMMONDS and E. H. AHRENS, JR. Usefulness of chromic oxide as an internal standard for balance studies in formula-fed patients and for assessment of colonic function. *J. Clin. Invest.* 47:127, 1968.

18. DINOSO, V. P., H. MESHKINPOUR, S. H. LORBER, G. GUTIERNEZ and W. Y. CHEY. Motor responses of the sigmoid colon and rectum to exogenous cholecystokinin and secretin. *Gastroenterology* 65:438, 1973.

19. DRASAR, B. S., and M. J. HILL (ed.). *Human Intestinal Flora*. London, New York, San Francisco, Academic Press, 1974, Chap. 1, p. 9.

20. DRASAR, B. S., A. G. RENWICK and R. T. WILLIAMS. The role of the gut flora in the metabolism of cyclamate. *Biochem. J.* 129:881, 1972.

21. EASTWOOD, M. A., N. FISHER, C. T. GREENWOOD and J. B. HUTCHINSON. Perspectives on the bran hypothesis. *Lancet* 1:1029, 1974.

22. EASTWOOD, M. A., and D. HAMILTON. Studies on the adsorption of bile salts to nonabsorbed components of diet. *Biochem. Biophys. Acta* 152:165, 1968.

23. EASTWOOD, M. A., J. R. KIRKPATRICK, W. D. MITCHELL, A. BONE, and T. HAMILTON. Effects of dietary supplements of wheat bran and cellulose on faeces and bowel function. *Br. Med. J.* 4:392, 1973.

24. FANTUS, B., and W. FRANKL. The mode of action of bran. I. Effect of bran upon composition of stools. *J. Lab. Clin. Med.* 26:1774, 1941.

25. FANTUS, B., N. HIRSCHBERG and W. FRANKL. The mode of action of bran. II. Influence of size and shape of bran particles and of crude fiber isolated from bran. *Rev. Gastroenterol.* 8:277, 1941.

26. FANTUS, B., O. WOZASEK and F. STEIGMANN. Studies on colon irritation: effect of bran. *Am. J. Dig. Dis.* 8:298, 1941.

27. FINDLAY, J. M., M. A. EASTWOOD and W. D. MITCHELL. The physical state of bile acids in the diarrhoeal stool of ileal dysfunction. *Gut* 14:319, 1973.

28. FINDLAY, J. M., W. D. MITCHELL, M. A. EASTWOOD, A. J. B. ANDERSON and A. N. SMITH. Intestinal streaming patterns in cholerrhoeic enteropathy and diverticular disease. *Gut* 15:207, 1974.

29. FINDLAY, J. M., A. N. SMITH, W. D. MITCHELL, A. J. B. ANDERSON and M. A. EASTWOOD. Effects of unprocessed bran on colon function in normal subjects and in diverticular disease. *Lancet* 1:146, 1974.

30. FORTH, W., W. RUMMEL and H. GLASNER. Zur resorption-shemmenden Wirkung von Gallensauren. *Naunyn-Schmiedeberg's Arch. Exp. Pathol. Pharmakol.* 254:364, 1966.

31. GREGOR, O., R. TOMAN and F. PRUSOVA. Gastrointestinal cancer and nutrition. *Gut* 10:1031, 1969.

32. HARVEY, R. F., E. W. POMARE and K. W. HEATON. Effects of increased dietary fibre on intestinal transit. *Lancet* 1:1278, 1973.

33. HILL, M. J. The effect of some factors on the fecal concentration of acid steroids, neutral steroids and urobilins. *J. Pathol.* 104:239, 1971.

34. HILL, M. J., and V. C. ARIES. Faecal steroid composition and its relationship to cancer of the large bowel. *J. Pathol.* 104:129, 1971.

35. HILL, M. J., B. S. DRASAR, V. ARIES, J. S. CROWTHER, G. HAWKSWORTH and R. E. O. WILLIAMS. Bacteria and the aetiology of cancer of the large bowel. *Lancet* **1**:95, 1971.

36. HINTON, J. M., J. E. LENNARD-JONES and A. C. YOUNG. A new method for measuring gut transit using radio-opaque markers. *Gut* **10**:842, 1969.

37. HOELZEL, F. The rate of passage of inert materials through the digestive tract. *Am. J. Physiol.* **92**:466, 1930.

38. HOFMANN, A. F. The syndrome of ileal disease and the broken enterohepatic circulation: cholerrheic enteropathy. *Gastroenterology* **52**:752, 1967.

39. HOFFMAN, K. Untersuchungen über die Zasammensetzung der Stuhlflora während eines lang-dauernden Ernährung-sversuches mit kohlenhydratreicher, mit fettreicher und mit enveibrucher Kost. *Zbl. Bakt. I. Abt. Orig.* **192**:500, 1964.

40. HOPPERT, C. A., and A. J. CLARK. Effect of long-continued consumption of bran by normal men. *J. Am. Diet. Assoc.* **18**:1, 1942.

41. HOPPERT, C. A., and A. J. CLARK. Bran muffins and normal laxation. *J. Am. Diet. Assoc.* **18**:524, 1942.

42. HOPPERT, C. A., and A. J. CLARK. Digestibility and effect on laxation of crude fiber and cellulose in certain common foods. *J. Am. Diet. Assoc.* **21**:157, 1945.

43. HYDEN, S. The recovery of polyethylene glycol after passage through the digestive tract. *K. Lantbr. Högsk. Annir.* **22**:411, 1956.

44. KIRWAN, W. O., and A. N. SMITH. Gastrointestinal transit estimated by an isotope capsule. *Scand. J. Gastroenterol.* **9**:763, 1974.

45. KIRWAN, W. O., A. N. SMITH, A. A. McCONNELL, W. D. MITCHELL and M. A. EASTWOOD. Action of different bran preparations on colonic function. *Br. Med. J.* **2**:187, 1974.

46. KREULA, M. S. Absorption of carotene from carrots in man and the use of the quanti-tative chromic oxide indicator method in the absorption experiments. *Biochem. J.* **41**:269, 1947.

47. LEVEILLE, G. A., and H. E. SAUBERLICH. Mechanism of the cholesterol-depressing effect of pectin in the cholesterol-fed rat. *J. Nutr.* **88**:209, 1966.

48. MANNERS, M. J., and D. E. KIDDER. Polyethylene glycol as marker in piglet diets with a high dry-matter content. *Br. J. Nutr.* **22**:515, 1968.

49. McCANCE, R. A., and E. M. WIDDOWSON. Old thoughts and new work on breads white and brown, *Lancet* **2**:205, 1955.

50. McCONNELL, A. A., M. A. EASTWOOD and W. D. MITCHELL. Physical characteristics of vegetable foodstuffs that could influence bowel function. *J. Sci. Food Agric.* **25**:1457, 1974.

51. MEKHJIAN, H. S., and S. F. PHILLIPS. Perfusion of the canine colon with unconjugated bile acids. *Gastroenterology* **59**:120, 1970.

52. MEKHJIAN, H. S., S. F. PHILLIPS and A. F. HOFMANN. Conjugated bile salts block water and electrolyte transport by the human colon. *Gastroenterology* **54**:1256, 1968 (abstract).

53. MEKHJIAN, H. S., S. F. PHILLIPS and A. F. HOFMANN. Colonic secretion of water and electro-lytes induced by bile acids: Perfusion studies in man. *J. Clin. Invest.* **50**:1569, 1971.

54. MITCHELL, W. D., and M. A. EASTWOOD. Faecal bile acids and neutral steroids in patients with ileal dysfunction *Scand. J. Gastroenterol.* **7**:29, 1972.

55. MITCHELL, W. D., J. M. FINDLAY, R. J. PRESCOTT, M. A. EASTWOOD and D. B. HORN. Bile acids in the diarrhoea of ileal resection. *Gut* **14**:348, 1973.

56. MOORE, W. E. C., E. P. CATO and L. V. HOLDEMAN. Anaerobic bacteria of the gastrointesti-nal flora and their occurrence in clinical infections. *J. Infect. Dis.* **119**:641, 1969.

57. PAINTER, N. S. Diverticular disease of the colon. *Br. Med. J.* **3**:475, 1968.

58. PAINTER, N. S. *Diseases of the Colon, Rectum and Anus, B. C. Morson (ed.). London, Heineman 1969, p. 201.*

59. Painter, N. S., A. Z. ALMEIDA and K. W. COLEBOURNE. Unprocessed bran in treatment of diverticular disease of the colon. *Br. Med. J.* **1**:137, 1972.

60. PAINTER, N. S., and D. P. BURKITT. Diverticular disease of the colon: A deficiency disease of Western civilization. *Br. Med. J.* **2**:450, 1971.

61. PAYLER, D. K., E. W. POMARE, K. W. HEATON and R. F. HARVEY. The effect of wheat bran on intestinal transit. *Gut* **16**:209, 1975.

62. PORTER, J. R., and L. F. RETTGER. Influence of diet on the distribution of bacteria in the stomach, small intestine and cecum of the white rat. *J. Infect. Dis.* **66**:104, 1940.

63. PORTMAN, O. W., and P. MURPHY. Excretion of bile acids and β-hydroxysterols by rats. *Arch. Biochem. Biophys.* **76**:367, 1958.

64. REDDY, B. S., and E. L. WYNDER. Large bowel carcinogenesis: Fecal constituents of populations with diverse incidence rates of colon cancer. *J. Natl. Cancer Inst.* **50**:1437, 1973.

65. RENWICK, A. G., and R. T. WILLIAMS. The fate of cyclamate in man and other species. *Biochem. J.* **129**:869, 1972.

66. RITCHIE, J. A., S. C. TRUELOVE and G. M. ARDRAN. Propulsion and retropulsion in the human colon demonstrated by time-lapse cinefluorography. *Gut* **9**:735, 1968.

67. SHIELDS, R., J. HARRIS and M. W. DAVIES. Suitability of polyethylene glycol as a dilution indicator in the human colon. *Gastroenterology* **54**:331, 1968.

68. SLESINGER, E. G. Observations on diverticulitis *Lancet* **1**:1325, 1930.

69. SPRIGGS, E. I., and O. A. MARPER. Multiple diverticula of the colon. *Lancet* **1**:1067, 1927.

70. STREICHER, M. H., and L. QUIRK. Constipation: Clinical and roentgenologic evaluation of the use of bran. *Am. J. Dig. Dis.* **10**:179, 1943.

71. THWE, M., R. MACRAE, W. D. MITCHELL, W. O. KIRWAN and M. A. EASTWOOD. Unpublished results.

72. WALKER, A. R. P., B. F. WALKER and B. D. RICHARDSON. Bowel transit times in Bantu population. *Br. Med. J.* **3**:48, 1970.

73. WALLER, S. Differential measurement of small and large bowel transit times in constipation and diarrhoea: A new approach. *Gut* **16**:372, 1975.

74. WHITBY, L. G., and D. LANG. Experience with the chromic oxide method of fecal marking in metabolic balance investigations on humans. *J. Clin. Invest.* **39**:854, 1960.

75. WILKINSON, R. Polyethylene glycol 4000 as a continuously administered nonabsorbable fecal marker for metabolic balance studies in human subjects. *Gut* **12**:654, 1971.

76. WILLARD, J. H., and H. BOCKUS. Clinical and therapeutic status of cases of colonic diverticulosis seen in office practice. *Am. J. Dig. Dis.* **3**:589, 1936.

77. WILLIAMS, R. D., and W. H. OLMSTED. The effect of cellulose, hemicellulose and lignin on the weight of stool. *J. Nutr.* **11**:433, 1936.

78. WOODBURY, J. F., and F. KERN, JR. Fecal excretion of bile acids: A new technique for studying bile acid kinetics in patients with ileal resection. *J. Clin. Invest.* **50**:2531, 1971.

79. WYNDER, E. L., and T. SHIGEMATSU. Environmental factors of cancer of the colon and rectum. *Cancer* **20**:1520, 1967.

9

Epidemiology of Bowel Disease

Martin A. Eastwood, Jenny Eastwood, and Michael Ward

I. Introduction

There is a group of bowel diseases for which no obvious infective cause is known. To distinguish these diseases from such conditions as dysentery or amebiasis, they are defined as noninfective bowel diseases. These diseases include hemorrhoids, appendicitis, diverticular disease, cancer of the colon, colitis, and Crohn's disease. Noninfective bowel diseases show considerable difference in prevalence throughout the world. Such observations suggest that if a disease varies in its prevalence, then the disease is not inevitable; if a disease is not inevitable, then it is preventable.

The common diagnosis for a clinical finding will vary from country to country; e.g., a physician in South India feeling a mass in the right lower abdomen will first consider amebiasis, in North India tuberculosis, in Saigon an appendix abscess, and in San Francisco a carcinoma.

In this chapter an outline will be made of the epidemiology of diseases of the bowel, with particular reference to diet. Our classification groups those bowel diseases in which it is thought that a deficiency of dietary fiber coincides with their occurrence, e.g., hemorrhoids, appendicitis, constipation, diverticular disease, and cancer of the bowel, and secondly, those diseases which coincide with an excess of dietary fiber or where polysaccharides may be implicated in the etiology. The latter group includes volvulus of the colon[19] and inflammatory

9

MARTIN A. EASTWOOD ● Consultant Physician, Wolfson Gastrointestinal Laboratories, Department of Medicine, Western General Hospital, Edinburgh, Scotland.   JENNY EASTWOOD ● Community Medicine Medical Officer, Lothian Health Board, Edinburgh, Scotland. MICHAEL WARD ● Wolfson Gastrointestinal Laboratories, Western General Hospital, Edinburgh, Scotland.

bowel diseases.[45] The incidence of all these diseases shows considerable variation with geographical location and the passage of time.

Two parts of the colon are principally affected by noninfective diseases. The rectum and sigmoid colon are the prime sites for ulcerative colitis, diverticular disease, polyps, and carcinoma. Even constipation is now recognized to be a distal colonic problem.[42] The cecum is the locus for carcinoma, appendicitis, Crohn's disease, and in some Eastern communities, diverticulosis. The intermediate transverse colon is rarely affected.[91] Such selectivity in sites affected and geographical incidence has given clues to the etiology of these diseases. The extremely low incidence of noninfective bowel disease in Africa has been frequently contrasted with the high incidence in Europe, North America, Japan, and Hawaii. Various hypotheses have also been constructed to explain the changes in prevalence of colonic diseases that may have occurred during this century.

II. Considerations in the Epidemiology of Bowel Disease

A. Age, Sex, Race

Diverticular disease, cancer, and constipation are predominantly diseases of the elderly. Ulcerative colitis shows a bimodal distribution, being very common in the young and also having another peak incidence in the sixth decade. This suggests two different etiological factors.[27] Appendicitis occurs most frequently in the young.[10] In communities in Africa where these diseases are rare, only a small proportion of the population live to be elderly. In some areas in Africa, 50% of children die before the age of 5 years and half the population are children.[101] Such a population ravaged by disease will be exposed to selective factors, which in the survivors may protect against subsequent disease. The sex biases in the prevalence of colonic disease may, in part, reflect the longevity of women and in part more subtle clues to etiology. The male/female ratio varies both for particular diseases and in different parts of the world.

B. Geographical Considerations and Migration

Many races, e.g., African,[35] some Indians,[107] and Japanese,[80] appear to be free of noninfective bowel disease. There is a temptation to ascribe to racial characteristics the differences between high and lower incidence areas. However, when nationals of a country with a low incidence migrate, they and their children assume the enhanced susceptibility to bowel disease of nationals of their new country [35,78] (Figure 1). There is little information to indicate any benefit

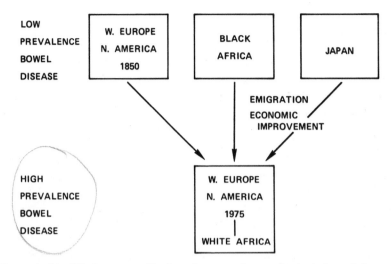

Figure 1. Identified groups with changing prevalence of colonic bowel disease.

derived by moving from a country with a high incidence of bowel disease to one with a low incidence. It is unlikely, however, that migrants would become less prosperous as a result of migration. The possible implication of wealth and sophistication of societies in the etiology of these diseases is supported by the white population of Africa having a higher incidence of bowel disease of the noninfective type than the indigenous African races.[35] In all studies of colonic disease, Jews appear to be peculiarly at risk, irrespective of where they live.[155,116]

C. Coexisting Diseases

Coexisting diseases may be subdivided into diseases which are linked, those with a potentiating effect, and those which may be protective.

Diseases linked with ulcerative colitis include the arthropathies, conjunctivitis, and colonic cancer.[121] The risk of conversion of the polyps in familial polyposis to a malignant form is a similar close relationship.[121]

Potentiating conditions may include the undoubted relationship between diverticular disease and colonic cancer with coronary heart disease in some communities.[130]

The effect of one disease on the susceptibility of an individual to a second disease is not well understood. The presence of Burkitt's lymphoma in African communities is paralleled by the incidence of malarial infection.[29] There are areas where tuberculosis has reached epidemic proportions, as in nineteenth-century Britain, at which time colonic carcinoma and diverticular disease were rareties.[31] Hookworm, dysentery, and schistosmiasis are endemic in many communities and may affect susceptibility to colonic carcinoma and diverticular disease. In two studies it was shown that 9 and 15% of two series of patients with colorectal carcinoma had bilharzial lesions of the colon, which contrasts with 20% in unselected autopsies.[122]

D. Dietary Differences and Changes

There are indications that the actual level of nourishment experienced by populations may have an influence of their susceptibility to particular diseases. The half of the human race that is either undernourished or malnourished[101] rarely develops carcinoma of the bowel or diverticular disease.

Because of the variability in the incidence of colonic disease throughout the world and the effect of migration on colonic disease susceptibility, diet is frequently suggested as an important etiological factor.[2,33] Communities in Africa which have a low incidence of noninfective diseases of the bowel in general eat a diet that is predominantly vegetable. Coincidental with the freedom from noninfective bowel disease is the problem of malnutrition, aggravated by tuberculosis, malaria, hookworm, and diarrhea leading to kwashiorkor or marasmus. The problems of ignorance, taboos, and seasonal food shortages are compounded by a staple diet that is protein-deficient, high in carbohydrate, but less than 10% fat. The African diet is 80% carbohydrate and contrasts with the European and North American diet, which is less than 50% carbohydrate.[101] The Masai alone have a more adequate protein intake.[101]

The common carbohydrate sources occurring in Africa are yams, millet, and sorghum. There then occurs an apparent transition in carbohydrate source coinciding with increasing sophistication of society through cassava or maize to rice and finally to bread or wheat products.[101]

The protein content of the carbohydrate sources is as follows:

Yams . . . . 2% protein
Potatoes . . . . 2% protein
Plaintains or bananas   . . . . low in protein
Cassava . . . . 1% protein
Maize . . . . 10% protein, but low in
                                       lysine and tryptophan
Wheat . . . . 9% protein in English wheat
                                       13% protein in Canadian wheat

This carbohydrate diet may be supplemented by legumes[136] or less frequently by fish and meat, which are generally rich in protein but low in fat. The intake of legumes varies from 11 g/head/day in Liberia to 100 g/head/day in Chad, an average range of between 30 and 80 g/head/day. This diet contrasts with the Western diet, where fat and protein of animal origin predominate and wheat is the chief carbohydrate source. Wheat is the world's most cultivated crop, and 300,000 metric tons of wheat per year are eaten by half the world population.[13]

Throughout history white fine flour has been prized.[58,103] The pericarp or testa consists of cellulose and hemicelluloses. Much emphasis has been placed on the production of white bread by the removal of the surrounding bran (epidermis, pericarp, or endocarp), which may contain aleurone which is rich in vitamins from the endosperm. Many diseases, including bowel disease of epidemic proportions, have been attributed to the removal of this fibrous material from the commonly used flour in some countries.

It has been suggested that the adoption of roller milling of wheat for the manufacture of all white flour in the late nineteenth century produced a dramatic fall in the fiber content, and that the flour previously produced contained almost 100% of the original grain, although removal of bran by sifting had been common practice since ancient time.[58] Coincidental with changes in the milling process, there was a change in Britain from the use of predominantly English wheat to North American wheat. There are differences in protein content between the two kinds of wheat, and there may well be differences in the hemicellulose and other polysaccharide constituents. In the late nineteenth century, fine flour (55% extraction) was provided only for confectionery in Britain.

The nutritive value of wheat must be seen in relation to patterns of the diet as a whole.[13] In prosperous countries like Great Britain and the United States, wheat products provide on the average less than 30% of the total calories, and even in economically poor countries the contribution of wheat rarely exceeds 60%. There has been a steady decline in flour consumption in Britain since in the last century (1840–1880), when the estimations of intake ranges from 260 to 600 g/day, to a contemporary figure of 180 g/day.[36,58] It has been suggested that the consumption of fruits and vegetables has remained constant or has even decreased, although such figures must be less reliable than flour figures because of the garden as a source of these foods.[58] In the United Kingdom there has been an increase in sugar consumption since 1840 of from 10 to 20 lb/head/year.[12] At the turn of the century the consumption was 90 lb/head/year and is now 120 lb/head/year.[40,151] This steady increase in consumption may be halted or even reversed by the recent sharp increase in the retail price of sugar.

The world production of sugar was 8 million tons in 1900, 20 million tons in 1945, and has risen to 70 million tons in 1970.[36] The highest sugar consumption per capita is in Iceland with 150g/head/day. Ireland, Holland, Denmark, and England eat more than 135g/head/day, whereas Rumania, Albania, Greece, Portugal, and Bulgaria eat less than 50 g/head/day.[151] It is worth noting that fat and sugar consumption have increased together, so that since the beginning of the century the total energy in European and North American countries derived from sugar and fat has increased from less than 33% to 58%.

It is often thought that the nineteenth-century diet was a pure diet, which slowly became adulterated as well as depleted, e.g., the removal of bran from flour.[35] Food adulteration was rife in the nineteenth century to an extent that is inconceivable today.[36] Beer was often matured with sulfuric acid. Tea was diluted with the leaves of ash, sloe, and elder matured on copper plates. It is estimated that in 1850, 4 million pounds of bogus tea were sold in comparison with 6 million pounds of tea imported by the East India Company to Britain. Bakers added 4 ounces of alum per 140-lb bag of flour to achieve whiteness. Ground peas were added to dilute the flour, as was gypsum, ground Derbyshire stone, and even flint. By 1863 "seconds" flour (sieved) was commonly used for breadmaking. One study has shown that in only one out of 370 cases was wholemeal flour used for bread. On the other hand, bread was an important source of nutrition for much of the population, who ate between 1 and 1½ lb/head/day. The advent of roller milling of flour in the late nineteenth century coincided with legislation, since which the adulteration of food has declined.

The comparison of turn-of-the-century Britain with African society extends beyond a freedom from cancer of the bowel, diverticular disease, and so on. Rowntree showed that in York 28% of the population of his time lived in poverty. In 1900 in Britain, then the richest country in the world and head of the largest empire, 20% of the population could be expected to have a pauper's funeral; 50% of children would grow up in poverty and be underweight. One out of every six babies of working-class parents would die before the age of 12 months. In 1904, in a survey in Leeds, 33% of children were undernourished and 50% had rickets. In the many rural districts, farm workers lived entirely on potatoes and bread. In the recruitment campaign for the Boer War, 40% of volunteers were rejected on health grounds. For millions of soldiers in the 1914–1918 war, army rations represented a higher standard of diet than they had previously known.[36] Table I shows the changes in food intake during the present century. It is against this background that the need for care is appreciated before isolating any one factor and attributing to it the entire responsibility for a particular disease entity.

III. Artifacts in Interpretation

Accurate age-specific incidence rates of diseases in primitive peoples are not available. Presentation at hospital reflects local customs, e.g., to allow the elderly sick to die at home rather than in hospital.[115] Hospital reports in India, for example, suggested a male predominance of rheumatic heart disease that was

TABLE I. Annual Consumption/Head in United Kingdom

Food	1903–1913 (lb)	1924–1928 (lb)	1934 (lb)	1964 (lb)
Butter	16	78	25	19
Cheese	7	9	10	10
Fruit	61	91	115	97
Margarine	6	12	8	17
Meat	135	134	143	117
Potatoes	208	194	210	180
Sugar	79	87	94	66
Vegetables	60	78	98	104
Wheat flour	211	198	197	181
Eggs (number of eggs)	104	120	152	240

aSource: J. Burnett, *Plenty and Want*. Pelican, 1968 (Ref. 36).

contrary to Western experience. Yet a house-to-house survey showed that women and children and the poor had a considerable incidence of rheumatic heart disease but did not normally come to hospital.[15] This demonstrates the need for studies in greater depth than perhaps are being carried out.

The relatively infrequent appearance of old people in disease surveys in Africa will to an extent be an artifact that is due to migrant populations' having both a rural and urban home.[115] Also, there is a differential use of medical services, some diseases being regarded as being better treated by Western medicine and others by African witch doctors. It has been shown that 50% of patients who come to Mpiol Hospital, Bulawayo, Rhodesia, had already visited a witch doctor. Witch doctors also visited hospital inpatients and treated them in the hospital bathroom during visiting hours.[115]

The problem also exists of obtaining an accurate diagnosis. This is in fact only likely to be a problem in differentiating between Crohn's disease and ulcerative colitis.

Implicit in certain hypotheses of bowel disease is the importance of stool weight as an index of susceptibility.[34] There are undoubted differences in stool weight between African communities and Western communities, more than 500 g/day for rural Africans contrasting with approximately 100 g/day for the British population. In Africa there is not only a high intake of vegetable dietary fiber but also a high incidence of parasitic infestation and apparently a constant use of laxatives and enemas.[22,122] Epidemiological information on changes in stool weight in Western communities over the past century is not readily available. The limited and presumably selective data for stool weight recorded in the United States and Great Britain between 1840 and 1906 give figures that are somewhat similar to those of modern Western-society diets, i.e., between 90 and 150 g/day.[58,70]

Because rectosigmoid cancer, polyps, diverticular disease, and hemorrhoids are diseases of the distal bowel and are associated with fecal stasis, great attention has been given to transintestinal transit times.[34] Problems in estimating transit time arise from shapes and specific gravity of the markers used. A commonly used method is that of Hinton, who used radiopaque plastic markers. An analysis of the reproducibility of measurements of intestinal transit times measured by Hinton's markers shows that even under control conditions a difference of 46 hr is necessary for a significance between individuals on the same trial, or 19 hr for comparison of the means of groups of six individuals.[58]

IV. Hemorrhoids

Hippocrates attributed hemorrhoids to bile and phlegm. Hyams and Philpott suggested that etiological factors in the development of hemorrhoids were hormonal factors, inflammation, infection, constipation, the erect stance, vascular stasis, lack of exercise, poor diet, straining by defecation, the physical stance at defecation and, with age, the loss of connective tissue and elasticity of the rectum. However, in their epidemiological investigations of hemorrhoids, they showed only that older patients have a greater prevalence of piles.[89] On the other hand, Burkitt attributes hemorrhoids entirely to lack of fiber in the diet and the problem of strain to pass a small inspissated stool.[32]

V. Appendicitis

The prevalence of appendicitis in African communities is very small.[38] Burkitt goes so far as to say that it is entirely a disease of Europeans and North Americans and is found only in those Africans who have adopted the Western style of life.[31] Ashley[10] has commented on the age-specific incidence of acute appendicitis in that it is a disease of the young. He suggested that it is restricted to a carrier subpopulation and that there are two etiological factors, one of which is the dietary factor, and the second the peculiar susceptibility of a proportion of the population to these dietary changes. In addition, he notes an increased incidence of acute appendicitis in the spring and therefore suggests an allergen or a virus as an alternative cause.

VI. Constipation, Intestinal Stasis, and Consequences

Constipation is difficult to define. It is a subjective symptom and may vary from passing insufficient stool, or insufficient stool for needs that may be real or imagined, to situations where there is a passage of only one stool annually. Implicit in some of the hypotheses for the etiology of colonic diseases is intestinal stasis. This has received much attention during the eighteenth, nineteenth, and twentieth centuries. Arbuthnott Lane[100] and Metchnikoff both believed that an abnormal retention of feces in any one part of the gastrointestinal tract resulted in toxins contaminating the entire system with resultant diminished resistance to disease. Such stasis was thought to lead to ulceration of the duodenum and stomach and also to cause a propensity to carcinoma of the pancreas and stomach. Burne[37] in 1840 suggested that habitual constipation is capable of producing many disorders including irritation of the sexual organs in man and many very serious affections of the sexual organs in women, including irregular menstruation. He suggested that if we neglect evacuation of the lower bowel, we allow foul and filthy matter to be retained in the body. Constipation, he said, was due to an inattention to the calls of nature and misplaced sense of delicacy, the want of proper conveniences, the civilized life, and sedentary habits and occupations. Little has been added to these causative elements in the past 100 years.

In 1911 Earle[57] wrote of constipation that it could be readily seen that from recent great advances in the preparation of food, there is little of the indigestible or offal left, so that by these refinements of civilization, constipation is induced by the consistency and refined character of the food. In order to prevent the nitrogenous portion of our food from packing to form a firm sciballous mass as it is inclined to do, it should be taken in conjuction with generous amounts of vegetables and fruits, which contain a large amount of cellulose. Cellulose is indigestible in the intestinal tract of man, except by microorganisms in the large intestine which are capable of breaking up cellulose to a limited extent. Consequently this type of digestive process furnishes a liberal amount of indigestible material, which separates the nitrogenous waste matter, which acts as a sponge to retain water, and which has an influence on the stimulation of the afferent nerves of the intestinal mucosa. Our food should therefore consist of a liberal amount of coarse vegetable, fruit and peel, and bread that contains husks.

Goodwin in 1909[70] wrote that the average output of feces was 120–180 g/day but that the causes of constipation were hereditary, that females were troubled more than males, that lack of exercise and reading at stool (a pernicious habit) were two important causes of constipation, and that an important treatment was vegetable diet. Arbuthnott Lane in 1909,[100] however, propounded that the large intestine was the cesspool of the body. He stated that gallstones, degenerative cystic changes in the breast, and carcinoma of the breast were due to intestinal stasis. Other conditions associated with intestinal stasis included sterility in women and the presence of toxic material in the circulation, with diminished resistance to tuberculosis, rheumatoid arthritis, gout, and gastric and duodenal ulcers.

An important approach to constipation, however, was that of Burkitt, Walker, and Painter[34] who showed, in a study of transit time of food residues in 1200 people from all over the world, that the more refined the diet, the smaller the stool and the slower the transit time and that there was an inverse relationship between daily stool weight and transit time.

VII. Diverticular Disease

A. Pathology

In Europe and North America diverticular disease always affects the sigmoid colon.[121] More proximal involvement only occurs when the sigmoid is also involved.[94,125,128] The rectum is never involved. It has long been suspected that thickening and shortening of the longitudinal muscle is secondary to increased luminal pressure. The muscle abnormality in diverticular disease is the most consistent and striking abnormality.[46,118,129] Edwards[60] suggested that the onset and progression of diverticulosis of the intestine is due to muscular forces arising from the bowel wall itself, not musculature failure, or to congenital factors. Diverticula of the colon are pouches of the mucous membrane projecting through the circular muscle layers of the bowel to the percolic fat.[55] The majority of the diverticula pass through the bowel wall at weak points in the circular muscle layer at the point where the main blood vessels pass through the muscle wall.[121]

Etiological factors that have been suggested are age and constipation, muscle degeneration secondary to age, disturbances of the neuromuscular system, and hyperemia and dilatation of vessels.[8,34,43,46,61,129] The distribution of diverticular disease in Western communities is almost 90% a left-sided phenomenon, whereas in Japan 60% of diverticula are found in the cecum; however, in Japanese living in Hawaii 40% of diverticula are found in the cecum.[82,88,126,128,138]

B. Age, Sex, Race

Diverticular disease affects a substantial proportion of the older age groups of Western populations; and this amounts to more than 30% of persons over the age of 60.[48,82,108,109,128,129,138] It is important to note that only 10% of these will experience symptoms.[125] The incidence of diverticular disease in Africa is very low. Throughout the world diverticula are uncommon in non-European stock, and where they do occur it appears to be a right-sided diverticulosis rather than rectosigmoid diverticulosis. This suggests that a common factor exists in the genesis of diverticula engendered by two disparate etiological factors or that there are distinct and different etiological factors.[126,138] In Chinese and Filipinos in the United States and in native Hawaiians, 15% of diverticula were present in the right colon, compared with Americans, in whom 95% of diverticula occur solely in the left colon.

Manousos showed that in normal subjects who underwent a barium enema in Oxford[108] the incidence of diverticula was 7.6% under the age of 60 but 35% over the age of 60. However, in Australia Dearlove found, based on 7000 barium enemas, a 7% rate over the age of 40, but few or none were found under that age.[48] Cleland and others claim a low diverticulosis rate in Australians compared with the high incidence in English of common stock.[41]

The average age of onset of diverticulosis for Scots is 68 years; for Europeans in Singapore and Fiji, 60 years; for Chinese and Nigerians, 54 years; and for Fijians living in Fiji, 32 years.[97] Edwards in London, from 1925 to 1938, found an incidence of 12% diverticula in 2000 postmortems, with most of these over the age of 40 and with an incidence that increased with age.[60]

In Japanese living in Hawaii, the peak incidence for diverticulosis of the ascending colon is between 50 and 70 years, and for the descending colon the peak is 70 years, with 40% of the diverticula being found in the right colon.[126,138] In Malta the age range is 20–90 years, with maximum prevalence between 60 and 80 years and the maximum effect being in the descending colon.[128] Likewise in Finland, both types of cases are found in individuals over 60 years. A female predominance appears to exist only by virtue of the greater longevity of females.

C. Geographical Considerations

Sources of data include hospital admission rates and barium enema and postmortem data, each source having its peculiar selectivity. Despite this there are areas of minimal incidence, e.g., Nigeria, Uganda, Ethiopia, Kuwait, and Korea.[18,125] The condition is rare in non-Europeans and is 40-fold more common for Europeans living in the tropics and 80-fold more common in places of high prevalence like Scotland.[97] The admission rates for diverticular disorders vary widely in the hospital regions of Scotland, England, and Wales. Rates per 100,000 population range from 21 to 66 for diverticular disease. The number of cases admitted in Scotland is double the numbers for England and Wales. In the north of Scotland the rate is double that of Scotland as a whole.[58,59] Diverticulosis is significantly (three times) more common in south Sweden than in south Finland, suggesting unsuspected etiological subtleties.[94]

D. Coexisting Diseases

It has been suggested that the irritable colon syndrome is a preliminary stage of colonic diverticulosis. The most common age for patients with irritable bowel syndrome was 46 years compared with 64 years for diverticular disease.[141] It is suggested that there is a slow transition from irritable bowel disease to diverticular disease. In a follow-up of patients with irritable bowel disease, it was shown that 24% developed diverticulosis and that this is twice the expected rate.[83] Yet irritable bowel syndrome is widespread throughout the world in countries where diverticulosis is rare, e.g., South India and Iran. It may well be that irritable bowel disease is a response of the bowel to a physical stress which is luminal and that this will vary from country to country. Many psychiatrists advocate emotional stress and anxiety as a cause of irritable bowel syndrome, suggesting a nice interplay between moral and vegetable dietary fiber.[63,87] Others have suggested that there is a distinct trend toward an increased incidence of diverticulosis with increased atheroma of the aorta.[142] Some have felt that ischemia is an important element in the development of diverticulosis because of anoxia and weakening of the bowel.[46,142] This is an interesting point as diverticulosis is commonest in an area of the bowel (rectosigmoid) which is particularly susceptible to ischemia, i.e., at the junction of areas supplied by the middle colic and middle rectal arteries. In other studies a coincidence of gallstones and hiatus hernia was found in 30% of subjects with diverticular disease.[82]

E. Dietary Influences

A clear relationship has been suggested between diet and colonic diverticulosis,[125] but as a time course of 40 years is being suggested, it is difficult to refute or substantiate the hypothesis. Undoubtedly a higher proportion of children in contemporary African and Asian societies are growing to maturity and old age than in the nineteenth century. The diet of affluence is 50% sugar and fat in providing calories, and a bread is eaten that is deficient in cereal fiber. Strong arguments have been developed that an excess of dietary sugar is responsible.[40] This theory relies on a coincidence between the increasing use of sugar in our dietaries and the increasing prevalence of diverticulosis. It is easy to dismiss such hypotheses with equivalent correlations between improbable and obviously irrelevant factors. Yet a deluge of sucrose to the gastrointestinal tract might result in some sucrose reaching the colon with subsequent metabolism of the sucrose by bacteria.

The alternative theory is that Africans who eat a high-fiber diet (yams, plantains, cassava, maize) do not get diverticular disease[31,32,35] in contrast to whites in Europe and North America who eat a fiber-deficient diet (high-extract wheat). The association is strengthened by the fact that patients with symptoms are helped by taking cereal bran.[59] In addition, studies have shown that many patients with diverticular disease have an increased pressure in the rectum when measured by manometric methods.[125] On treating these individuals with cereal bran, the pressure in the bowel diminishes. Because of this, it has been postulated that the reason for the development of the diverticula is a deficiency in the diet causing the development of a high pressure, which could only be relieved by a high-fiber-containing diet.[125] However, therapeutic value does not necessarily mean a true deficiency.[58]

The importance of these theories is to emphasize that diverticular disease is not inevitable, and that careful considerations of numerous factors will assist in the eradication of this disorder. The pressure and cereal-fiber theory depends on a demonstration of increased pressure in the rectosigmoid area.[125] Such an increase undoubtedly occurs in many patients with diverticular disease. Yet there are little or no data on colonic pressures in normal individuals of any age. There are no pressure results for comparison in racial groups who are free of this condition, e.g., Africans and Asians, so that the high colonic pressures found in patients with symptomatic diverticular disease may be related to the genesis of the symptom not the cause. It must here be recalled that from the population believed to have diverticulosis, only one in ten individuals complains of symptoms.

Also, we emphasize that just as there are considerable differences in the prevalence of diverticula internationally, there are considerable differences in the intake of vegetables between different countries, high in Italy and low in Sweden and Great Britain. The differences in prevalence of diverticular disease between England, Wales, and Scotland and between the regions of Scotland are to an extent paralleled by differences in fruit and vegetable intake.[59]

VIII. Cancer of the Bowel

A. Pathology

Cancer of the colon and rectum is one of the commonest forms of malignant disease.[52,155] The tumor is usually an adenocarcinoma.[121] Many epidemiologists separate rectal from colonic carcinoma, believing that this will unlock some etiological clues.[23,99] There is great discussion about the prevalence of rectosigmoid-junction carcinoma, but this is probably due only to problems of definition. There are undoubted differences in the ratio between the colonic and rectal carcinoma, but distortion may occur because of lack of definition of the point of transition from the rectum to the sigmoid colon.[79] In any survey, the variables include such anatomical variations as well as other variations.

In a large series of 15,000 cases in Birmingham, Warwickshire, studied by postmortem, 11% were found in the cecum, 5% in the ascending colon, 22% in the pelvic colon, and 35% in the pelvorectal junction.[147] In Los Angeles, 16% of carcinomas were found in the cecum and ascending colon, 6% in the transverse colon, and 55% in the sigmoid colon and rectum.[137] An important point in the diagnosis of carcinoma has been that a large number of cases were locatable by digital examination of the rectum. This number varies from 43% in Sao Paulo to 12% in New Orleans, but it has been suggested that the number of cases that can be diagnosed in this way is diminishing.[99] Between 1928 and 1967 at the Lahey Clinic the number of cases in the right colon increased from 7 to 22% and those in the rectosigmoid colon dropped from 80 to 62%.[44]

The nineteenth-century approach to carcinoma was that while there was a stimulus (trauma, inflammation) to the genesis of cancer, the cancer itself was regarded as an irregularity of nature.[84,135] The demonstration that neoplastic cells could be transferred to other organisms and that viruses and chemicals could induce cancer altered this somewhat fatalistic approach.[76] Some known inducers of cancer are chemical agents (exogenous and endogenous), physical agents (ionizing radiation), genetic predisposition, and viral cause. The only proven tumors of viral origin in man are warts, with other possibilities being sarcoma, leukemia, and Burkitt's lymphoma.[7] Most known chemical carcinogens affect the liver, skin, and bladder,[77] but there are chemicals that can cause carcinoma of the colon in animals, e.g., azoxymethane and cycadin.[139] Most discussions of the etiology of carcinoma mention diet as the most important cause. A virtually undiscussed possibility is that carcinoma of the colon is viral in origin.[72] For example, Epstein–Barr virus infections present in various ways,[62] and secondary factors become relevant; e.g., Burkitt's lymphoma is most common in areas where malaria is epidemic.[29] It is not impossible that the equally endemic worms and other intestinal infestations in Africa and Asia may have a protective effect against carcinoma. Another hypothesis is that European and North American society has been long exposed to tuberculosis, leaving a population equally susceptible to new diseases, as malaria leaves communities susceptible to Burkitt's lymphoma.

B. Age, Sex, Race

A working basis for consideration of carcinoma of the colon is that in low-risk countries colonic cancer, when it does occur, is concentrated in the cecum and ascending colon[79] in about 60% of cases. Most of the increases that occur are in patients up to the age of 50 or 55 years. However, when there is any substantial increase in incidence of cancer of the colon, it is in the sigmoid colon, particularly in men over the age of 55 years and in women at a late age. In low-risk populations there is a low male-to-female ratio. In high-risk populations the male-to-female ratio is unity. The rise in the ratio of sigmoid to cecal carcinoma between a low- and a high-risk community is steeper for males than for females. Cancer of the colon is one of the few forms of nonendocrine-linked cancer where there is not a marked male excess. It has been suggested that there is a difference in the male-to-female ratio for both colonic and rectal carcinoma before and after the menopause. Males respond more readily to changing conditions and demonstrate a more labile incidence than do females.

It would appear that the highest rate for carcinoma of the colon is in English-speaking countries.[139] Intestinal cancer appears to be relatively more common among whites and blacks in the United States than among any of the African nationals, in whom it is certainly one of the rare forms of malignant disease.[30,47,95] In Israel it has been shown that the incidence of both colonic and rectal cancer is significantly higher among European-born Jews than among those born in Asia or Africa. It is suggested that the disease appears to be more common in the Jewish population than in other inhabitants of New York.[155] American blacks have a higher incidence of colonic cancer than Africans,[2,28,30] and it is also higher for Japanese living in the United States than for those in Japan.[26,78,153] Black people in the United States have an incidence of carcinoma of the colon of 70/100,000; Caucasians living in Hawaii, 68/100,000, and Japanese living in Hawaii, 66/100,000; Japanese living in Japan and black Rhodesians have an incidence of 12 and 18/100,000, respectively; black South Africans, 11/100,000, and Nigerians, 6/100,000.

Environment does appear to have an effect on incidence that is stronger than race[52] (Figure 2). There are variations in the colon/rectum ratios in various situations from 0.9:1 in Danes to 4.2:1 in Mexicans living in Texas.[79] Tumors of the alimentary tract are extremely rare in Africans, the incidence of mucoid and differentiated carcinoma in North America and Britain being 10%, whereas in Africa it is 30%, and lymphoreticular tumors are also more common than elsewhere.[95,122] White people living in Africa have an incidence of carcinoma of the bowel which is greater than that of the indigenous races.[35] For most African nationals neither rectal nor colonic cancer accounts for more than 3% of their total cancer.[122]

It has been argued that one reason for the reduced incidence of cancer in black Africa is that Africans have a much shorter life expectancy.[115] A comparison is made of the age-specific survival rates of Africans and Norwegians (Table II), and a much higher survival rate is seen among the Norwegians.[47] It is important to note that in Africans there is a decrease in age-specific incidence rates in older age groups which is in marked contrast to European and North American experience.[47] Black Americans have a carcinoma rate which is five times that of Ugandans, but of course black Americans in general enjoy greater longevity.[2] The apparent anomaly in age-related incidence rates may be due to a diminishing risk or to the fact that, because of the very high mortality in the young African, only very robust individuals live to be at risk. Neoplastic lesions

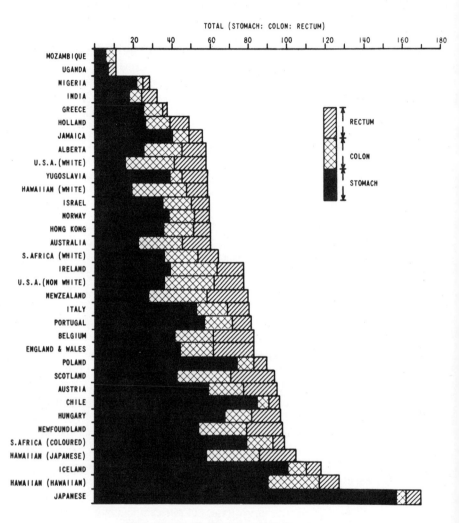

Figure 2. Cancer of stomach, colon, and rectum in various countries showing varying predominance internationally of each type of cancer. Prepared from original data by R. Doll, *Br. J. Cancer* **23**:1, 1969.

TABLE II. Comparison of Age-Specific Survival Rates
for Ugandans and Norwegians[a]

Age	Kyadondo, Uganda, 1959		Norway, 1956	
	Male	Female	Male	Female
45–49	100	100	100	100
55–59	47	53	81	86
65–69	24	30	50	58
75–79	12	13	27	34

[a]Source: J. N. P. Davies, J. Knowelden, and B. A. Wilson. Incidence rates of cancer in Kyadondo County, Uganda. *J. Natl. Cancer Inst.* **35**:818, 1965.

in the colon and anorectal region in black people affect a wide age range, 23–91 years. The disease occurs largely in the young and with a considerable female-to-male preponderance of 3.5:1.[95,122]

C. Social Class

It is difficult to identify antecedents to colorectal cancer.[28] There are familial risks; e.g., if a parent has carcinoma of the rectum, then the risk for the children is three to four times greater than that in the populations at large.[104–106] The socioeconomic status of patients with colonic cancer appears to be somewhat higher than that of patients with rectal carcinoma.[153,154] This compares with gastric cancer, which has an inverse relation to social class and familial aggregation; persons with blood group A are also at greater risk of gastric cancer. In New York there was seen to be an association between obesity and carcinoma and also between cigar smoking and carcinoma. These characteristics must be secondary factors.[80,81] A study was made of six families with a high incidence of carcinoma of the colon, in whom 20% of individuals at risk developed a colonic adenocarcinoma. Seven hundred twenty-five members covering four to five generations were identified. It was shown that 173 developed carcinoma and 81 adenocarcinoma, suggesting that carcinoma of the colon is transmitted in a pattern consistent with autosomal-dominant factors.[104]

However, in an overwhelming majority of patients the most common malignancies are caused by nongenetic factors. City dwellers are at greater risk of developing colon and rectal carcinomas than rural dwellers in the United States, Denmark, Finland, Norway, and, to a lesser extent, England.[81]

D. Geography and Migration

The highest rate of cancer of the bowel and rectum is found in North America and New Zealand, the rate in Scandinavia, England, and Wales is midway, and that in Japan, Bombay, Singapore, and the Caribbean Islands is lowest.[52] There is a sixfold variation. Cancer incidence has a different age distribution in different parts of the world so that cohort effects may be demonstrated. Because of the cultural factors, an age distribution of 35–64 has been used internationally. Figure 2 shows the incidence of carcinomas of the stomach, rectum, and bowel in different parts of the world. It would seem that in many parts of the world, e.g., the United States and Great Britain, carcinoma of the large intestine is a very important neoplasia. Other countries, however, have an incidence of carcinoma of the stomach which dwarfs that of carcinoma of the bowel.

It is noteworthy that in certain countries neither carcinoma of the bowel nor of the stomach is found in significant numbers.[107] In the United States in 1948 there were 33,694 deaths from carcinoma of the colon, 16% of all malignant neoplasia.[137] In 1964, 41,763 persons died, although this amounted to only 14% of cancer deaths.[155] In the United States the incidence is highest in the northern states and lower in the southern states.[81] Within England and Wales deaths from carcinoma of the colon and rectum, 5500 deaths per annum for rectal carcinoma and 9000 for carcinoma of the colon,[121] are second only to lung cancer, with gastric cancer a close third. However, within Africa, carcinoma of the colon represents a small proportion of all carcinomas; for instance, in Ibidan only 1% of all cancer is colorectal, and it is very rare to find carcinoma of the intestine in Uganda,[122] whereas in Egypt 60% of gastrointestinal cancer is colorectal.

There are clues to the etiology of carcinoma of the bowel given by data on migrants from areas of low to high prevalence. The incidence of bowel cancer is low in Africans but significant in American blacks.[2,30] Immigrants to the United States have colonic cancer rates that are equivalent to those of United States-born citizens. The rate of stomach cancer in United States immigrants is, however, equivalent to that of their country of origin. The incidence of bowel disease in Japan is low, although Japanese living in the United States have an incidence of colonic cancer that is equivalent to that of other United States inhabitants, whereas the rectal cancer rate is equivalent to that of Japanese living in Japan.[154] It is interesting that the cancer mortality for first- and second-generation Japanese in Hawaii differs in that the rate of carcinoma of the stomach is less than that for Japanese living in Japan.[26]

It would appear that migrants from low-incidence countries to high-incidence countries are exposed to a colonic carcinogen in the environment which rapidly changes the colonic cancer rate in these immigrants; this change in rate is totally unlike the situation for carcinoma of the stomach.[23,53,81] Some would say this was due to their change of diet, and it may be so, but what is unexpected is that their fecal transit times were not significantly different.[68] This suggests that it may be the chemical composition of the fecal content of the bowel that is all-important[30] and not the fecal volume, which has been a much favored explanation.

E. Coexisting Diseases

The most important condition coexisting with carcinoma of the colon is adenomatous polyps.[121] These are benign; i.e., the growth from the intestinal mucous membrane does not advance or grow beyond the confines of the mucous membranes. Malignancy implies a breach of these confines and after this a threat to the life of the host. Benign polyps are usually adenomatous in type. It has been suggested that there is a genetic predisposition to adenomatous polyps.[106]

The number of intestinal adenomatous polyps in any one person may vary from one to many thousands, although to have more than 20 polyps is rare. The presence of one polyp indicates the tendency to develop more. The distribution coincides with the distribution of carcinoma, i.e., in the rectum and sigmoid colon region. These polyps are regarded by many as precancerous. This is because they are often found in the base of a frank carcinoma, but the cellular modification to neoplasia is not inevitable.[121] About one in three of all operative specimens for cancer of the colon or colon and rectum contains benign polyps in addition to the cancer. This is indirectly suggestive evidence for an etiological relationship between benign neoplastic polyps and carcinoma.[120]

The age distribution of patients diagnosed as having adenoma and carcinoma is the same, with the peak age 57 years for benign polyps and 62 years for carcinoma. It must be stressed, however, that in populations where there is little or no carcinoma of the rectum and colon, as in indigenous Africans, polyps are equally uncommon.[30] There is, however, considerable risk of carcinoma developing in families with familial multiple polyposis of the colon,[121] but this is a different condition from benign adenomatous polyposis which only sometimes progresses to malignancy and is a common condition in much of Europe and North America.

Ulcerative colitis also has a predisposition to carcinomatous change. There is an annual incidence of carcinoma in ulcerative colitis of 6.5/100,000 in Great Britain. Only 1% of carcinoma of the colorectal region has its origin in ulcerative colitis. The neoplastic change in extensive ulcerative colitis of more than 10 years duration is, however, 10%.[117,119] In 1908 Edwards of St. Marks Hospital, London, said that rectal disease was more frequent in civilized people because of sedentary habits, improper feeding, constipation, and the abuse of purgatives.[61] Tumors of the lower alimentary tract are rare in Africans.[2,30] Some authors stress the widespread use of herbal and proprietory laxatives in African communities.[122] Fifty-seven percent of South African Bantu are said to use aperients and 37% enemas,[22] so it is not surprising to learn that Africans pass three to four bowel motions per day, particularly in view of the high incidence of parasites, which must of themselves certainly affect the bowel.[122]

Burkitt has discussed the role of constipation and fecal arrest resulting in raised intraluminal pressure and prolonged exposure of the lower alimentary mucosa to carcinogenic substances in the feces,[33] suggesting that enhanced carcinogenesis leads to polyps and cancer. In no survey of patients with carcinoma of the colon and rectum, however, has constipation been a significant factor.[20,153,155] Constipation is more common in women, yet the female-to-male ratio for the incidence of these diseases is less than unity. Neither previous appendectomy[155] nor the use of laxatives, with the possible exception of liquid paraffin, has been found to have an association with cancer of these sites.[20] In animal experiments liquid paraffin has never been found to be oncogenic.

There are data to support the theory that carcinoma of the bowel is related to an enhanced fecal bile acid excretion.[85,86] Other studies, however, show that bile acids inhibit sodium and water reabsorption from the colon, so that an enhanced stool output in carcinoma because of the laxative effect of bile acid would be expected, not the reverse as suggested by the constipation hypothesis. This paradox is increased in complexity by the association between ischemic heart disease mortality and carcinoma of the rectum and colon, which has been described internationally. Rose and his colleagues looked at 90 cases of colonic carcinoma.[130] The international mortality rate for coronary heart disease showed a relationship of $r = -0.78$, but in the patients studied the serum cholesterols were found to be low. Such a low serum cholesterol did not apply to carcinoma of other gastrointestinal sites including the rectum and anal canal. In many larger studies, e.g., of Japanese,[153] the serum cholesterol in patients with carcinoma of the rectum and colon was shown not to differ from control subjects.

Hill and his colleagues[86] extended this suggestion that cholesterol metabolism and colonic carcinoma were interrelated. When there is an excess of serum cholesterol, it is converted to bile acids, which pass to the colon. This is associated with a propensity to neoplasia. The fecal bile acid concentration in patients with carcinoma was 6.8 mg/g dry weight, and in the controls with nonmalignant gastrointestinal disease, it was 4.5 mg/g dry weight. Among patients with large-bowel carcinoma, 82% had fecal bile acid concentrations of over 6 mg/g, and 82% of cancer patients had nuclear dehydrogenating bacteria in comparison to 43% of controls. When these factors were combined, 70% of cancer patients had a high fecal bile acid and nuclear dehydrogenating bacterial content of the stool compared with 9% of the comparison group.[86]

Aries and his colleagues[9] have discussed the role of bacteria in the etiology of cancer of the colon and rectum. In their comparison of English and Ugandan feces, the English feces were found to have more bacteroides and bifido bacteria and fewer streptococci than the Ugandan stools, raising the possibility of more carcinogens being produced by these anaerobic bacteria.

The paradox developing is that while cardiologists are trying to diminish serum cholesterol and increase the fecal bile acid excretion to diminish ischemic heart disease, the possibility is now emerging that carcinoma of the colon and rectum may be related to enhanced fecal bile acids. Burkitt et al.[34] showed the importance of transit time along the intestinal tract in the etiology of colonic disease. They studied the transit time and the passage of stool of 12,000 people all over the world and related this to incidence of colonic disease. They found that a refined diet decreased stool weight and transit time. They also suggested an inverse relationship between an individual's stool weight and transit time and his susceptibility to colonic disease.

This suggestion is at variance with findings of a study of transit time of the population of Hawaii by Glober and his colleagues,[68] which showed that the mean colonic transit time in the white Hawaiians was 54 ± 28 hr, whereas in the Japanese it was 31 ± 14 hr. The incidence of colonic cancer was equivalent in all of the groups studied. Data from intestinal transit times are, however, notoriously unreliable.[58]

There is an association between carcinoma of the colon and carcinoma of the breast (0.8), and both correlate with affluence, high-animal-fat content of the diet, and the availability of motor vehicles.[53] There may be a familial etiological factor, with an excess of first-degree relatives with cancer of the colon being identified compared with the controls.[104,106]

F. Dietary Influences

The chance of a patient developing colonic cancer is proportional to the incidence of colonic cancer in the area in which he resides rather than in that in which he was born.[79,81] Because of the relative absence of colonic disease in Africans, the high incidence in Europe and North America, and the rather rapid effects of migration on the incidence of colonic disease, epidemiologists formerly sought a dietary variable. Infection is rarely mentioned as an etiological factor in colonic disease, even though, as discussed in the introduction to this section, viral infection is an important possibility as a cause of other tumors. The increase in the incidence of colonic diseases since the nineteenth century could also be significant.

Available data do not establish a major environmental factor or factors,[28] although dietary patterns may accord with the distribution of bowel cancer. This association is greater for cancer of the colon than for cancer of the rectum.

Dietary information on specific food items obtained in retrospective histories is of little value in determining the influence of diet in cancer of the large bowel.[154] Diet is, however, the fulcrum of the fiber hypothesis for African rural communities. These communities are established, easily identified populations who eat a high-fiber, low-fat, low-protein diet and who have no diagnosed cancer of the bowel.[30] Low-dietary-fiber, high-fat, high-protein dietaries are seen to coexist with a high incidence of colonic cancer. The argument then proceeds to suggest that the return of an adequate amount of cereal fiber to the diet would abolish constipation, use of laxatives (also seemingly used by Africans), appendicitis, diverticular disease, and cancer of the bowel. Yet the African diet is not a cereal-fiber-containing diet.

The cause of the apparent confusion is perhaps the use of the term fiber, an archaic term, used in legislation about animal feedingstuffs to prevent adulteration of food.[45] It could be compared with the use of nitrogen values as a basis for explaining enzyme reactions. The requirement is to identify the composition of items in African diets, e.g., constituent polysaccharides.

Drasar and Irving have looked at the consumption of various fiber-containing foods and found no correlation between diet containing potatoes, pulses, nuts, raisins, vegetables, and fruits, with carcinoma of the colon. The only marginally significant correlation was with cereals at – 0.30.[54]

Undoubtedly Westernization of the diet with an increased standard of living parallels the rate of colonic cancer.[78,139,153,155] There is more colonic cancer with an increase of fat and animal protein in the diet and a closer correlation than with dietary sugar and carbohydrate. Dietary changes may alter the balance and type of the colonic bacteria by control of the supply of substrate and hence the flora acting upon it, e.g., bile acids. The bacteria in turn may dictate how potential carcinogens are converted to innocuous or oncogenic materials.[23] Among potential carcinogens are tryptophan, tyrosine, lycine, argenine, and lecithin, which may be converted to nitrosamine. N-Nitroso compounds present in bacon, smoked fish, and mushrooms may be converted to carcinogens.[98] Reducing agents are protective against the conversion of potential carcinogens to actual carcinogens. Sausages and other prepared foods have a high ascorbic acid content and in some dietaries are even the principal source of ascorbic acid, so this reducing agent may have an unremarked protecting effect.

Workers whose experience and data are based on Japan suggest that dietary fat is all important.[155] Japanese in Japan who have a low incidence of carcinoma of the bowel obtained only 12% of their calories from fat, whereas in the United States it is 40%. It has been shown[155] that there is a higher proportion of professional and technical workers among male colonic cancer patients than among controls. Drasar and Irving have looked at the food items based on Food and Agricultural Organization data and found a correlation of carcinoma of the bowel with total dietary fat of 0.8, with animal protein of 0.8, and with animal fat of 0.78, each of which correlates better than any socioeconomic variable. Dietary fiber based on vegetables being 3% fiber does not relate to colonic cancer nor do sugar or sweets ($r = 0.29$).[53,54] Japanese colonic cancer patients have diets that are lower in rice but contain more fruit and milk, and hence fat content, than control subjects, and yet the diet of patients with rectal cancer is indistinguishable from that of controls.[153]

Berg and Howell were attracted by the relationship between bowel cancer and beef consumption.[14] They developed a mathematical description which suggests that many of the factors of the colon are equivalent to those in the rectum. They suggest beef as a factor which influences rectal carcinoma only and which accounts for the difference in the susceptibility of the two sites. In Scotland 19% more beef is eaten than in England, and there is 19% more carcinoma of the colon and rectum. Similarly, in New Zealand, Uruguay, and Argentina, a large amount of beef is eaten, and this coincides with an incidence of carcinoma higher than elsewhere in the world. Observations of this sort, however, lack the precision of association found between carcinoma of the lung and cigarette smoking.

Bjelke suggests that processed meat and coffee are etiologically important in carcinoma of the colon.[17] Voisin wrote a dissertation on the link between health and the mineral balance of the soil, remarking that cancer is more frequent where soils are deficient in copper.[145] Figure 3 is an attempt to draw these various theories together. The substrate is of obvious importance, irrespective of whether it is endogenous or exogenous in origin, and the predominant presence of reducing or oxidizing agents from hydrolyzed protein on fat or carbohydrate will have an important influence on bacterial behavior (see Chapter 4). The problem of carcinogenesis probably devolves on bacterial activity, whether or not the substrates presenting themselves to the colon are converted to innocent or carcinogenic substances. Fiber will also be important because of the surface effects it provides and because of its water-holding capacity and hence water-diluting effects.

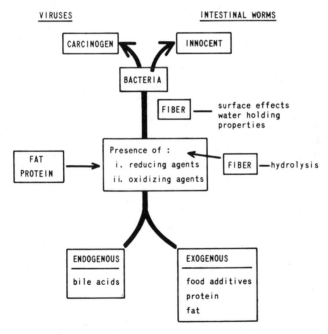

Figure 3. Influences on the metabolism of substrate of exogenous and endogenous origin by bacteria: (a) the presence of reducing agents provided by hydrolysis of fats, proteins, or fiber; (b) provision of surface effects by fiber or dilution of the colonic contents by water-holding capacity of fiber. These various influences could dictate whether the end product of this metabolism is carcinogenic or innocent.

IX. Volvulus of the Colon

Volvulus of the sigmoid colon is the commonest cause of colonic obstruction in India,[5,74,75] East Africa,[132] Scandinavia,[25,127] and Eastern Europe,[25] although it is extremely rare in the United States and Britain. The obstruction is caused by the distal bowel twisting on its mesentery. It is found more commonly after the age of 50 and is reported twice as frequently in males as in females. Ghavarni and Saidi,[66,131] from Shiraz in Iran, showed that in a population whose staple diet was rice, barley, wheat, and other vegetable matter, volvulus was very common, particularly in the sigmoid colon, although diverticulosis of the colon was not found. Similar data are available from Gulati et al.[74] in Delhi, who showed that the commonest cause of acute intestinal obstruction was volvulus of the sigmoid colon. The incidence is high in Hindus, who are vegetarians, and less in Muslims and Sikhs, who are nonvegetarians. It is interesting that in Nazi-occupied countries during World War II, when the fiber content of the diet increased, this was paralleled by an increased incidence of volvulus of the colon.

X. Inflammatory Bowel Disease

The term "inflammatory bowel disease" could be taken, in the broadest sense, as applying to any condition of the small or large bowel in which the classical inflammatory response of living tissues to any injury is seen. The cause of this response may be infective, ischemic, chemical, or traumatic, but by common usage the term is restricted to two conditions of unknown etiology: ulcerative colitis and Crohn's disease.

A. Pathology

Ulcerative colitis involves the colon and only the colon. The pathological effect is continuous in the bowel, with the distal part or the rectum always involved and with a variable proximal spread up to and possibly including the whole colon. Microscopically, it is primarily a mucosal and submucosal disease with confluent shallow ulcers.

Crohn's disease may involve any part of the alimentary tract from the mouth to the anus and is characteristically discontinuous or segmental. When the small bowel is affected, functional derangement may produce a variable degree of malabsorption. Microscopically, all the layers of the bowel wall are involved; that is to say, it is "transmural." It is characterized by ulcers, which progress to fissures or sinuses that may involve neighboring structures. Granulomata, collections of multinucleate giant cells and lymphocytes, are found in the bowel wall and in the regional lymph nodes in 60–70% of the cases.

When the colon alone is the site of inflammation, the differential diagnosis between the two diseases is sometimes difficult, with resultant epidemiological problems.[102] The symptoms and other features of the diseases indicate that ulcerative colitis is a more homogeneous entity than Crohn's disease and that the latter may have more than one etiology.[90]

As with many diseases of unknown etiology, there is no shortage of experimental data and hypotheses.[49,51,150] These fall into three broad groups: (1) infective, (2) immune aberration, and (3) ingested substances. For the present discussion mention will be made only of some points relevant to group 3.

In Crohn's disease, for instance, it has been recognized for many years that if the continuity of the bowel is changed so that there is diversion or bypass of the inflamed area, the disease activity in the bypassed area frequently heals. Unfortunately the disease continues at other sites that remain in contact with the luminal contents and recurs in the bypassed areas if continuity is restored.[124] It seems possible to achieve a similar result "medically" by total emptying of the alimentary tract, with provision for metabolic requirements by intravenous feeding with protein hydrolysates, hypertonic carbohydrate solutions, and fat emulsions.[6,65,144]

B. Age, Sex, Race

The age distribution of patients with ulcerative colitis is very constant in reports from widespread parts of the world. A careful mathematical study of the phenomenon has been made,[27] and this shows a bimodal distribution, with one group having an onset at the age of about 20 years and a later clinically and probably genetically distinct group arising in the 50s. These authors suggest that this distribution is not consistent with the primary cause of ulcerative colitis being an extrinsic factor such as an organism or dietary factor, as they suggest that these would not be constant in different parts of the world. They put forward the possibility that ulcerative colitis is confined to individuals with a genetic predisposition in whom the attack is initiated by random events. An alternative explanation of the bimodal distribution is that two separate disease entities exist.[64] Explanations are further complicated by the necessity of excluding ischemic colitis in the older group. A similar bimodal age distribution is found in Crohn's disease.[73] There is a peak incidence in the 20s and 30s and a second peak incidence in some areas in the 70s and 80s. Most of the recent increase in Crohn's disease can be explained by the appearance of this older female colonic peak.

The female-to-male ratio of ulcerative colitis is remarkably constant at 1.5:1 in British and American studies.[49] This is in contrast to the sex ratio for Crohn's disease which varies from study to study from 0.5:1 female-to-male in Baltimore[116] to 1.9:1 in the northeast of Scotland.[96] These findings are perhaps further evidence of the heterogeneous nature of Crohn's disease contrasting with the homogeneous nature of ulcerative colitis.

Epidemiological clues with regard to race have been sought in areas where different races share a common environment. For example, in the United States Jews seem particularly prone to develop ulcerative colitis and Crohn's disease.[1]

In the Baltimore study Jews had approximately 4.5 times the incidence of ulcerative colitis of non-Jews, and black Americans had a much lower incidence than whites.[116] However, Jews do not seem to suffer unduly in other areas of the world,[69] and ulcerative colitis does not seem to be particularly common in Israel. When it does occur, it is more common in Jews of Western origin.[16] A disproportionate number of this subgroup of Jews in the United States might account for at least part of the enhanced incidence, as this subgroup has an increased occurrence of familial disorders. In a study of two racial groups in New Zealand, the prevalence of ulcerative colitis among Europeans in New Zealand was 41/100,000 compared with 2.5/100,000 for Maoris.[149]

The findings relating to Crohn's disease are similar to those for ulcerative colitis, with a higher risk for Jews compared with non-Jews and for whites compared with nonwhites in the United States.[116]

C. Geographical Considerations

Inflammatory bowel disease is most commonly encountered in northern Europe and North America. Britain has a particularly high incidence of ulcerative colitis, which reaches 6.5/100,000, and of Crohn's disease, which reaches 3.4/100,000.[50] In north Europe, the incidence of ulcerative colitis varies from 0.5/100,000 to 2.6/100,000[69] and of Crohn's disease from 0.3/100,000[67] to 2.5/100,000; however, ulcerative colitis is still approximately twice as common in England.[114] In southern Europe, ulcerative colitis and Crohn's disease are less commonly reported and probably have a true lower incidence. In Baltimore, the incidence was found to be 3.5/100,000 for ulcerative colitis and 1.8/100,000 for Crohn's disease.[116] Ulcerative colitis has been reported from Israel,[16] India, Africa, Egypt, and Brazil,[69] and although accurate estimates are not possible, the disease is obviously uncommon in these areas. Small series have also been reported from New Zealand, southeast Australia, and Japan;[69] however, nearly all the reported experience of Crohn's disease originates from Europe and North America, and it would seem that there has been an increase in incidence of this disease in the last 20 years.[64,67,114,123,134]

In Nottingham between 1958 and 1971, the incidence increased fivefold from 0.7/100,000 to 3.6/100,000.[114] Similar figures were available for England as a whole, with a threefold increase; the increase was particularly pronounced for the colonic type of Crohn's disease. Data from Sweden and from two separate areas of Scotland indicate an increase that is particularly marked in females more than 50 years old.[96,123,134] These widespread changes in incidence rates would seem to be true increases, not simply due to a higher diagnostic rate, and suggest a changing environmental factor.

D. Coexisting Diseases

Patients with ulcerative colitis may have relatives with Crohn's disease; more commonly the reverse is true.[11,93] Also the patient with inflammatory bowel disease or his relatives may be affected by ankylosing spondylitis.[110,112,113]

An overlapping genotype for the three conditions has been suggested as an explanation of this association.[111] In addition, patients with ulcerative colitis have an undue liability to develop malignant disease of the colon.

E. Familial Studies

Many instances have been reported of the familial occurrence of inflammatory bowel disease. Many of these have been isolated reports, which makes interpretation difficult, but support has recently come from controlled studies.[21,133] In a recent review it has been estimated that between 15 and 20% of subjects with inflammatory bowel disease have affected relatives.[92]

F. Dietary Influences

Allergy to normal dietary components has long been considered as a possible provoking factor in inflammatory bowel disease.[4] Allergy to milk protein has been suggested among other possible dietary factors. Some patients with ulcerative colitis have circulating antibodies to milk proteins,[71,140,152] and although these results have not always been substantiated,[56] a proportion of patients do seem to benefit from exclusion of milk from their diet.[143] This may in fact be related to a lactase deficiency.[3] Patients with Crohn's disease do not seem to be intolerant of milk and do not have milk antibodies, but some benefit seems to be obtained from parenteral "nonallergenic diets" which are elemental diets of monosaccharides, amino acids, and fatty acids.[146]

It is, of course, possible that the causes of the disease are in food additives. It is said that there are more than 10,000 well-defined synthetic chemicals added to the diet, and that on the average inhabitants of the Western world eat about 2 kg of additives per year in the form of flavoring, stabilizers, antioxidants, etc. It is said that more than 40,000 tons of surface activators were added to food in the United States in 1965.[39]

It is perhaps finally worth mentioning the special case of the additive carrageenan. This is an extract of seaweed, consisting mainly of sulfated polysaccharides, which is used as a food emulsifier. When degraded by acid hydrolysis, a product is produced which has been used in the treatment of peptic ulcers. Extensive studies by Watt and Marcus in 1973[148] showed that in some animals, notably guinea pigs, colonic ulceration occurs. No case of ulcerative colitis has been reported in any patient given degraded carrageenans. There are defects in this experimental system as a model of ulcerative colitis, but it does provide an important insight into the possibly toxic additives in a Western "processed" diet. These might be compared with the adulterants of previous centuries.

XI. Summary

There has been a change in incidence of bowel disease attributable to the differences in environment between the nineteenth and twentieth centuries. Changes are also attributed to geographical variables. The changing nature of these diseases suggests that there is some variable factor which, if eliminated, would be of value in countries where there is a disease of particularly high incidence. Epidemiologists have suggested that protein, fat, and sugar intake are all important. Other relevant factors are malnourishment and infestation with parasites. A case has been made for the importance of vegetable dietary fiber in the etiology of hemorrhoids, appendicitis, constipation, diverticular disease, and cancer of the bowel. A depletion in the vegetable fiber content of the diet is believed by some to be important, and this case has been examined. In another group, an excess of vegetable dietary fiber may be causative of disease, namely volvulus. Also, the potentially toxic effects of other dietary additives need to be more closely examined.

References

1. ACHESON, E. D. The distribution of ulcerative colitis and regional enteritis in U.S. veterans with particular reference to the Jewish religion. *Gut* **1**:291, 1960.
2. ADETAYO GRILLO, I., L. F. BOND and W. W. EBONG. Cancer of the colon in Nigerians and American Negroes. *J. Natl. Med. Assoc.* **63**:357, 1971.

3. ANALFIN, D., and P. HOLT. Lactase deficiency in ulcerative colitis, regional enteritis and viral hepatitis. *Am. J. Dig. Dis.* **12**:81, 1967.
4. ANDERSON, A. F. R. Ulcerative colitis an allergic phenomenon. *Am. J. Dig. Dis.* **9**:91, 1942.
5. ANDERSON, D. A. Volvulus in Western India. *Br. J. Surg.* **44**:132, 1956–57.
6. ANDERSON, D. L., and H. W. BOYCE. Use of parenteral nutrition in the treatment of advanced regional enteritis. *Am. J. Dig. Dis.* **18**:633, 1973.
7. ANDREWES, C. *Viruses and Cancer*. London, Weidenfeld & Nicolson, 1970.
8. ARFWIDSSON, S. Pathogenesis of multiple diverticula of the sigmoid colon in diverticular disease. *Acta Chir. Scand., Suppl.* **342**, 1954.
9. ARIES, V., J. S. CROWTHER, B. S. DRASAR, M. J. HILL and R. E. O. WILLIAMS. Bacteriology and the etiology of cancer of the large bowel. *Gut* **10**:334, 1969.
10. ASHLEY, D. J. B. Observations on the epidemiology of appendicitis. *Gut* **8**:533, 1967.
11. ATWELL, J. D., H. L. DUTHIE and J. C. GOLIGHER. The outcome of Crohn's disease. *Br. J. Surg.* **52**:966, 1965.
12. AYKROYD, W. R. Sugar in history. *In: Sugars in Nutrition*, H. L. Sipple and K. W. McNutt (eds.). New York, Academic Press, 1974, p. 3.
13. AYKROYD, W. R., and J. DOUGHTY. Wheat in human nutrition. *F.A.O. Nutritional Studies*, No. 23, Rome, Italy.
14. BERG, J. W., and M. A. HOWELL. The geographic pathology of bowel cancer. *Cancer* **34**:807, 1974.
15. BERRY, J. N. Prevalence survey for chronic rheumatic heart disease and rheumatic fever in Northern India. *Br. Heart J.* **34**:143, 1972.
16. BIRNBAUM, D., J. J. GREEN and G. KALLNER. Ulcerative colitis among the ethnic groups in Israel. *Arch. Intern. Med.* **105**:843, 1960.
17. BJELKE, E. Colon cancer and blood cholesterol. *Lancet* **1**:1116, 1974.
18. BOHRER, S. P., and E. A. LEWIS. Diverticula of the colon in Ibadan, Nigeria. *Trop. Geogr. Med.* **26**:9, 1974.
19. BOTSFORD, T. W., S. J. HEALEY and F. VEITH. Volvulus of the colon. *Am. J. Surg.* **114**:900, 1967.
20. BOYD, J. T., and R. DOLL. Gastrointestinal Cancer and the use of liquid paraffin. *Br. J. Cancer* **8**:231, 1954.
21. BRADER, V., E. WEEKE, J. H. OLSON, P. ANTHONISEN and P. RIIS. A genetic study of ulcerative colitis. *Scand. J. Gastroenterol.* **1**:49, 1966.
22. BREMNER, C. G. Ano rectal disease in the South African Bantu. I. Bowel habit and physiology. *S. Afr. J. Surg.* **2**:119, 1964.
23. BRITISH MEDICAL JOURNAL. Leader: "Diet and Colonic Cancer." **1**:339, 1974.
24. BRITISH MEDICAL JOURNAL. Leader: Faecal Fibre Fortunes," **2**:580, 1975.
25. BRUUSGAARD, C. Volvulus of the sigmoid colon and its treatment. *Surgery* **22**:466, 1947.
26. BUELL, P., and J. E. DUNN, JR. Cancer mortality among Japanese Issei and Nisei of California. *Cancer* **18**:656, 1965.
27. BURCH, P. R. J., F. T. DE DOMBAL and C. I. WATKINSON. Etiology of ulcerative colitis. II. A new hypothesis. *Gut* **10**:277, 1969.
28. BURDETTE, W. J. Identification of antecedents to colorectal cancer. *Cancer* **28**:51, 1971.
29. BURKITT, D. P. *"African Lymphoma," Racial and Geographical Factors in Tumour Incidence*. A. A. Shivas (ed.). Edinburgh University Monograph No. 2, 1967.
30. BURKITT, D. P. Epidemiology of cancer of the colon and rectum. *Cancer* **28**:3, 1971.
31. BURKITT, D. P. Cancer and other noninfective diseases of the bowel. Epidemiology and possible causative factors. *Rend. Gastroenterol.* **5**:33, 1973.
32. BURKITT, D. P. Epidemiology of large bowel disease: The role of fibre. *Proc. Nutr. Soc.* **32**:145, 1973.
33. BURKITT, D. P. An epidemiologic approach to cancer of the large intestine. *Dis. Colon Rectum* **17**:456, 1974.

34. BURKITT, D. P., A. R. P. WALKER and N. S. PAINTER. Effect of dietary fibre on stools and transit times and its role in the causation of disease. *Lancet* **2**:1408, 1972.
35. BURKITT, D. P., A. R. P. WALKER and N. S. PAINTER. Dietary fiber and disease. *JAMA* **229**:1068, 1974.
36. BURNETT, J. *Plenty and Want*. London, Penguin, 1968.
37. BURNE, J. *A Treatise on the Causes and Consequences of Habitual Constipation*. London, Longman, Orme, Brown Green Longman, 1840.
38. CARAYON, A. Supplementary survey of appendicitis in Africans (1000 cases). *Soc. Med. Afr. Noire, Lang Franc* **13**:696, 1968.
39. CARSTENSEN, J., and E. POULSEN. Food additives and food contaminants. *In: Regional Enteritis—Crohn's Disease*. Skandia International Symposium. Stockholm, Nordiska Bokhandeln's Förlag, 1971.
40. CLEAVE, T. L., G. D. CAMPBELL and N. S. PAINTER. *Diabetes, Coronary Thrombosis and the Saccharine Disease* 2d ed., Bristol, John Wright & Sons, *2nd Edition* 1969.
41. CLELAND, J. B. Incidence of diverticulosis. *Br. Med. J.* **1**:579, 1968.
42. CONNELL, A. M. Physiology of the colon. *In: Management of Constipation*, F. A. Jones and E. W. Godding (eds.). Oxford, Blackwell, 1972.
43. CONNELL, A. M. Applied physiology of the colon: Factors relevant to diverticular disease. *In: Clinics in Gastroenterology,* A. N. Smith (ed.). London, Saunders, 1975, Vol. 4, p. 23
44. CORMAN, M. L., N. W. SWINTON, D. D. O'KEEFE and M. C. VEIDENHEIMER. Colorectal carcinoma at the Lahey clinic 1962–1966. *Am. J. Surg.* **125**:434, 1973.
45. CUMMINGS, J. H. Dietary fibre, *Gut* **14**:69, 1973.
46. DAVID, V. C. Diverticulosis and diverticulitis with particular reference to the development of diverticula of the colon. *Surg. Gynae. Obstet.* **56**:375, 1933.
47. DAVIES, J. N. P., J. KNOWELDEN and B. A. WILSON. Incidence rates of cancer in Kyadondo County, Uganda. *J. Natl. Cancer Inst.* **35**:789, 1965.
48. DEARLOVE, T. P. Diverticulitis and diverticulosis with a report on a rare complication. *Med. J. Aust.* **1**:470, 1954.
49. DE DOMBAL, F. T. Ulcerative colitis: Epidemiology and aetiology course and prognosis, *Br. Med. J.* **1**:649, 1971.
50. DE DOMBAL, F. T. Symposium on Crohn's disease: Epidemiology and natural history. *Proc. R. Soc. Med.* **64**:161, 1971.
51. DE DOMBAL, F. T., P. R. J. BURCH and G. WATKINSON. Aetiology of ulcerative colitis. I. A review of past and present hypothesis. *Gut* **10**:270, 1969.
52. DOLL, R. The geographical distribution of cancer. *Br. J. Cancer* **23**:1, 1969.
53. DRASAR, B. S., and D. IRVING. Environmental factors and cancer of the colon and breast. *Br. J. Cancer* **27**:167, 1973.
54. DRASAR, B. S., and D. IRVING, Fibre and cancer of the colon. *Br. J. Cancer* **28**:462, 1973.
55. DRUMMOND, H. Sacculi of the large intestine with special reference to their relations to the blood vessels of the bowel wall. *Br. J. Surg.* **4**:407, 1917.
56. DUDEK, B., H. H. SPIRO and W. R. THAYER. A study of ulcerative colitis and circulating antibodies to milk proteins. *Gastroenterol.* **49**:544, 1965.
57. EARLE, S. T. Diseases of anus, rectum and sigmoid. Philadelphia and London, Lippincott, 1911.
58. EASTWOOD, M. A., N. FISHER, C. T. GREENWOOD and J. B. HUTCHINSON. Perspectives on the bran hypothesis. *Lancet* **1**:1029, 1974.
59. EASTWOOD, M. A., and W. D. MITCHELL. The place of vegetable fibre in diet. *Br. J. Hosp. Med.* **12**:123, 1974.
60. EDWARDS, H. C. *Diverticula and Diverticulitis of the Intestine*. Bristol, J. Wright & Sons, 1939.
61. EDWARDS, S. *Diseases of the Rectum and Sigmoid Colon*. London, Churchill, 1908.

62. EPSTEIN, M. A., and B. G. ACHONG. Various forms of Epstein–Barr virus infection in man—established facts and a general concept. *Lancet* **2**:836, 1973.
63. ESLER, M. D., and K. J. GOULSTON. Levels of anxiety in colonic disorders. *N. Engl. J. Med.* **288**:16, 1973.
64. EVANS, J.G., and E. D. ACHESON. An epidemiological study of ulcerative colitis and regional enteritis in the Oxford area. *Gut* **6**:311, 1965.
65. FISCHER, J. E., G. S. FOSTER, R. M. ABEL, W. M. ABBOTT and J. A. RYAN. Hyperalimentation as primary therapy for inflammatory bowel disease. *Am. J. Surg.* **125**:165, 1973.
66. GHAVARNI, A., and F. SAIDI. Pattern of colonic disorders in Iran. *Dis. Colon Rectum* **12**:462, 1969.
67. GJONE, E., O. M. ORNING and J. MYREN. Crohn's disease in Norway 1956–1963. *Gut* **7**:372, 1966.
68. GLOBER, G. A., K. L. KLEIN, J. O. MOORE and B. C. ABBA. Bowel transit times in two populations experiencing similar colon cancer risks. *Lancet* **2**:80, 1974.
69. GOLIGHER, J. C., F. T. DE DOMBAL, J. McK. WATTS, and G. WATKINSON. In: *Ulcerative Colitis*. London, Balliere, Tindall and Cassell, 1968.
70. GOODWIN, S. *Constipation and Intestinal Obstruction*. Philadelphia, Saunders, 1909.
71. GRAY, J. Antibodies to cow's milk in ulcerative colitis. *Br. Med. J.* **2**:1265, 1961.
72. GREGORY, J. E. *Pathogensis of Cancer*, 2d ed. Pasadena, California, Fremont Foundation, 1952.
73. GRIMLEY EVANS, J. The epidemiology of Crohn's disease. In: *Clinics in Gastroenterology—Crohn's Disease*, B. N. Brooke (ed.). London, Saunders, 1972, p. 335.
74. GULATI, S. M., N. K. GROVER, N. K. TAGORE and O. P. TANEJA. Volvulus of the sigmoid colon in Delhi, India. *Dis. Colon Rectum* **17**:219, 1974.
75. GUPTA, S., and M. P. VAIDYA. Volvulus of the sigmoid colon. *Ind. J. Surg.* **31**:569, 1969.
76. GYE, W. E., and W. J. PURDY. *The Cause of Cancer*. London, Cassell, 1931.
77. HADDOW, A. The chemical and genetic mechanisms of carcinogenesis. I. Nature and mode of action. In: *The Physiopathology of Cancer*, F. Homburger and W. H. Fishman (eds.). London, Cassells, 1953.
78. HAENSZEL, W., J. W. BERG, M. SEGI, M. KURIHARA, M. and F. B. LOCKE. Large-bowel cancer in Hawaiian Japanese. *J. Natl. Cancer Inst.* **51**:1765, 1973.
79. HAENSZEL, W., and P. CORREA. Cancer of the colon and rectum and adenomatous polyps in a review of epidemiologic findings. *Cancer* **28**:14, 1971.
80. HAENSZEL, W., and P. CORREA: Cancer of the large intestine: Epidemiologic findings. *Dis. Colon Rectum* **16**:371, 1973.
81. HAENSZEL, W., and E. A. DAWSON. A note on mortality from cancer of the colon and rectum in the United States. *Cancer* **18**:265, 1965.
82. HAVIA. T. Diverticulosis of the colon. *Acta Chir. Scand.* **137**:367, 1971.
83. HAVIA, T., and R. MANNER. The irritable colon syndrome—a follow up study with special reference to the development of diverticula. *Acta Chir. Scand.* **137**:569, 1971.
84. HEMMETER, J. C. *Diseases of the Intestine*. London, Rebman, 1901.
85. HILL, M. J. Bacteria and the etiology of colonic cancer. *Cancer* **34**:815, 1974.
86. HILL, M. J., B. S. DRASAR, R. E. O. WILLIAMS, T. W. MEADE, A. G. COX, J. E. P. SIMPSON, and B. C. MORSON. Faecal bile acids and clostridia in patients with cancer of the large bowel. *Lancet* **2**:535, 1975.
87. HISLOP, I. G. Psychological significance of irritable colon syndrome. *Gut* **12**:452, 1971.
88. HUGHES, L. E. Post mortem survey of diverticular disease of the colon. *Gut* **10**:336, 1969.
89. HYAMS, L., and J. PHILPOTT. An epidemiologic investigation of haemorrhoids. *Am. J. Proctol.* **21**:177, 1970.
90. JONES-HYWEL, J., J. E. LENNARD-JONES, B. C. MORSON, M. CHAPMAN, M. J. SACKIN, P. H. A. SNEATH, C. C. SPICER, and W. I. CARD. Numerical toxonomy and discriminant analysis applied to nonspecific colitis. *Q. J. Med.* **42**:715, 1973.

91. KENT, S. J. S. Diverticulosis of the transverse colon. *Br. Med. J.* **2**:219, 1973.
92. KIRSNER, J. B., Genetic aspects of inflammatory bowel disease. *In: Clinics in Gastroenterology—Genetics of Intestinal Disorders*, R. B. McConnell (ed.). London, Saunders, 1973.
93. KIRSNER, J. B., and J. A. SPENCER. Familial occurrences of ulcerative colitis, regional enteritis and iliocolitis. *Ann. Intern. Med.* **59**:133, 1963.
94. KOHLER, R. The incidence of colonic diverticulosis in Finland and Sweden. *Acta Chir. Scand.* **126**:148, 1963.
95. KOLADE, S. O., E. B. CHUNG, J. E. WHITE and L. D. LEFFALL, JR. Neoplastic lesions of the colon and anorectum in blacks. *J. Natl. Med. Assoc.* **65**:142, 1973.
96. KYLE, J. An epidemiological study of Crohn's disease in N.E. Scotland. *Gastroenterology* **61**:826, 1971.
97. KYLE, J., A. O. ADESOLA, L. F. TINCKLER and J. DE BEAUX. Hospital admission rates to major teaching hospitals in Fiji, Singapore, Nigeria, N.E. Scotland. *Scand. J. Gastroenterol.* **2**:77, 1967.
98. LANCET. Leader: "Environmental Nitrosamines." **2**:1243, 1973.
99. LANCET. Leader: "Beyond the Examining Finger." **2**:1185, 1974.
100. LANE, W. A. *The Operative Treatment of Chronic Constipation*. London, Nisbet & Col, 1909.
101. LATHAM, M. C. *Human Nutrition in Tropical Africa*. Rome, Italy, F.A.O., 1965.
102. LENNARD-JONES, J. E. Differentiation between Crohn's disease, ulcerative colitis and diverticulitis. *In: Clinics in Gastroenterology—Crohn's Disease*, B. N. Brooke (ed.). London, Saunders, 1972, p. 367.
103. LEVITICUS, Chapter 23, v. 17.
104. LOVETT, E. Heredity and bowel disease. *Gut* **15**:345, 1974.
105. LYNCH, H. T., H. GUIRGIS, M. SWARTZ, J. LYNCH, A. J. KRUSH and A. R. KAPLAN. Genetics and colon cancer. *Arch. Sug.* **106**:669, 1973.
106. LYNCH, H. T., and A. J. KRUSH. Heredity and adenocarcinoma of the colon. *Gastroenterology* **53**:517, 1967.
107. MALHOTRA, S. L. Geographical distribution of gastrointestinal cancers with special reference to causation. *Gut* **8**:361, 1967.
108. MANOUSOS, O. N., S. C. TRUELOVE and K. LUMSDEN. Prevalence of colonic diverticulosis in the general population of Oxford area. *Br. Med. J.* **3**:762, 1967.
109. MANOUSOS, O. N., G. VRACHLIOTIS, G. PAPAEVANGELOU, E. DETORAKIS, P. DORITIS, L. STERGIOU and G. MERIKAS. Relation of diverticulosis of the colon to environmental factors in Greece. *Dig. Dis.* **18**:174, 1973.
110. MCBRIDE, J. A., M. J. KING, A. G. BAIKIT, G. P. CREAN and W. SIRCUS. Ankylosing spondylitis and chronic inflammatory disease of the intestine. *Br. Med. J.* **2**:483, 1963.
111. MCCONNELL, J. B. Genetics of Crohn's disease. *In: Clinics in Gastroenterology—Crohn's Disease*, B. N. Brooke (ed.). London, Saunders, 1972, p. 321.
112. MCCONNELL, R. B. Inflammatory bowel disease and ankylosing spondylitis. Birth defects. Original Article Ser. **8**:42, 1972.
113. MCRAE, I. F., and V. WRIGHT. Ulcerative colitis and sacro-ileitis: A family study. *Ann. Rheum. Dis.* **29**:559, 1970.
114. MILLER, D. S., A. C. KEIGHLEY and M. J. S. LANGMAN. Changing patterns in the epidemiology of Crohn's disease. *Lancet* **2**:691, 1974.
115. MITCHELL, H. F. Sociological aspects of cancer rate surveys in Africa. *In: Tumours of the Alimentary Tract in Africans*, J. F. Murray (ed.). Monograph No. 25. Washington, National Cancer Inst., 1967.
116. MONK, M., A. I. MENDELEFF, C. J. SIEGEL and A. LILLIENFELD. An epidemiological study of ulcerative colitis and regional enteritis among adults in Baltimore. II. Social and demographical features. *Gastroenterology* **56**:847, 1969.

117. MORGAN, C. N. Malignancy in the inflammatory diseases of the large intestine. *Cancer* **28**:41, 1971.
118. MORSON, B. C. The muscle abnormality in diverticular disease of the colon. *Proc. R. Soc. Med.* **56**:798, 1963.
119. MORSON, B. C. Cancer in ulcerative colitis. *Gut* **7**:425, 1966.
120. MORSON, B. C. The polyp–cancer sequence in the large bowel. *Proc. R. Soc. Med.* **67**:451, 1974.
121. MORSON, B. C., and I. M. P. DAWSON. *Gastrointestinal Pathology*. Oxford, Blackwell Scientific Publication, 1972.
122. MURRAY, J. F. (ed.). *Tumours of the Alimentary Tract in Africans*. Monograph No. 25. Washington, U.S. National Cancer Inst., U.S. Department of Health, Education and Welfare, 1971.
123. NORLEN, B. J., U. KRAUSE and L. BERGMAN. An epidemiological study of Crohn's disease. *Scand. J. Gastroenterol.* **5**:385, 1970.
124. OBERHELMAN, JR., H. A., S. KOHATSU, K. B. TAYLOR and R. M. KIVEL. Diverting ileostomy in the surgical management of Crohn's disease of the colon. *Am. J. Surg.* **115**:231, 1968.
125. PAINTER, N. S., and D. P. BURKITT. Diverticular disease of the colon, a 20th century problem: *In: Clinics in Gastroenterology*, A.N. Smith (ed.). London, Saunders, 1975, Vol. 4, p. 3.
126. PECK, D. A., R. LABART and V. G. WAITE. Diverticular disease of the right colon. *Dis. Colon Rectum* **11**:49, 1968.
127. PETERSON, H. I. Volvulus of the cecum. *Ann. Surg.* **166**:296, 1967.
128. PODESTA, M. T., and J. L. PACE. The incidence of diverticula of the colon in Malta based on radiology. *Rend. Gastroenterol.* **5**:168, 1973.
129. RODKEY, G. V., and C. E. WELCH. Diverticulosis of the colon—evaluation in concept and therapy. *Clin. N. Am.* **45**:1231, 1965.
130. ROSE, G., H. BLACKBURN, A. KEYS, H. L. TAYLOR, W. B. KANNALL, O. PAUL, D. D. REID and J. STAMLER. Colon cancer and blood cholesterol. *Lancet* **1**:181, 1974.
131. SAIDI, F. High incidence of intestinal volvulus in Iran. *Gut* **10**:838, 1969.
132. SHEPHERD, J. J. Treatment of volvulus of sigmoid colon, a review of 425 cases. *Br. Med. J.* **1**:280, 1968.
133. SINGER, H. C., J. A. D. ANDERSON, H. FRISCHER and J. B. KIRSNER. Familial aspects of inflammatory bowel disease. *Gastroenterology* **61**:423, 1971.
134. SMITH, I. S., S. YOUNG, G. GILLESPIE, J. O'CONNOR and J. R. BELL. Epidemiological aspects of Crohn's disease in Clydesdale 1961–1970, *Gut* **16**:62, 1975.
135. SNOW, H. L. *Clinical Notes on Cancer, its Etiology and Treatment*. London, Churchill, 1883.
136. STANTON, W. R. *Grain Legumes in Africa*. Rome, F.A.O., 1966.
137. STEINER, P. E. *Cancer, Race and Geography*. Baltimore, Williams and Wilkins, 1954.
138. STEMMERMANN, G. N., and R. YATAI. Diverticulosis and polyps of the large intestine: a necropsy study of Hawaiian Japanese. *Cancer* **31**:1260, 1973.
139. STEWARD, H. L. Geographic pathology of cancer of the colon and rectum. *Cancer* **28**:25, 1971.
140. TAYLOR, K., and S. TRUELOVE. Circulating antibodies to milk proteins in ulcerative colitis. *Br. Med. J.* **2**:924, 1961.
141. THIJN, C. J. P. Irritable colon: A preliminary stage of colonic diverticulosis. *Radiol. Clin. Biol.* **42**:468, 1973.
142. TROWELL, H., N. S. PAINTER and D. P. BURKITT. Aspects of the epidemiology of diverticular disease and ischemic heart disease. *Am. J. Dig. Dis.* **19**:864, 1974.
143. TRUELOVE, S. Ulcerative colitis provoked by milk. *Br. Med. J.* **1**:154, 1961.
144. VOGEL, C. M., T. R. CORWIN and A. E. BAUE. Intravenous hyperalimentation. *Arch. Surg.* **108**:460, 1974.

145. Voisin, A. *Soil Grass and Cancer.* London, Crosby Lockwood & Son, 1959.
146. Voitk, A. J., V. Echave, J. A. Feller, K. A. Brown and F. N. Gurd. Experience with elemental diet in the treatment of inflammatory bowel disease—is this primary therapy? *Arch. Surg.* **101**:329, 1973.
147. Waterhouse, J. A. H. *Cancer Handbook of Epidemiology and Prognosis.* Edinburgh, Churchill Livingstone, 1974.
148. Watt, J., and R. Marcus. Experimental ulcerative disease of the colon in animals. *Gut* **14**:506, 1973.
149. Wigley, R. D., and B. P. McLaurin. A study of ulcerative colitis in New Zealand showing a low incidence in Maoris. *Br. Med. J.* **2**:228, 1962.
150. Willougby, J. M. T. *The Small Intestine,* B. C. Creamer (ed.). London, Heinemann, 1974, Chap. 9, p. 146.
151. Wretlind, A. World sugar production and usage in Europe. *In: Sugars in Nutrition,* H. L. Sipple and K. W. McNutt (eds.). New York, Academic Press, 1974.
152. Wright, A., and S. Truelove. Circulating antibodies to dietary proteins in ulcerative colitis. *Br. Med. J.* **2**:142, 1965.
153. Wynder, E. L., T. Kajitana, S. Ishikawa, H. Dodo and A. Takano. Environmental factors of cancer of the colon and rectum. II. Japanese epidemiological data. *Cancer* **23**:1210, 1969.
154. Wynder, E. L., and B. Reddy. Studies of large bowel cancers: Human leads to experimental application. *J. Natl. Cancer Inst.* **50**:1099, 1973.
155. Wynder, E. L., and T. Shigematsu. Environmental factors of cancer of the colon and rectum. *Cancer* **20**:1520, 1967.

10

Gastrointestinal Diseases and Fiber Intake with Special Reference to South African Populations

Alexander R. P. Walker

I. Introduction

South Africa occupies about half a million square miles. The main populations are Negroes (16 million), Whites (4 million), Coloureds (Eur-African-Malays, 2 million), and Indians (1 million). Negroes may be observed in all stages of transition of diet and manner of life, from those pursuing a primitive and frugal existence in remote country regions to those dwelling in large cities, segments of whom are accustomed to a sophisticated manner of life and possess cars and employ servants. Extremes in socioeconomic state, although less wide, also prevail with the other populations.

Changes, in varying degree, in patterns of diseases and diet have occurred in the last half century or more. However, the changes as seen in South African populations are encountered throughout the world. Because of their universality, they will first be described generally to provide an appropriate background against which the changing situations in South African populations may be better appreciated.

ALEXANDER R. P. WALKER ● Medical Research Council, Human Biochemistry Research Unit, South African Institute for Medical Research, Johannesburg, South Africa.

II. General Pattern of Changes in Disease and Diet

In many parts of the world there has been a dramatic alteration in disease pattern from one predominantly of infections to one of degenerative diseases. The alteration is evident when (1) our ancestors are compared with ourselves, (2) rural are compared with town populations in developing countries, and (3) persons in developing countries are compared with those who have settled in Western countries.

During transition from primitiveness to sophistication, the mortality rate of the young has fallen and expectation of life at birth has greatly increased. This gratifying situation masks the fact that at middle age and thereafter, the gain in expectation of life has been disappointingly small; indeed, it has barely improved in the last three centuries. Among the elderly, therefore, the change in pattern of disease has brought no amelioration of mortality.

Etiologically, the conditions and diseases that have become more common are almost wholly of environmental origin. Which factors have changed so markedly such that at present two thirds of all deaths in Western populations are due to coronary heart disease, cerebral vascular disease, and cancer? The most plausible factor, of course, is diet, although other changes must not be overlooked, e.g., in physical activity, cigarette smoking, atmospheric pollution, and stresses linked with urbanization and rise in privilege.

In the three contexts enumerated, which dietary changes have occurred or are occurring? Calorie intake increases. Total intakes of protein and fat rise slightly, but variably. Total carbohydrate intake falls; less bread and other cereal foods are eaten, and they tend to be refined in character. Sugar intake becomes doubled or more. There are changes in intakes of mineral salts and vitamins; intakes of some nutrients rise; others fall. Bulk-forming capacity of diets unequivocally decreases. Although there are increases in the consumption of vegetables and fruits, their bulk-forming capacity does not compensate for that derived formerly from lightly milled cereal products. It is the fiber* in cereals that is the primary regulatory factor in transit time of bowel content, frequency of defecation, and weight of feces passed *per diem*. Fiber ingestion too, has the capacity to alter the chemical and microbiological makeup of feces.

Of the dietary changes depicted (discussed in detail for South African populations), which component or components can be specifically or particularly blamed for the tremendous changes in disease pattern, with or without modification from one or more nondietary factors?

Historically, alarm over the changing pattern of diseases began before the end of the last century. In Britain, Brunton,[17] Treves,[81] and MacEwen,[52] and

*Throughout this contribution the term fiber is used in the orthodox sense, i.e., as determined analytically by the alkali and acid digestion procedure.

later McCarrison,[58] blamed refinement of foods in general. In the United States, Charles Mayo[56] referring to appendicitis, maintained that "a soil has become engendered and has become susceptible to diseases by changes of food that have developed within a few decades." In South Africa, associations with changes in food habits, especially fiber depletion, were pointed out by Walker[87,89] with respect to changes in gastrointestinal conditions or diseases (bowel motility, appendicitis, gallstones), diabetes, atherosclerosis, and coronary heart disease. In Britain, Cleave et al.,[23] Burkitt,[19,20] Painter et al.,[63] and Trowell[84] have considerably advanced and enlarged the hypothesis of the incrimination of food refinement and fiber depletion to include the emergence and increasing prevalence of many other diseases: obesity, diverticular disease, colonic cancer, varicose veins, hemorrhoids, and hiatus hernia. Burkitt[19] particularly has amassed data that show associated changes in the prevalence, as well as remarkable interrelations in the emergence, of the diseases mentioned. He emphasizes that geographically and chronologically, without exception, all populations accustomed to diets high in fiber have negligible or low prevalences of those diseases.

For the changes in disease pattern, Burkitt and associates attach blame principally to fiber depletion.[19,20,64,66,84,87,89] Cleave[23,24] believes that refined carbohydrates, especially sugar, as well as fiber depletion, are mostly implicated. Yudkin[101] maintains that increased consumption of sugar is the chief responsible factor. More recently, Reddy and Wynder[67] and Drasar and Hill,[29] who have confined their attention to the relationship between diet and colonic cancer, aver that the rise in fat intake is primarily involved in the emergence of the disease.

The foregoing provides the background for a discussion of gastrointestinal diseases in South African populations and the bearing of fiber intake on their causation and possible prevention.

III. Diets of South African Populations

A. Diets of Negroes

The diet of Negroes varies.[88] Among dwellers in the Homelands and to a lesser extent among workers on farms, the traditional diet with local modifications is still followed. Broadly, maize, mainly lightly milled, is still the staple, supplemented in parts with "kaffir corn" (*Sorghum vulgare*), millet, and wheat products. Additional foods include dried peas and beans, groundnuts, pumpkin, "kaffir melon" and other vegetables, fruits, and wild greens (*m'fino, morogo*). Consumption of fermented cereal products (*marewu*, "kaffir beer") varies greatly. Meat is consumed irregularly and milk usually in small quantities in

TABLE I. Dietary Intakes of Venda and Cape Town
Negro Males, Ranging in Age from 20 to 50 Years Old (in grams)

Diet element	Venda rural males[50]	Venda urban males[50]	Cape Town bachelors[54]
Calories (in kcal)	3420	3200	4240
Protein	116	126	125
Fat	65	52	96
Carbohydrate	590	555	721
Fiber	25	5	10

season. Rural Negroes purchase varying, although increasing, amounts of sugar, tea, coffee, soft drinks, condensed milk, and tinned fish.

In towns, domestic servants eat much the same foods as their employers; maize-meal porridge, however, is still popular. Some groups of men, especially the gold and coal miners, are well catered for by industrial concerns. However, the vast majority of urban Negroes buy their own food and eat a partially Westernized diet. Bread and maize products, usually refined in character, are major sources of energy. Sugar intake is increasing. Meat is eaten fairly regularly, often one or more times a day, and more milk is taken. Municipally prepared kaffir beer made from maize and kaffir corn is popular with men and to a lesser extent with women.

For the bulk of older children and adults, the diet is probably adequate in calories and gross protein, low in animal protein and fat, high in carbohydrate and fiber (more especially in rural areas), low in calcium, usually or frequently high in iron, and borderline or low (with exceptions in some groups) in most of the vitamins.

Examples of dietary intakes, from a study on Venda Negro males and on Negro bachelors in Cape Town, are shown in Table I.

The dietary pattern of Negroes substantially is that of the forebears of Whites living one or two centuries ago and is similar to that of many other developing populations: it is, moreover, the dietary pattern of indigent rural peasant populations in Europe and elsewhere. Thus, the pattern of diet of Negroes is common to the huge bulk of the world's population.

B. Diet of Indians

Diets differ among constituent groups of Indians.[88,93] Moslems usually are nonvegetarians, and partake of all common foods save pork. Some Hindus are vegetarian, others are nonvegetarian. About half of their calorie intake is derived from carbohydrate, 25–40% from fat, and about 10–15% from protein. Carbohydrate is largely supplied by rice, bread, potatoes, and sugar. Fat is derived from *ghee* (produced by heating butter and removing the sediment by filtering through cloth) and to a lesser extent from vegetable oils. Milk, pulses, and cereals are chief sources of protein for Hindu vegetarians. Mutton, chicken, eggs, pulses, and cereals are main sources of protein for Hindu nonvegetarians. Consumption of beef is forbidden by their religion. Additionally, spices, chilies, garlic, and other flavorings are frequent ingredients in everyday dishes. In a group of Indian male students in Durban, mean intakes were: calories, 2043; protein, 64 g; fat, 122 g; carbohydrate, 415 g; fiber intake was not given.[8]

C. Diet of Coloureds

Compared with that of Whites, the diet of Coloureds includes less protein (especially animal protein), less fat (especially animal fat), more carbohydrate (mainly from wheat and maize), and less fruit.[16] Fiber intake is slightly more than that of Whites. The more prosperous segments of Coloureds consume a diet similar to that of Whites.

D. Diet of Whites

Local Whites consume the same diet as Whites in Britain[43] and other Western countries. In a group of South African middle-class urban adult males, mean intakes were: calories, 2564; protein 99 g; fat 109 g; carbohydrate, 256 g; fiber intake was not given.[73] Fiber intakes are believed to be similar to those in Britain, but consumption of fruit, because of its readier availability, is higher among local Whites. In the past, their diet was similar to that consumed in Britain.

As to changes in fiber intake, in Britain a century ago lightly milled cereals formed a large proportion of the diet. The average daily consumption of bread was about 600 g per head[14]; moreover, much oatmeal porridge was eaten by the poorer classes. About 1880, a change occurred in the character of the national loaf.[30] Whereas formerly it was made mainly from stone-ground flour, the introduction of roller mills resulted in the production of a bran-free white flour, yielding a loaf that quickly won popularity. By about 1890, replacement seems to have been largely effected: at any rate by 1906 it was complained by some that the old flour was almost unobtainable.[13] A progressive reduction also took place in the amount of bread eaten: in 1881, daily consumption averaged about 450 g per head; by 1904–1909, it had fallen to about 300 g.[62] Nowadays it is about 200 g.[43] These combined changes resulted in a marked decrease in cereal fiber intake. Although there is controversy over the magnitude of the fall,[31] there is no doubt, as Trowell[82,83] has shown, that a marked reduction did occur. Dietary changes in the United States resembled those in Britain (Table II).

TABLE II. Dietary Intakes in Britain Among the Poorer Classes (in grams)[a]

Diet element	Low income, 1920[65]	High income, 1975[43]	Low income, 1975[43]
Calories (in kcal)	2560	2490	2660
Protein	76	131	122
Fat	73	126	126
Carbohydrate	384	279	327

[a]Fiber intakes are not known with accuracy.

IV. Pattern of Gastrointestinal Diseases in South African Populations

Commencing with the mouth, the following diseases will be considered: (1) dental caries, (2) bowel motility and constipation, (3) appendicitis, (4) diverticulitis, (5) colonic cancer, and (6) miscellaneous—peptic ulcer, gallstones, ulcerative colitis, irritable bowel syndrome; these will be referred to briefly since requisite information is meager.

For each disease, information will be given on its epidemiology in the four ethnic groups and, where possible, on changes in frequency compared with the past. This will be followed by discussion of the etiology and of possible means of the alleviation or prevention of the disease by dietary means, with special reference to an increased intake of fiber.

A. Dental Caries

1. Epidemiology

Dental caries, unorthodoxly, is included since it is one of the diseases of civilization that has greatly increased simultaneously with the dietary changes described. Among primitive and developing populations, excellent teeth are the rule, although certainly deterioration occurs with urbanization.[9] In young Negroes of 16–17 years, in country areas compared with Johannesburg, mean DMF (decayed-missing-filled) scores of about 1.5 and 2.5, respectively, have been noted. In other groups of the same age in Johannesburg, data were: Coloureds, 6.5; Indians, 7.5; Whites, 10.0.[70]

2. Etiology

The enormous deterioration of teeth found in Whites and other sophisticated populations is usually accepted as being due almost wholly to the rise in intake of sugary and other soft carbohydrate foods.[97] This view is highly plausible; moreover, considerable experimental evidence may be cited in support. So firmly grounded is the belief that it has almost completely overwhelmed consideration of the role of other food components, including the decreasing intake of fibrous foods. The sugar hypothesis has limited validity. (1) Urban South African Negro high school pupils consuming a mean of about 100 g sugar *per diem* have excellent teeth.[70] (2) Until 1940, White children in the Island of Lewis, Scotland, had excellent teeth (DMF about 2) despite a high sugar intake,[44] similar in level to that of White children in Johannesburg who have a mean DMF score of 10.[70] (3) In Britain, observations on young children in 1929 and 1943 revealed a very marked fall in caries prevalence.[59] Yet during the interval (at least until 1940), a national sugar intake rose. (4) During World War II, the caries situation in children improved in all European countries[9,53]; the change was associated with a fall in sugar intake and a rise in consumption of fiber contained in cereals, vegetables, and fruits. Yet, in Germany, dental caries also decreased, but in spite of a *rise* in sugar intake.[95] This suggests that alterations other than decrease in sugar intake were instrumental in effecting a reduction in caries. (5) Short-term studies on young children showed that a considerable rise in intake of sugar or sugar-containing foods did not raise caries scores significantly.[47] Conversely, several short-term studies have demonstrated that contrasting intakes of sugar and sugar-containing foods are associated with only slight differences in DMF scores, by one to two units.[35] Accordingly, it is thought that the role of intake of other dietary components, including fiber-containing foods, should be investigated by epidemiological and other studies. From the foregoing, it is clear that considerable caution is required in attaching specific blame to particular nutrients or foods, not only as causes of dental caries, but of other diseases to be described.

B. Bowel Motility: Constipation

1. Epidemiology

Defecation Frequency. Hippocrates[1] was aware that a diet which contained whole-grain cereal products yielded bulky stools passed regularly and with ease, and that a diet low in such foodstuffs resulted in less frequent passage of hard feces. Hippocrates thought that defecation should be twice or thrice a day, the frequency usual among African Negroes and similarly placed populations. Among rural Negro children aged 10–12 years, a daily frequency of 2.8 was found. In town children, the figure decreased to 2.3. Frequencies were slightly lower in older children and adults.[90] For urban pupils of 16–20 years, figures of 1.7 and 2.1 were obtained.[86,91,94] Concerning adults, Bremner[10] questioned such at Baragwanath Hospital (2500 beds), Johannesburg, and obtained a figure of

2.0 motions *per diem*. Groups of Indian, Coloured, and White pupils aged 16–20 years had defecation frequencies of 1.5, 1.7, and 1.1, respectively.[94] The latter figure, for Whites, approximates that reported for Whites elsewhere.[25]

Transit Time. As noted by McCance et al.[57] and others, transit time is usually faster in diets high in fiber-containing foods compared with those that are low in these foods. Using carmine as a marker, data for schoolchildren were found to be: rural Negroes, 8–10 hr; urban Negroes, 10–15 hr; Whites, 20–30 hr.[86,91]

Amount of Feces. Diets high, compared with those low, in fiber-containing foods lead to the voiding of greater amounts of feces, often voluminous and unformed.[26,57,86] Mean weights range from 30 to 70 g dry feces *per diem* in rural Negro adults to 20–25 g in White adults.[91]

Inadequate Bowel Motility or Constipation. Unfortunately, purgation is practised in all ages, past and present, in all countries and cultures.[22] There are two motivations: cleansing and the relief of constipation. Hence, in both past and present, information on bowel movement as evoked by the diet consumed is unsatisfactory. In South Africa, constipation, as reflected by use of purgatives, is uncommon in rural Negroes, commoner in urban Negroes, and very common among Whites.[91] It must be appreciated that Negroes tend to regard themselves as constipated should they have one, not two or three, daily motions; hence connotations of constipation in different populations are not the same.

Bowel Motility in the Past. In Britain, before changes occurred in bread composition and consumption, habitual constipation was certainly common.[22,34,53] It occurred mostly in women, particularly of the upper class, among whom it was fashionable to be "delicate"; to achieve this delectable state, a "skimpy diet" was consumed which amounted to "chronic starvation."[17,22] Among rural people, whose diet, in addition to cereal foods, usually included a large amount of vegetables, constipation was less common. Who heard of a peasant or laboring man being constipated?[34] However, the important issue is, did the dietary changes already described increase the occurrence of the condition? Several authorities blamed fine wheaten bread for the disorder. At the 1910 Annual Meeting of the British Medical Association, Goodhart,[38] introducing a discussion on chronic constipation, "affirmed that with advancing civilization, aperients would always be with us, but lamented the change from the occasional pill of our forefathers to the excess of the present day." In 1938, Sir Heneage Ogilvie[61] considered that "stasis is an almost inevitable sequel of the modern diet, with the over-milling of cereals and the reduced ingestion of cellulose." These views suggest, but do not prove, that constipation may have become more common. In recent times, constipation has been variously described as "more or less habitual," "of wide occurrence," "the national curse," and "the bane of the British people."

In the United States, Riley[71] related how "the mass of the food of our fathers and grandfathers . . . was subjected to the simplest and most necessary processing only," and then referred to the changes in diet when Benjamin Harrison was President. In 1928, the editor of the *Journal of the American Medical Association*, in an article on "bran as a laxative," stressed "the imperative need that millions of people feel for something that will assist in the regulation of the bowel."[2] In that country even 30 years ago there was an annual expenditure of 100 million dollars on purgatives.[79] Nowadays, expenditure is far higher.[26,49]

2. Etiology

Constipation, or inadequate bowel motility, is due principally to insufficient bulk-forming capacity of the habitual diet.[6] Can eating more cereal-fiber-containing foods provide an effective remedy? Improvement certainly occurs in measure when persons change from eating white to brown bread or when a supplement of bran is taken.[6,24,26,57,87,92] Locally, in long-term studies in which White adults changed from consuming a diet that included 450 g white bread to one including 450 g whole-grain bread, mean transit time (using carmine) decreased from 28 to 19 hr, frequency of defecation rose from 1.0 to 1.6 times *per diem*, and the amount of feces (dry) voided daily, 29 to 41 g.[91,92] McCance et al.[57] reported similar observations. As an example from wartime, in Eire, when whole-grain bread became the only bread available nationally, the sale of Epsom salts for laxative purposes fell by several tons per month.[74]

C. Appendicitis

1. Epidemiology

In South Africa, information on appendectomy prevalence was secured on 15,317 16- to 20-year-old pupils and students in the four ethnic groups.[94] Data were also obtained on fiber intake, frequency of defecation, and transit time of digesta. Among students of 18–20 years, appendectomy was very uncommon in rural Negroes (0.5%) and periurban Negroes (0.9%), slightly more common in urban Negroes (1.4%), but far more common in Whites (16.5%); prevalences in Coloured and Indian groups were low (1.7 and 2.9%). Rural Negroes had a far larger fiber intake, greater frequency of defecation, and much shorter transit times; yet in the ethnic groups in urban areas, despite wide differences in appendectomy prevalence, data on these variables were not dissimilar. Corresponding studies were made on 1325 White pupils in institutional homes; their diet was less sophisticated in respect to fiber, sugar, and fat intakes than that of the general population. Pupils had slightly greater defecation frequency and lower transit time; their appendectomy incidence was only 23% of that of an appropriate control group. Short[78] reported similar findings.

How valid are these data? Among African Negroes, numerous reports indicate the disease to be rare.[18,94] The best confirmatory evidence concerns Negro laborers recruited from areas in southern Africa and employed on the gold mines. The great majority are 19–25 years old. They are fed on an adequate diet, which includes large amounts of lightly milled cereal products, legumes, and vegetables. One mining group has a labor force of 100,000. Taking into account the wartime appendectomy incidence in the United States Navy[85] as well as probability data for this age group in New Zealand,[51] roughly 850 appendectomies would be expected annually from a White group of the same size as that of the Negroes. Yet in 1970, among the latter only 22 were operated upon.[94]

Concerning the data on Whites, in Johannesburg and Pretoria, large cities with medical schools, mean appendectomy rates for students were 18.2 and 13.0% respectively. Yet among students at Potchefstroom and Grahamstown, small country towns, rates were much the same, 19.4 and 16.7% respectively. Universities in the latter towns include a large proportion of students brought up in country areas. Using data on the age distribution of appendectomy,[5,99] calculations indicate that the prevalence of the operation per *total* population of South African Whites is about 17%, a proportion similar to that for England,[5] about 14%, and for New Zealand,[51] about 16%. Hence, standard of judgement as to whether or not to perform appendectomy is deemed the same in different parts. Accordingly, the appendectomy data on the four South African ethnic groups of students are believed to be valid.

Appendicitis among Whites in the Past. The disease, even when making allowance for the change in nomenclature (typhlitis, perityphlitis), was rare in England before 1880. A spectacular rise took place soon afterwards.[18,78] Watson,[96] from figures obtained at St. Bartholomew's and St. Thomas' Hospitals, as well as certain large provincial hospitals, reported an increase of about 600% between 1876 to 1880 and 1896 to 1900. Other authorities accepted that a considerable rise did, in fact, occur.[12] Present prevalence of the disease has already been given.[94] Mortality from the operation, previously high, had fallen to "near vanishing point" by 1946.[60]

2. Etiology

That the chief predisposing cause is depletion of cellulose or fiber from the diet was put forward and developed by Short[78] and later by Burkitt.[18] It is believed that a fiber-depleted diet can result in the formation of fecaliths and also excessive segmentation of the appendix, either of which may obstruct the appendix lumen. Obstruction might raise the intraluminal pressure sufficiently to devitalize the appendicular mucosa and allow bacterial invasion. Aschoff [47] had similar ideas. It is interesting to note that Fantus et al.[36] observed filling and emptying ("flushing out") movements radiographically when studying the effect of bran supplements on the appendix of healthy adults. As to the precipitating cause of appendicitis, it is not known.[18,78]

Effect of Changes in Diet on Appendicitis Prevalence.   Firstly, regarding populations forced to make dietary changes by war conditions, in Russia after World War I, privations were so intense that food was limited to about half a kilogram of "black" bread daily, together with a little soup made from coarse vegetables. Simultaneously to the diet taking on greater coarseness, appendicitis almost disappeared.[78] In Switzerland during World War II, an average of about 500 g brown bread (made from 90% extraction flour) and about 400 g vegetables (other than potatoes) were rationed per head. While this coarse diet was consumed, appendicitis became rare.[37] In the Channel Islands during the German occupation, white bread was replaced by whole-grain bread, and the consumption of vegetables, always abundant, often reached very high levels. Appendicitis incidence fell considerably.[7] Constipation, a common complaint before the occupation, almost completely disappeared: An early effect of liberation was widespread constipation, attributed to the reintroduction of white bread.

Next, as to communities of people under confinement, also forced to make dietary changes, Briscoe[11] in 1912 reported appendicitis to be very rare in asylums and similar institutions in England and Wales. Spencer[80] in 1938 noted one death in Portland during a 10-year period. In 1940 at Johannesburg jail, accommodating 500 White prisoners, there had been only four cases of appendicitis in the previous 20 years. At the periods indicated, the diet consumed in such institutions was simpler and higher in bulk-forming capacity than that to which the inmates were previously accustomed.

Comment.   A diet high in bulk-forming capacity is therefore consistent with a low incidence of appendicitis; moreover, evidence indicates that a change from a refined diet to one less refined is accompanied by a reduced incidence of the disease.

Young and Russell[100] considered that "under conditions of modern life, it becomes improbable that a sufficient change in dietary habits will be introduced to influence to an appreciable degree the incidence of appendicitis." The much lower incidence in White children in homes suggests that only a small increase in everyday fiber intake, with accompanying increase in bulk-forming capacity, may considerably reduce proneness to the disease.

D. Diverticular Disease

1. Epidemiology

Diverticular disease is rare in primitive and emerging populations; it was rare in Whites in Western countries until about 1920.[19,20,63,64] In South Africa, diverticulitis is unknown in rural Negroes. In Soweto, Johannesburg, where over a million Negroes reside, three cases were detected from 580 barium enema studies made at Baragwanath Hospital (2500 beds) in 1972. At Coronation Hospital (500 beds), Johannesburg, where Coloureds and Indians attend, data on diverticulitis frequency are not available. However, available information is that the disease is uncommon in Coloureds; moreover while it occurs among Indians, the frequency is far lower than that in South African Whites.[39] Among the latter, diverticulitis has been stated to be present in about a third of middle-aged and elderly persons as diagnosed from barium enema investigations.[46] In Western countries, diverticulitis has been stated to be the commonest disease of the colon, in one study afflicting 7.6% of patients under 60 years of age and 34.6% of those over 60.[64]

2. Etiology

The major advances in this field are due to Painter and coworkers,[63,64] who regard the disease as caused by fiber depletion. They have demonstrated that the addition of bran to the diet relieves symptoms even when the disease is well established. The protective function of fiber is explained on mechanical grounds, namely, that fiber affects the consistency of the fecal stream and hence the pressures that the colon has to generate to propel its contents. A recent study by Latto[48] revealed that diverticulitis patients ingested less fiber than did controls.

E. Colonic Cancer

1. Epidemiology

Cancer of the colon is either unknown or very rare in primitive and developing populations. Even among South African Negroes born in large cities, the disease remains rare; in Johannesburg no significant rise has occurred since 1963.[72] In the past, colonic cancer was uncommon in Western populations;[69] it now ranks next to lung cancer in men and breast cancer in women, accounting for about 12–14% of all cancers, and about 3% of all deaths.[28]

2. Etiology

The cause of the disease undoubtedly is environmental. Its epidemiology suggests that changes in the pattern of diet are responsible. It is highly likely that the incriminating factor or factors are not exogenous, but are noxious metabolites produced in the bowel, arising from interactions between digesta and intestinal flora.[4]

Of the alterations in intake of food components, which change or changes are most likely to be implicated?

Hypotheses. (1) According to Czech workers,[40] there is a close correlation between national intakes of protein and colonic cancer mortality rates; hence, level of intake of this component may have etiological significance. (2) Yudkin[101] regards cancer of the bowel as one of the results of the increase in sugar intake. (3) Cleave[24] blames the increase on consumption of refined carbohydrate foods, especially sugar, and, to a lesser extent, the associated fall in fiber intake. (4) Reddy and Wynder[67] in New York and Drasar, Hill, and associate workers[29] in London, consider increase in fat consumption to be the chief etiological factor. (5) Fiber depletion is regarded as the primary predisposing factor by Burkitt, Walker, and others.[19,20]

The protein, fat, and sugar hypotheses claim support, at least in part, from close associations between national intakes of specific food components and colonic cancer mortality rates. Adherents of the fiber hypothesis have not used this procedure, principally because although intakes of protein, fat, and sugar in different populations are known with some certainty, this is not the case with fiber intakes.[83,84] Yet despite this uncertainty, the London workers,[29] from their correlation coefficient studies, have concluded that level of fiber intake is almost irrelevant in the causation of colonic cancer.

Apart from the foregoing reservation, it must be recognized that correlation studies of the above type have contributed little information of value on the etiology of degenerative diseases. The mortality rate from coronary heart disease is ten times that of colonic cancer. Yet despite intensive correlation coefficient and other studies, there is still vehement controversy over which food component bears chief responsibility. In another view of the situation, the Scottish people have a very high mortality rate from colonic cancer and a mortality rate from coronary heart disease far higher than that of, say, the French; but no one has seriously contemplated that the Scottish people consume far more fat, sugar, coffee, and soft water and far less crude fiber. Nevertheless, despite the limitations of correlation studies, such is the lethality of colonic cancer that all leads should be adequately explored, whether they concern fat,[29,67] beef,[41] or other foodstuffs.

3. Laboratory Studies

The investigations undertaken have principally concerned fat and fiber intakes. (1) Drasar, Hill and coworkers[29] have examined feces from populations prone and nonprone to colonic cancer. They noted that feces from the former contained much higher concentrations of bile acids and sterols; further, feces from the two groups differed microbiologically. They also reported contrasts between the makeup of feces from groups of persons with and without cancer of the colon. Experimentally, they found that halving the intake of fat approximately halved the fecal concentrations of bile salts and sterols. (2) Reddy and Wynder[67] have shown that Americans consuming an average diet had greater fecal concentrations of bile acids and sterols and excreted greater amounts than vegetarians and other groups whose diet, *inter alia*, contained less fat and more fiber than the average. Both these groups of workers, therefore, attach great etiological importance to the rise in fat intake that has occurred within the last century. The general belief is that high concentrations of these metabolites (bile acids and sterols) in the digesta interact with the associated intestinal microflora to yield substances that have carcinogenic properties. (3) In South Africa, Antonis and Bersohn[3] fed diets containing varying amounts of fiber and fat to groups of Negro and White prisoners. They found that increasing the fiber intake, as well as lowering the fat intake, decreased the concentrations of fecal bile acids and sterols. In their studies the effect of increasing fiber intake was greater than that from decreasing fat intake. (4) Eastwood et al.[32] reported that the ingestion of 16 g bran increased fecal bulk and at the same time lowered fecal concentration of bile acids. (5) Walker[90,91] carried out closely similar studies on South African Negro children, and obtained the same pattern of results. He also noted a similar lowering effect on bile acids concentration when groups of children consumed a large supplement of oranges, five to seven *per diem*, containing about 4 g fiber. Studies (3) to (5) thus demonstrate that addition of fiber-containing foods also significantly reduces fecal concentrations of bile acids and sterols.

Preventive Measures. Assuming that the principal mechanism of carcinogenesis in the colon stems from high fecal concentrations of bile acids and sterols, then it becomes imperative to lower these concentrations. This can be accomplished by lowering the fat intake or by raising the fiber intake, or by both procedures. Together, these changes tend toward the diet consumed by less sophisticated populations among whom colonic cancer is rare or of low frequency.

A second preventive measure would be to limit the time of contact of noxious material with bowel mucosa. The importance of contact and transit times would be diminished were intestinal contents totally homogeneous. Our investigations on chemical composition in different parts of a length of stool, together with other studies, have indicated that fecal material is not homogeneous. Presumably, therefore, noxious material occurs in pockets of digesta, in which case contact and transit times should not be prolonged. While the role of these times in promoting the development of colonic cancer is controversial, it is altogether premature, in view of the very inadequate state of knowledge on the whole subject, to conclude that is unimportant.[29,42]

What are the chances of Western populations making worthwhile changes in diet to lessen the risk of developing colonic cancer? Hill[42] has stated "mere halving of the daily fat intake to 50 to 60 g *per diem* would result in a much reduced fecal bile acid concentration while still leaving a very acceptable diet." The likelihood of this suggestion being adopted as a practical measure is out of the question. Firstly, even slum populations in Glasgow 30 years ago ingested 70–80 g fat *per diem*.[65] Secondly, people will not even reduce their fat intake (and thereby reduce the palatability of their diet) to ward off coronary heart disease, a killer ten times greater than colonic cancer. Equally, of course, even if the fiber hypothesis is verified by further research, extremely few would be prepared to revert to diets of the past, which included eating large amounts of lightly milled bread and porridge.

If, however, the desirable bowel *milieu interieur* (chemical and microbiological) could be achieved simply by a daily sprinkling of a tablespoonful or so of bran on breakfast porridge or cereal food, without any other change in diet, then this measure conceivably could become acceptable and be adopted by many. Still more so would this be the case should ease of bowel movement follow and, moreover, should additional health benefits accrue in respect to other gastrointestinal diseases. Pomare and Heaton[66] have emphasized that consumption of bran must be regarded as restoring to the diet a component which was formerly present. The addition is physiological, i.e., commensurate with earlier dietary habits. A supplement of, say, 25 g bran, which contains about 3 g fiber, is included in 125 g wholemeal or four to six slices of wholemeal bread.

What we urgently need to know, of course, are the minimum changes in diet and the minimum changes in bowel *milieu interieur* that might lead to a fall in the incidence of colonic cancer. But how can this knowledge be attained when the period of development of the disease is so very long, and there is no period of secondary prevention as there is with coronary heart disease? The answer that supporters of the fiber hypothesis would give is that if by dietary means appendicitis incidence could be reduced, it would confidently be expected that consumption of the same diet, if sustained, would effectively reduce the incidence of colonic cancer. Epidemiologically, the latter occurs very rarely until appendicitis becomes of common occurrence. It has already been indicated that the dietary changes involved need not be large, since Short[78] in Bristol, and we in Johannesburg[94] have shown that among children in homes where the diet consumed is less sophisticated, the incidence of appendicitis may be far lower than that in the average child population. Accordingly, the accurate defining of the diet in homes or similar institutions where this lower incidence demonstrably prevails should be one of the subjects for future research.

F. Miscellaneous Diseases

A number of additional diseases have either emerged or significantly increased with the changes in diet described. These diseases include peptic ulcer, gallbladder disease, ulcerative colitis, and irritable bowel syndrome.

Among rural Negroes, all these diseases are unknown or extremely rare. However, they occur among Negroes in large centers of population, although emerging at different time periods. For example, peptic ulcer is now certainly common,[55] whereas the first cases of irritable bowel syndrome in Negroes in Johannesburg have just been reported.[76] All the diseases are more common in Coloureds and Indians than in urban Negroes, but are less common than in the White population.

Undoubtedly, changes in diet (fiber depletion associated with a high intake of refined carbohydrate foods) bear a large measure of responsibility. But present knowledge is insufficient to assess the extent of the role of fiber intake, except in the case of irritable bowel syndrome, where a diet high in bulk-forming capacity has been shown to be beneficial.[76]

V. Discussion

When all the gastrointestinal diseases under review are taken together, it is clear that their amelioration is more likely to be forthcoming from increasing the intake of fiber-containing foods than by other dietary means.

Obviously, a considerable amount of further research work must be undertaken to throw additional light on the etiologies of the diseases discussed. There are several avenues of approach: (1) Roles of fiber, other than as a diluent or bulking agent, must be investigated. More short- and long-term experimental studies are required to learn of the effects of increasing or decreasing various dietary components on the makeup of feces.[77] Metabolites other than bile acids and sterols may be potentially carcinogenic. (2) Controlled clinical investigations should be carried out on "the efficacy of fiber-enriched diets in a wide variety of disorders including irritable bowel syndrome, 'prediverticular' disease, diverticulosis, refractory constipation and postvagotomy diarrhea. Such studies will be awaited with great interest since drug therapy of the above listed disorders is often unsatisfactory."[21] (3) As emphasized by many,[15,33] much more should be known of minority groups in Western populations, whose diet, metabolism, and disease pattern differ from the average. Populations would include children in homes where appendicitis incidence is significantly lower than the average. Other populations, of course, would be vegetarians, especially those accustomed to a diet which is really high in bulk-forming capacity. The vegetarian group studied by Reddy and Wynder[67] voided an average of only 20 g dry feces *per diem*, an amount associated with a low fiber intake, only 3–4 g.[27] Even in the general population, it should be possible, with perseverence, to select groups accustomed to consuming diets high in fiber and bulk-forming capacity, diets relatively low in fat, etc., and hence learn of their patterns of health in respect to proneness or nonproneness to the diseases under discussion. (4) Information must be kept up-to-date on the changing pattern of gastrointestinal diseases, due directly or indirectly to changes in diet, more especially as they prevail in primitive and developing populations. Research must also be undertaken to clarify experiences of particular populations: for example, Navajo Indians have a high fat intake, and some groups of Eskimos have a high-fat, low-fiber diet; yet neither population appears to be prone to colonic cancer.[68,75] Is it essential for high fat and low fiber intakes to lie within the pattern of a Western diet before noxious effects occur?

Finally, it must be remembered that, in the main, no matter how fruitful are the results of the research suggested, the diseases at issue are multifactorial in etiology. It is not reasonable therefore to expect that knowledge of the predisposing causes of these diseases will fully explain their times of emergence and prevalence, sex differences, individual susceptibilities, etc. It must also be kept in mind that in any population, there are individuals or groups of people who appear especially prone to develop particular diseases. Thus, Ashley[5] believes there is a subpopulation, about 12%, especially susceptible to appendicitis. In the general United States population Wilkins and Hackman[98] showed there to be two distinct classes of cholesterol converters and nonconverters. This difference, they believe, is possibly related to variations in risk level for bowel cancer.

VI. Summary

Comparing ourselves with our ancestors, and comparing urban with rural dwellers in developing countries, there are contrasting differences in patterns of disease and in patterns of diet and manner of life. In general, the pattern of disease changes from one chiefly of infections to one principally of degenerative diseases. In South Africa, there are four ethnic groups: Negroes, Coloureds, Indians, and Whites. These have been examined in relation to their increasing proneness to degenerative diseases of the gastrointestinal tract: dental caries, poor bowel motility and constipation, appendicitis, diverticulitis, colonic cancer, and certain other diseases. Examination of the changes in diet that occur as these diseases emerge and become more common suggests that the primary underlying or predisposing cause is the decrease in the intake of fiber-containing foods or the bulk-forming capacity of diet. Should the latter be restored, it would be reasonable to expect both a measure of protection and, in some conditions or diseases (inadequate bowel movement and diverticulitis), an ameliorative effect. Considerable further research is required to throw more light on the etiology of the diseases cited. Findings, however, are unlikely to explain fully their times of emergence, their current prevalences, or susceptibilities of individuals to the diseases.

References

1. ADAMS, F. *The Genuine Works of Hippocrates*. Baltimore, Maryland, Williams and Wilkins, 1939.
2. AMERICAN MEDICAL ASSOCIATION, JOURNAL OF. Bran as a laxative. Editorial. *JAMA* **90**:206, 1928.
3. ANTONIS, A., and I. BERSOHN. The influence of diet on fecal lipids in South African White and Bantu prisoners. *Am. J. Clin. Nutr.* **11**:142, 1962.
4. ARIES, V., J. CROWTHER, B. S. DRASAR, M. J. HILL and R. E. O. WILLIAMS. Bacteria and the etiology of cancer of the large bowel. *Gut* **10**:334, 1969.
5. ASHLEY, D. J. B. Observations on the epidemiology of appendicitis. *Gut* **8**:533, 1967.
6. AVERY JONES, F., and E. W. GODDING. *Management of Constipation*. Oxford, Blackwell, 1972, p. 38.
7. BANKS, A. L., and H. E. MAGEE. Effects of enemy occupation on the state of health and nutrition in the Channel Islands. *Mon. Bull. Min. Hlth. Lond.* **4**:184, 1945.
8. BOOYENS, J., and V. M. DE WAAL. The food intake, activity pattern and energy expenditure of male Indian students. *S. Afr. Med. J.* **43**:344, 1969.
9. BOWEN, W. H. Dental caries. *Arch. Dis. Child.* **47**:849, 1972.
10. BREMNER, C. G. Anorectal disease in the South African Bantu. 1: Bowel habit and physiology. *S. Afr. J. Surg.* **2**:119, 1964.
11. BRISCOE, J. F. Appendicitis in asylums. *Lancet* **2**:175, 1912.
12. BRITISH MEDICAL JOURNAL. The alleged epidemicity of appendicitis. Editorial. *Br. Med. J.* **1**:815, 1906.
13. BRITISH MEDICAL JOURNAL. Bread and milk. Editorial. *Br. Med. J.* **1**:759, 1906.
14. BRITISH MEDICAL JOURNAL. Brown bread versus white. Editorial. *Br. Med. J.* **2**:752, 1937.
15. BRITISH MEDICAL JOURNAL. Leading Article. Diet and colonic cancer. *Br. Med. J.* **1**:339, 1974.
16. BRONTE-STEWART, B., A. KEYS and J. F. BROCK. Serum cholesterol, diet, and coronary heart disease. *Lancet* **2**:1103, 1955.
17. BRUNTON, T. L. Constipation. *In: Allbutt's System of Medicine*, T. C. Albutt (ed.). New York, MacMillan, 1902, Vol. III, Chap. 10, p. 696.
18. BURKITT, D. P. The etiology of appendicitis. *Br. J. Surg.* **58**:695, 1971.
19. BURKITT, D. P. Some diseases characteristic of modern Western civilization. *Br. Med. J.* **1**:274, 1973.
20. BURKITT, D. P., A. R. P. WALKER and N. S. PAINTER. Dietary fiber and disease. *JAMA* **229**:1068, 1974.
21. BURKITT, D. P., A. R. P. WALKER and N. S. PAINTER. Effect of dietary fiber on stools and transit times and its role in the causation of disease. *In: The Digestive System. Year Book of Medicine 1973*. Chicago, Year Book Medical Publishers, 1973, p. 452.
22. CHEADLE, W. B. Pathology and treatment of chronic constipation in childhood, and its sequel, atony and dilatation of the colon. *Lancet* **2**:1063, 1886.
23. CLEAVE, T. L., G. D. CAMPBELL and N. S. PAINTER. *Diabetes, Coronary Thrombosis, and the Saccharine Disease*. Bristol, Wright, 1966.
24. CLEAVE, T. L. *The Saccharine Disease*. Bristol, Wright, 1974.
25. CONNELL, A. M., C. HILTON, G. IRVINE, J. E. LENNARD-JONES and J. J. MISIEWICZ. Variation of bowel habit in two population samples. *Proc. R. Soc. Med.* **59**:11, 1966.
26. CUMMINGS, J. H. Progress report: Laxative abuse. *Gut* **15**:758, 1974.
27. DIAO, E. K., and F. A. JOHNSTON. The fecal residue from a normal diet of known composition. *Gastroenterology* **33**:605, 1957.
28. DOLL, R. The geographical distribution of cancer. *Br. J. Cancer* **23**:1, 1969.

29. DRASAR, B. S., and M. J. HILL. *Human Intestinal Flora*. New York, Academic Press, 1974.
30. DRUMMOND, J. C., and A. WILBRAHAM. *The Englishman's Food*. London, Jonathan Cape, 1939.
31. EASTWOOD, M. A., N. FISHER, C. T. GREENWOOD and J. B. HUTCHINSON. Perspectives on the bran hypothesis. *Lancet* **1**:1029, 1974.
32. EASTWOOD, M. A., J. R. KIRKPATRICK, W. D. MITCHELL, A. BONE and T. HAMILTON. Effects of dietary supplements of wheat bran and cellulose on faeces and bowel function. *Br. Med. J.* **4**:392, 1973.
33. ELLIS, F. R., and T. A. B. SANDERS. Diet and colonic cancer. *Br. Med. J.* **2**:505, 1974.
34. EWART, W. Discussion on chronic constipation and its treatment. *Br. Med. J.* **2**:1045, 1910.
35. FANNING, E. A., T. GOTJAMANOS and N. J. VOWLES. Dental caries in children related to availability of sweets at school canteens. *Med. J. Aust.* **1**:1131, 1969.
36. FANTUS, B., G. KOPSTEIN and H. R. SCHMIDT. Roentgen study of intestinal motility as influenced by bran. *JAMA* **114**:404, 1940.
37. FLEISCH, A. Nutrition in Switzerland during the war. *Schweiz. Med. Wschr.* **76**:889, 1946.
38. GOODHART, J. F. The treatment of chronic constipation. *Lancet* **2**:468, 1910.
39. GRIEVE, S. Personal communication.
40. GREGOR, O., R. TOMAN, and F. PRUSOVA. Gastrointestinal cancer and nutrition. *Gut* **10**:1031, 1969.
41. HAENSZEL, W., J. W. BERG, M. SEGI, M. KURIHARA and F. B. LOCKE. Large-bowel cancer in Hawaiian Japanese. *J. Natl. Cancer Inst.* **51**:1765, 1973.
42. HILL, M. J. Steroid nuclear dehydrogenation and colon cancer. *Am. J. Clin. Nutr.* **27**:1475, 1974.
43. HOUSEHOLD FOOD CONSUMPTION AND EXPENDITURE: 1970 and 1971. Rep. Natl. Food Sur. Comm. London, H. M. Stat. Office, 1973.
44. KING, J. D. Dental disease in the Island of Lewis. *Med. Res. Counc. Spec. Rep. Ser. No. 241.* London, H. M. Stat. Office, 1940.
45. KING, J. D., M. MELLANBY, H. H. STONES and H. N. GREEN. The effect of sugar supplements on dental caries in children. *Med. Res. Counc. Spec. Rep. Ser. No. 288.* London, H. M. Stat. Office, 1955.
46. KLOPPERS, P. J. Personal communication.
47. LANCET. The pathology of appendicitis. Editorial. *Lancet* **1**:311, 1933.
48. LATTO, C. Personal communication.
49. LLOYD, C., and J. CROOKS. Are we overconsuming? *World Health*, April 16, 1974.
50. LUBBE, A. M. Study of rural and urban Venda males: Dietary evaluation. *S. Afr. Med. J.* **45**:1289, 1971.
51. LUDBROOK, J., and G. F. S. SPEARS. The risk of developing appendicitis. *Br. J. Surg.* **52**:856, 1965.
52. MACEWEN, W. The function of the caecum and appendix. *Br. Med. J.* **2**:873, 1904.
53. MAGEE, H. E. Activities in nutrition of the Ministry of Health in England during the war. *Proc. Nutr. Soc.* **5**:211, 1947.
54. MANNING, E. B., J. I. MANN, E. SOPHANGISA and A. S. TRUSWELL. Dietary patterns in urbanised Blacks. A study in Guguletu, Cape Town, 1971. *S. Afr. Med. J.* **48**:485, 1974.
55. MARKS, I. N. Gastrointestinal disorders. *In: Clinical Medicine in Africans in Southern Africa*, G. D. Campbell, Y. K. Seedat, and G. Daynes (eds.). London, Churchill Livingstone, 1973.
56. MAYO, C. H. The appendix in relation to, or as the cause of, other abdominal diseases. *JAMA* **83**:592, 1924.
57. McCANCE, R. A., K. M. PRIOR and E. M. WIDDOWSON. A radiological study of the rate of passage of brown and white bread through the digestive tract of man. *Br. J. Nutr.* **7**:98, 1953.
58. McCARRISON, R. Deficiency disease with special reference to gastrointestinal disorders. *Br. Med. J.* **1**:822, 1920.

59. MELLANBY, M., and H. COUMOULOS. Teeth of 5-year old London schoolchildren. *Br. Med. J.* **2**:565, 1946.

60. METROPOLITAN LIFE INSURANCE COMPANY. The pathology of appendicitis. *Stat. Bull. Metrop. Life Insur.* **28**:7, 1947.

61. OGILVIE, W. H. The changing ground of surgery. *Br. Med. J.* **1**:1193, 1938.

62. ORR, J. B. *Food, Health and Income*. London, MacMillan, 1936.

63. PAINTER, N. S., A. Z. ALMEIDA and K. W. COLEBOURNE. Unprocessed bran in treatment of diverticular disease of the colon. *Br. Med. J.* **2**:137, 1972.

64. PAINTER, N. S., and D. P. BURKITT. Diverticular disease of the colon, a twentieth century problem. *Clin. Gastroenterol.* **4**:3, 1975.

65. PATON, D. N., and L. FINDLAY. Poverty, Nutrition and Growth. *Med. Res. Counc. Spec. Rep. Ser. No. 101*. London, H. M. Stat. Office, 1926.

66. POMARE, E. W., and K. W. HEATON. Alteration of bile salt metabolism by dietary fiber (bran). *Br. Med. J.* **4**:262, 1973.

67. REDDY, B. S., and E. L. WYNDER. Large bowel carcinogenesis: Fecal constituents of populations with diverse incidence rates of colon cancer. *J. Natl. Cancer Inst.* **50**:1437, 1973.

68. REICHENBACH, D. D. Autopsy incidence of diseases among Southwestern American Indians. *Arch. Pathol.* **84**:81, 1967.

69. RENNAES, S., and E. W. OSTBERG. Cancer ventriculi mortality in different population groups. *Br. J. Cancer* **9**:7, 1955.

70. RETIEF, D. H., P. E. CLEATON-JONES and A. R. P. WALKER. Dental caries and sugar intake in South African pupils of 16 to 17 years in four ethnic groups. *Br. Dent. J.* **138**:463, 1975.

71. RILEY, R. H. The health department and the food of the people. *JAMA* **138**:333, 1948.

72. ROBERTSON, M. A., J. S. HARINGTON and E. BRADSHAW. The cancer pattern in Africans at Baragwanath Hospital, Johannesburg. *Br. J. Cancer* **25**:377, 1971.

73. ROSSOUW, D. J., J. J. FOURIE, L. E. VAN HEERDEN and F. M. ENGELBRECHT. A dietary survey of free-living middle-aged White males in the Western Cape. *S. Afr. Med. J.* **48**:2528, 1974.

74. SAUNDERS, J. C. Wholemeal and laxatives. *Lancet* **1**:516, 1944.

75. SCHAEFER, O. Medical observations and problems in the Canadian Arctic. *Can. Med. Assoc. J.* **81**:386, 1959.

76. SEGAL, L., and J. A. HUNT. The irritable bowel syndrome in the urban South African Negro. *S. Afr. Med. J.* **49**:1645, 1975.

77. SHERLOCK, P. Diet, bile acids and colon cancer. Selected Summaries. *Gastroenterology* **66**:163, 1974.

78. SHORT, A. R. *The Causation of Appendicitis*. Bristol, Wright, 1946.

79. SMITH, A. E. Bowel and body versus laxatives. *Hygeia* **23**:414, 1945.

80. SPENCER, A. M. Aetiology of acute appendicitis. *Br. Med. J.* **1**:227, 1938.

81. TREVES, F. Perityphlitis. *In: System of Medicine*, T. C. Allbutt. (ed.). New York, MacMillan, 1897, Vol. III, Chap. 4, p. 879.

82. TROWELL, H. C. Bran hypothesis. *Lancet* **2**:54, 1974.

83. TROWELL, H. C. Dietary fiber, coronary heart disease and diabetes mellitus. *Plant Foods for Man.* **1**:11, 1973.

84. TROWELL, H. C. Dietary fiber, ischaemic heart disease and diabetes mellitus. *Proc. Nutr. Soc.* **32**:151, 1973.

85. UNITED STATES NAVAL MEDICAL BULLETIN. Editorial. Mortality in appendicitis. *Natl. Med. Bull. Wash.* **49**:1180, 1949.

86. WALKER, A. R. P. The effect of recent changes of food habits on bowel motility. *S. Afr. Med. J.* **21**:590, 1947.

87. WALKER, A. R. P. Dietary fiber and the pattern of diseases. Editorial. *Ann. Intern. Med.* **80**:663, 1974.

88. WALKER, A. R. P. Nutritional, biochemical, and other studies on South African populations. *S. Afr. Med. J.* **40**:814, 1966.
89. WALKER, A. R. P. Some aspects of nutritional research in South Africa. *Nutr. Rev.* **14**:321, 1956.
90. WALKER, A. R. P. Studies on the effects of high crude fiber intake on transit time and the absorption of nutrients in South African Negro schoolchildren. *Am. J. Clin. Nutr.* **28**:1161, 1975.
91. WALKER, A. R. P. Unpublished work.
92. WALKER, A. R. P., F. W. FOX and J. T. IRVING. Studies in human mineral metabolism. The effect of bread rich in phytate phosphorus on the metabolism of certain mineral salts with special reference to calcium. *Biochem. J.* **42**:452, 1948.
93. WALKER, A. R. P., S. D. MISTRY and H. C. SEFTEL. Studies in glycosuria and diabetes in non-White populations of the Transvaal, Part II. Indians. *S. Afr. Med. J.* **37**:1217, 1963.
94. WALKER, A. R. P., B. D. RICHARDSON, B. F. WALKER and WOOLFORD. Appendicitis, fiber intake and bowel behaviour in ethnic groups in South Africa. *Postgrad. Med. J.* **49**:243, 1973.
95. WANDELT, S. Statistical survey of the relation between sugar consumption and dental caries. *Ernahrungs-Umschau* **15**:302, 1968 (*Nutr. Abstr. Rev.* **39**:573, 1969).
96. WATSON, C. The alleged epidemicity of appendicitis. *Br. Med. J.* **1**:947, 1906.
97. WHO SCIENTIFIC GROUP. The Etiology and Prevention of Dental Caries. *World Health Org. Tech. Rep. Ser. No. 494.* Geneva, WHO, 1972.
98. WILKINS, T. D., and A. S. HACKMAN. Two patterns of neutral steroid conversion in the feces of normal North Americans. *Cancer Res.* **34**:2250, 1974.
99. WRIGHT, R. B. Invalidism from acute appendicitis. *Lancet* **2**:475, 1963.
100. YOUNG, M., and W. T. RUSSELL. Appendicitis, a statistical study. *Med. Res. Counc. Spec. Rep. Ser. No. 233.* London, H. M. Stat. Office, 1939.
101. YUDKIN, J. *Pure White and Deadly.* London, Davis-Poynter, 1972.

11

The Effects of Dietary Fiber: Are They All Good?

Ian Macdonald

I. Introduction

The concept that a constituent of the diet that is not absorbed or even metabolized can influence metabolism is a notion alien to those whose interests lie in the study of the biochemistry of health and disease. The blotting paper effect of dietary fiber raises a series of intriguing questions, not only as to how this effect is brought about, but also as to what substances present in the gut are adsorbed in this way and how this adsorption and subsequent elimination affect the metabolic well-being of the individual. Currently there are very few answers to these and other related questions when applied to man, partly because this is a new area of investigation and partly because some of the desirable effects claimed for fiber are only highlighted after many years of "fiber deficiency." Because there is little in the way of fact, there is an abundance of speculation on the role of dietary fiber in the prevention of disease, and some of the speculators display a fervor not often seen in science.

The current widespread interest in dietary fiber is not confined to those whose main interest is in medicine and nutrition; the imagination of the lay person has been captured by the belief that fiber is a "health" food. This attitude is possibly part of the generally held view that modern technology has some side effects that could be undesirable and that maybe modern food in the Western world can be too purified. It is, therefore, important to try to make an assessment of the current situation regarding the role of dietary fiber in health and disease

IAN MACDONALD ● Professor of Applied Physiology, Department of Physiology, Guy's Hospital Medical School, London, England.

that is based on fact. However, it must be appreciated that within a compara-
tively short time with the advent of new facts, the role of fiber will become
clearer.

II. The Evidence of Geographic Pathology

In 1970 Burkitt put forward the view that "there is reason to suspect a
common or related cause for diseases which are both associated with one another
in their geographical distribution and tend to occur together in the same
patients."[4] He pointed out that such diseases as diabetes, obesity, atherosclerosis, and noninfective bowel disease such as diverticulosis, adenomatous
polyposis, cancer, and ulcerative colitis are consistently rare or absent in rural
populations in developing countries and consistently common in all economically developed countries. Burkitt then postulated that the ingestible fiber found
in the food of the developing countries, which is removed in the processing of
sugar and white flour, was the factor responsible for the rarity of the diseases
listed above. This hypothesis was similar to that put forward earlier by Cleave
and Campbell[12] who chose to give the generic name "the saccharine diseases" to
those conditions associated with the consumption of refined carbohydrate either
as white flour or as sugar. Further support for this hypothesis has been
published,[5,7,8,10,11] and in addition, Trowell has emphasized the prophylactic
properties of dietary fiber in atherosclerosis.[41-43] The efficacy, in an uncontrolled
trial, of unprocessed bran in the treatment of diverticular disease of the colon was
reported in 1972.[33]

The evidence presented by these authors is circumstantial and as such is
capable of other interpretations; the evidence is more in the nature of an association than a relationship. It is difficult to obtain reliable figures for the incidence
of various diseases in rural areas in developing countries and the assumption by
Painter and Burkitt[31] that colonic diverticula are rare in the East has been chal-
lenged by Antia and Desai.[1] The latter authors also make the point that the high
incidence of coronary heart disease and cancer of the intestinal tract are in older
age groups and therefore seen in higher proportion in the West owing to the
greater longevity and better diagnostic resources found in the West. Further
queries on the purported rarity of certain diseases in developing countries have
also been raised.[23,40]

III. The Validity of Assumptions

In any hypothesis, it is not unreasonable to make assumptions, and the use
of teleology is acceptable; to persist when the assumptions are either incorrect or

of a second or third order weakens the hypothesis. Eastwood et al.[18] have pointed out that the plateau in the mortality from diverticular disease in the United Kingdom started in 1933, some 6 years before the wartime increase in flour extraction. This could be due to the introduction of roller-milled flour in 1880, 50 or so years earlier. However, in the mid-1950s the mortality from diverticular disease in women started to rise and currently is about twice that of men. As Eastwood et al. comment, this sudden appearance of great susceptibility in the female cannot be attributed to changes in milling in 1880–1890.

A decrease in intestinal transit time when fiber is added to the diet has been assumed to be advantageous in itself,[10] and from this a second-order assumption is made that slow transit not only allows potential carcinogens to be formed by colonic organisms but also allows increased contact time of the carcinogens with the mucosa of the colon.[3,6] However, doubt has been thrown on whether fiber does, in fact, decrease transit time in most people.[19,24]

It has been stated that the reduced incidence of atherosclerosis in developing countries is due to the high fiber content of the diet.[41,43] This evidence was an association, and when it was reported that in experimental animals on high-fiber regimens there was an increased excretion of bile salts,[30] there was a tendency to assume that fiber not only prevents atherosclerosis but does this by increasing the excretion of cholesterol. This assumption may be true, but it is not acceptable on the evidence so far presented, especially as studies in man, unlike those in chicks,[21] rabbits,[27] and rats,[15] have as yet failed to find a fall in serum cholesterol concentration on high-fiber diets.[13,16,26,28]

It is inherent in any hypothesis that casts its net as wide as does the fiber hypothesis that much of the hypothesis will have to be discarded as new facts are learned. This being so, it is essential that those aspects of the hypothesis which become untenable must be immediately discarded and not perpetuated. Failure to do so could lead to unwarranted therapy, as well as bring into disrepute a hypothesis that contains much that could well be useful.

IV. Proven Advantages of Dietary Fiber in Man

At the moment, it seems that the one clinical condition in which the fiber in the diet is of undoubted therapeutic value is diverticulosis. Intraluminal colonic pressure is high in diverticular disease,[2,32] and gut transit is prolonged;[20] fiber in the form of bran has been shown to reduce both intraluminal pressure and transit time.[20] The patients also have markedly fewer symptoms.[17,33] In addition, although it is not documented, the softer stools caused by an increased intake in the fiber almost certainly ease the discomfort of hemorrhoids and may well contribute to their improvement.

It is not correct, however, to assume that because an agent is therapeutic, the application of that agent, *sine qua non*, prevents the disease arising. It may be that a fiber-rich diet is prophylactic against diverticular disease and hemorrhoids, but to state that because the fiber-rich diet improves these conditions, its deficiency must therefore be a causative agent is not good logic. In fact Eastwood[17] has produced some reasoning to suggest fiber could induce colonic diverticula.

V. Some Possible Side Effects of Dietary Fiber

One of the advantages of white flour is that the removal of the husk (bran) means that much of phytate (inosital hexophosphate) is also removed. It may be that the possible harmful role of phytate in forming an insoluble calcium compound and hence leading to a deficient intake of calcium has been exaggerated, especially in the developed areas where calcium intake tends to be high. However, the recognition that trace amounts of metals such as zinc, chromium, magnesium, and copper are essential in the diet has led to the suggestion that these divalent metals may also form insoluble compounds with phytate.[35] If the intake of these trace elements is low and the small amount consumed is then made unavailable by the phytate in some fiber, a deficiency state may be produced. There is evidence that the presence of fiber in the form of whole wheat has led to iron, calcium, and zinc depletion in children in Iran.[23] It is of interest to note that the phytate in whole wheat can be destroyed by yeast fermentation.[34]

Other hazards of dietary fiber that have been suggested are volvulus and intussusception. It is considered that these abnormal postures of the small intestine may be the result of the bulk of its contents, caused by the excessive consumption of vegetable roughage.[9] It has been stated that fiber contains n-alkyl resorcinol and that this compound is a growth inhibitor in animals.[44] A recent report claims that fiber, especially wheat bran, contains both trypsin and chymotrypsin inhibitors.[29]

These observations obviously need to be substantiated, but they make the point that there are few substances that are consumed that are beneficial under all circumstances, and before recommending an increase in dietary fiber intake, it is necessary to examine its effects carefully and in more depth.

Southgate[37] has raised three theoretical considerations that must be taken into account when dietary fiber intake is increased. Firstly, he points out that many of the diets eaten in different parts of the world, which appear to differ in the amount of fiber they contain, also differ qualitatively in other respects, and it is important to consider the possible effects of the other changes before ascribing everything to dietary fiber. Secondly, Southgate raises the possibility that the decrease in transit time (if, indeed, this is a property of fiber) leaves less time for the processes of digestion and absorption. And thirdly, the physicochemical effects of fiber, such as water-binding, may reduce the rate of diffusion of the products of digestion towards the absorptive surface. Southgate also wonders whether the mechanical erosion of the mucosal surface of the gut leads to increased losses of endogenous material. Some of these considerations have received support from a study in men and women where it was found that increasing the intake of unavailable carbohydrate (fiber) resulted in a greater fecal loss of energy and in most instances, of nitrogen and fat.[38] Heaton[25] also believes that food fiber is an obstacle to energy intake, a point also taken up by Trowell.[43]

VI. Sources of Confusion

It has been stated on several occasions that the term dietary fiber means different things to different people, and an attempt has been made by Trowell[42] and Cummings[14] to accept a single definition. According to these authors, dietary fiber is the remnants of vegetable cell walls that are not hydrolyzed by the alimentary enzymes of man. Not only must there be international agreement on the definition of fiber, but, as was recently pointed out, it must be appreciated that not all dietary fiber has the same composition.[39] Indeed, some dietary fiber, e.g., wheat-bran, contains, apart from the fiber, several active nutritional ingredients, which themselves can and do influence metabolic activity.

There are statements in the medical literature that the fiber content of the diet has decreased, particularly in Britain since the advent of the roller milling in 1880–1890. A more precise assessment of the fiber content of the diet has shown that there has been no such decrease,[36] but that the nature of the fiber has changed. Thus, it is not only important to agree upon a definition of fiber, but it is likely to be clinically more useful if the various components of dietary fiber were studied in isolation. The high speculation-to-fact ratio in the field of dietary fiber could be reduced if careful research with fiber isolates could be carried out, preferably in man, and possibly in nonhuman primates also.

References

1. ANTIA, F. P., and H. G. DESAI. Colonic diverticula and dietary fiber. *Lancet* 1:814, 1974.
2. ARFWIDSSON, S., N. G. KOCK and L. LEHMANN. Pathogenesis of multiple diverticula of the sigmoid colon in diverticular disease. *Acta Clin. Scand. Suppl.* **342**:1, 1964.
3. ARIS, V., J. S. CROWTHER, B. S. DRASER, M. J. HILL and R. E. O. WILLIAMS. Bacteria and the etiology of cancer of the large bowel. *Gut* **10**:334, 1969.
4. BURKITT, D. P. Relationship as a clue to causation. *Lancet* **2**:1237, 1970.

5. BURKITT, D. P. Varicose veins, deep vein thrombosis and hemorrhoids: epidemiology and suggested etiology. *Br. Med. J.* **2**:556, 1972.

6. BURKITT, D. P. The importance of fibre in food. *Bull. Br. Nutr. Found.* **7**:29, 1972.

7. BURKITT, D. P. Some diseases characteristic of modern Western civilization. *Br. Med. J.* **1**:274, 1973.

8. BURKITT, D. P., and P. A. JAMES. Low residue diets and hiatus hernia. *Lancet* **2**:128, 1973.

9. BURKITT, D. P., C. L. NELSON and E. H. WILLIAMS. Some geographical variations in disease pattern in East and Central Africa. *E. Afr. Med. J.* **40**:1, 1963.

10. BURKITT, D. P., A. R. P. WALKER and N. S. PAINTER. Effect of dietary fibre on stools and transit times and its role in the causation of disease. *Lancet* **2**:1408, 1972.

11. CLEAVE, T. L. *The Saccharine Disease*. Bristol, J. Wright, 1974.

12. CLEAVE, T. L., and G. D. CAMPBELL. *Diabetes, Coronary Thrombosis and the Saccharine Disease*. Bristol, J. Wright. 1966.

13. CONNELL, A. M., C. L. SMITH and M. SOMSEL. Absence of effect of bran on blood lipids. *Lancet* **1**:496, 1975.

14. CUMMINGS, J. H. Dietary fibre. *Gut* **14**:69, 1973.

15. DEVI, K. S., and P. A. KARUP. Effects of certain pulses on the serum, liver and aortic lipid levels in rats fed a hypercholesterolenic diet. *Atherosclerosis* **11**:479, 1970.

16. EASTWOOD, M. A. Dietary fiber and serum lipids. *Lancet* **2**:1222, 1969.

17. EASTWOOD, M. A. Personal communication.

18. EASTWOOD, M. A., N. FISHER, C. T. GREENWOOD and J. B. HUTCHINSON. Perspectives on the bran hypothesis. *Lancet* **1**:1029, 1974.

19. EASTWOOD, M. A., J. R. KIRKPATRICK, W. D. MITCHELL, A. BONE and T. HAMILTON. Effects of dietary supplements of wheat bran and cellulose on feces and bowel function. *Br. Med. J.* **4**:392, 1973.

20. FINDLAY, J. M., A. N. SMITH, W. D. MITCHELL, A. J. B. ANDERSON and M. A. EASTWOOD. Effects of unprocessed bran on colon function in normal subjects and in diverticular disease. *Lancet* **1**:146, 1974.

21. FISHER, H., and P. GRIMINGER. Cholesterol-lowering effects of certain grains and of oat fractions on the chick. *Proc. Soc. Exp. Biol. Med.* **126**:108, 1967.

22. GOULSTON, E. Diet and diverticulitis. *Br. Med. J.* **2**:378, 1967.

23. HAGHSHENASS, M., M. MAHLOUDJI, J. G. REINHOLD and N. MOHAMMADI. Iron deficiency anaemia in an Iranian population associated with high intakes of iron. *Am. J. Clin. Nutr.* **25**:1143, 1972.

24. HARVEY, R. F., E. W. POMARE and K. W. HEATON. Effects of increased dietary fibre on intestinal transit. *Lancet* **1**:1278, 1973.

25. HEATON, K. W. Food fibre as an obstacle to energy intake. *Lancet* **2**:1418, 1973.

26. HEATON, K. W., and E. W. POMARE. Effect of bran on blood lipids and calcium. *Lancet* **1**:49, 1974.

27. KRITCHEVSKY, D., P. SALLATA and S. A. TEPPER. Experimental atherosclerosis in rabbits fed cholesterol-free diets. *J. Atheroscler. Res.* **8**:697, 1968.

28. LINDER, P., and B. MOLLER. Lignin: A cholestrol lowering agent? *Lancet* **2**:1256, 1973.

29. MISTUANAGA, T. Some properties of protease inhibitors in wheat grain. *J. Nutr. Sci. Vitaminol.* **20**:153, 1974.

30. MORGAN, B., M. HEALD, S. D. ATKIN and J. GREEN. Dietary fibre and sterol metabolism in the rat. *Br. J. Nutr.* **32**:447, 1974.

31. PAINTER, N. S., and D. P. BURKITT. Diverticular disease of the colon: a deficiency disease of Western civilization. *Br. Med. J.* **2**:450, 1971.

32. PAINTER, N. S., and S. C. TRUELOVE. The intraluminal pressure patterns in diverticulosis of the colon. *Gut* **5**:201, 1964.

33. PAINTER, N. S., A. Z. ALMEIDA and K. W. COLEBOURNE. Unprocessed bran in treatment of diverticular disease of the colon. *Br. Med. J.* **2**:137, 1972.
34. REINHOLD, J. G. Phytate destruction by yeast fermentation in whole wheat meals. *J. Am. Diet. Assoc.* **66**:38, 1975.
35. REINHOLD, J. G., K. NASR, A. LAHIMGARZADEH and H. HEDAYATI. Effects of purified phytate and phytate-rich bread upon metabolism of zinc, calcium, phosphorous and nitrogen in man. *Lancet* **1**:283, 1973.
36. ROBERTSON, J. Changes in the fibre content of the British diet. *Nature* **238**:290, 1972.
37. SOUTHGATE, D. A. T. Fibre and other unavailable carbohydrates and their effects on the energy value of the diet. *Proc. Nutr. Soc.* **32**:131, 1973.
38. SOUTHGATE, D. A. T., and J. V. G. A. DURNIN. Calorie conversion factors. An experimental reassessment of the factors used in the calculation of the energy value of human diets. *Br. J. Nutr.* **24**:517, 1970.
39. SPILLER, G. A., and R. J. AMEN. Research on dietary fiber. *Lancet* **2**:1259, 1974.
40. TELLEGREN, A. O. H. Prevention of deep vein thrombosis. *Br. Med. J.* **2**:467, 1972.
41. TROWELL, H. Ischemic heart disease and dietary fiber. *Am. J. Clin. Nutr.* **25**:926, 1972.
42. TROWELL, H. Crude fiber, dietary fiber and atherosclerosis. *Atherosclerosis* **16**:138, 1972.
43. TROWELL, H. Dietary fiber, ischemic heart disease and diabetes mellitus. *Proc. Nutr. Soc.* **32**:151, 1973.
44. WIERINGA, G. W. *Growth-inhibitory Substances in Rye. Publ. 156.* Wageninger, Holland, Inst. Storage and Processing of Agric. Produce, 1973.

Index

Adenocarcinoma, 219, 223
Agar, 18-19
Alginates, 18, 50, 115
Alphacel®, 154-155, 157, 163
Appendicitis
  incidence
    age, 208, 214
    economic status, 249
    geographical, 207, 214
    race, 249-250
  diet, 214, 243, 250-251
  etiology, 250
  fiber, 227, 234, 250-251, 257
Arabinans, 36
Arabinogalactans, 37, 42
Arabinoxylans, 41
Atherosclerosis, 264-265
  in animals, 174-175
Avicel®, 155, 158, 166

Bacteria, 121, 126, 139, 144, 146, 196
  and carcinogens, 123, 199, 226, 228
  fermentation by, 133, 138, 145
  and fiber, 120-121, 127, 154, 157
  growth of, 131, 136-137
  and HCl, 134
  metabolism by, 122
  methanogenic, 133
  in rumen, 143
Bile, 117, 172, 214
  acids
    adsorption by fiber, 117, 124-125, 203
    binding by fiber, 178-181, 190
    and cancer, 200, 226, 253, 256
    and cholesterol, 172
    in colon, 185, 199-200, 203
    fecal, 200, 203, 226, 253
    metabolism of, 122, 200

Bile (*cont'd*)
  salts, 11, 16, 163, 172, 178-181, 200
Bilirubin, 122, 124
Bran
  and bile acids, 180, 190, 200, 254
  and bile salts, 179
  and bowel *milieu interieur,* 255
  and colonic irritation, 188
  crude fiber fraction in, 3
  definition, 6
  and diverticular disease, 198, 203, 219
  as a laxative, 187-190, 249
  and lipid levels, 174
  and phytic acid, 20, 266
  properties, 114-116
  and stool weight, 187-192, 194, 203, 254
  and transit time, 191-192, 203
  water-holding capacity, 113-114, 182,
    190, 193
Bulk, 3-4, 152

Cancer (carcinoma), 126, 242
  of the colon
    and diet, 199, 210, 227, 243, 253
    etiology, 126, 199, 219-220, 225-229,
      252-255
    and fiber, 190, 199-203, 207, 227-229,
      234, 243, 252-255, 257
    incidence, 208, 210, 264
      age, 208, 220-223, 225
      economic status, 212, 223, 227
      geographical, 207, 212, 221-222
      race, 220, 252
      sex, 221, 252
    colorectal, 210, 219-220, 223
    of the rectum, 219-220, 223-225, 227-
      229
    rectosigmoid, 220